Applied Probability and Statistics (*Continued*)

BHAT · Elements of Applied Stochastic Processes
BOX and DRAPER · Evolutionary Operation: A Statistical Method for Process Improvement
BROWNLEE · Statistical Theory and Methodology in Science and Engineering, *Second Edition*
CHAKRAVARTI, LAHA and ROY · Handbook of Methods of Applied Statistics, Vol. I
CHAKRAVARTI, LAHA and ROY · Handbook of Methods of Applied Statistics, Vol. II
CHERNOFF and MOSES · Elementary Decision Theory
CHIANG · Introduction to Stochastic Processes in Biostatistics
CLELLAND, deCANI, BROWN, BURSK, and MURRAY · Basic Statistics with Business Applications
COCHRAN · Sampling Techniques, *Second Edition*
COCHRAN and COX · Experimental Designs, *Second Edition*
COX · Planning of Experiments
COX and MILLER · The Theory of Stochastic Processes
DANIEL and WOOD · Fitting Equations to Data
DAVID · Order Statistics
DEMING · Sample Design in Business Research
DODGE and ROMIG · Sampling Inspection Tables, *Second Edition*
DRAPER and SMITH · Applied Regression Analysis
ELANDT-JOHNSON · Probability Models and Statistical Methods in Genetics
GOLDBERGER · Econometric Theory
GUTTMAN, WILKS and HUNTER · Introductory Engineering Statistics, *Second Edition*
HAHN and SHAPIRO · Statistical Models in Engineering
HALD · Statistical Tables and Formulas
HALD · Statistical Theory with Engineering Applications
HOEL · Elementary Statistics, *Third Edition*
HUANG · Regression and Econometric Methods
JOHNSON and KOTZ · Distributions in Statistics
    Discrete Distributions
    Continuous Univariate Distributions-1
    Continuous Univariate Distributions-2
    Continuous Multivariate Distributions
JOHNSON and LEONE · Statistics and Experimental Design: In Engineering and the Physical Sciences, Volumes I and II
LANCASTER · The Chi Squared Distribution
LEWIS · Stochastic Point Processes
MILTON · Rank Order Probabilities: Two-Sample Normal Shift Alternatives
RAO and MITRA · Generalized Inverse of Matrices and Its Applications
SARD and WEINTRAUB · A Book of Splines
SARHAN and GREENBERG · Contributions to Order Statistics
SEAL · Stochastic Theory of a Risk Business
SEARLE · Linear Models
THOMAS · An Introduction to Applied Probability and Random Processes
WHITTLE · Optimization under Constraints
WILLIAMS · Regression Analysis

continued on back

Applied Probability and Statistics (*Continued*)
    WOLD and JUREEN · Demand Analysis
    WONNACOTT and WONNACOTT · Introduction to Econometric Methods
    YOUDEN · Statistical Methods for Chemists
    ZELLNER · An Introduction to Bayesian Inference in Econometrics

Tracts on Probability and Statistics
    BILLINGSLEY · Ergodic Theory and Information
    BILLINGSLEY · Convergence of Probability Measures
    CRAMÉR and LEADBETTER · Stationary and Related Stochastic Processes
    JARDINE and SIBSON · Mathematical Taxonomy
    KINGMAN · Regenerative Phenomena
    RIORDAN · Combinatorial Identities
    TAKÁCS · Combinatorial Methods in the Theory of Stochastic Processes

Sarah Byrd Askew
LIBRARY

WILLIAM PATERSON COLLEGE
WAYNE, NEW JERSEY

Statistical Inference under Order Restrictions

Statistical Inference under Order Restrictions

The Theory and Application of Isotonic Regression

R. E. BARLOW
University of California, Berkeley

D. J. BARTHOLOMEW and
J. M. BREMNER
University of Kent at Canterbury

H. D. BRUNK
Oregon State University, Corvallis

JOHN WILEY & SONS
London · New York · Sydney · Toronto

Copyright © 1972 John Wiley & Sons Ltd.

All Rights Reserved. No part of this publication may be reproduced, stored in a retrieval system, or transmitted, in any form or by any means, electronic, mechanical, photocopying, recording or otherwise, without the prior written permission of the Copyright owner.

Library of Congress Catalog Card Number: 74-39231

ISBN 0 471 04970 0

Preface

Some results dealing with statistical inference in the presence of order conditions have been rediscovered several times independently during the course of the past decade or two, in various contexts, in research in several disciplines. It seemed to us it was time to make work in this area accessible in a single book. As our title implies, we have aimed to present a broad spectrum of the subject ranging from its abstract mathematical foundations to the tables and formulas needed for applications. For convenience, we have given the most general mathematical discussion in the last chapter, but elsewhere theory and applications are intertwined. We hope that our readers, whatever their main interests, will find the interplay between these two aspects as illuminating as we do.

We have not thought it necessary to include a large number of numerical examples. They have been introduced whenever we thought their presence would help the exposition; but, as the reader will quickly discover, the arithmetical manipulations involved are mainly the routine calculation of means and sums of squares. It did not seem desirable to burden the text with more of these than necessary.

The selection of material naturally reflects our particular interest, and there is much interesting and important work which has not been developed fully in the text. We have indicated in "Complements" sections at the ends of the chapters where such further work as we are aware of can be found. These "Complements" contain also brief historical notes; and while we have tried to be accurate, inevitably some errors will have crept in, for which we apologize.

There are two rather obvious omissions. Apart from brief mention in Chapter 2, there is no discussion of Bayesian inference and no mention at all of interval estimation. In both cases, the reason for the omission is simple. At the present time, there is virtually no theory available; we hope that some of our readers will be inspired to fill these gaps.

The subtitle of the book is "The Theory and Application of Isotonic Regression", and we hope to have made clear in Chapter 1 the reason for the selection of this subtitle. The adjective "isotonic" is used as a synonym for "order-preserving"; order restrictions on parameters generally can be regarded as requiring that the parameter, as a function of an index, shall be isotonic with respect to a partial order on the index set. The term "regression" arises in the context of a typical problem: given a function, the problem is to fit it in the sense of least squares by an isotonic function (rather than, for example, by a linear function).

In an introductory section of Chapter 1, we mention a number of situations in which order restrictions play an essential role in statistical inference. An example in a regression context introduces in Section 1.2 the problem of least squares fit by a function defined on a finite set whose values are required to satisfy certain order restrictions. Properties of the solution are developed, and an algorithm is presented for its computation in the case of a linear or simple ordering. A formula is also given, as well as a graphical interpretation. The section closes with an example drawn from applications in psychology. The third section develops properties of isotonic regression used in subsequent chapters. It concludes with an example based on an application in social science. Section 1.4 introduces a wide class of extremum problems whose solutions are furnished by isotonic regression, using as illustration maximum likelihood estimation of ordered binomial parameters. The fifth section discusses problems dual to those in Section 1.4; one such is the "taut string" problem. Two examples drawn from operations research are included.

Chapter 2 deals with order-restricted estimation in general and estimation of a regression function known to be isotonic in particular. Estimation of ordered multinomial parameters is discussed in an example, and least squares estimation in a preference ranking situation in another. The third section states and justifies an elaboration of the algorithm mentioned above, as well as algorithms applying to more general partial orderings. Section 2.4 establishes the isotonic regression of sample means as maximum likelihood estimates of population means known to be ordered in a certain way, when the populations belong to a common exponential family. Section 2.5 is devoted to maximum likelihood estimation of stochastically ordered distributions.

Chapter 3 is devoted to the problem of testing the null hypothesis that k normal populations have equal means against the alternative that the means satisfy order restrictions. The likelihood ratio test is derived both for the case when the population variances are known and when they are unknown but equal. The null hypothesis distribution is obtained in each case and its density is shown to be a weighted sum of χ^2 or Beta densities respectively. The weights are probabilities which have an independent interest since they give the distribution of the number of distinct values among the maximum likelihood estimators of the means. They are discussed at some length in the latter part of Section 3.3. The power of the tests is investigated in Section 3.4 using exact and approximate methods. This investigation leads to conclusions about the possible optimality of the likelihood ratio test.

Chapter 4 contains some generalizations and extensions of the problem treated in the previous chapter. A one-sided test for the mean of a multivariate normal distribution is given in the first section and shown to be closely related to the likelihood ratio test of Chapter 3. A different class of tests for normal means based on contrasts is described in Section 4.2; their power is compared with the likelihood ratio test. An extension of the likelihood ratio method to distributions belonging to the exponential family is given in Section 4.3. The theory of distribution-free tests for ordered alternatives is given in the final section where it is shown that their asymptotic power compares very favourably with those based on normal theory.

Chapter 5 deals with the problem of estimating unimodal and monotone densities and failure rate functions. The maximum likelihood estimators under these restrictions are identified as isotonic regression estimators with respect to certain "basic" estimators. Strong consistency of the empirical distribution is used to prove strong consistency of the isotonic estimators. This is true in spite of the fact that the basic estimators corresponding to the maximum likelihood estimators are themselves *not* consistent. By modifying the basic estimators through the use of "wide windows" in Section 5.4, isotonic estimators are produced which improve asymptotically upon the maximum likelihood estimators. Asymptotic distributions for isotonic estimators of the generalized failure rate function are also given in Section 5.4. The generalized failure rate function includes the usual failure rate function and density as special cases. Section 5.5 considers isotonic estimators for distributions star-ordered with respect to a specified distribution. This subject is motivated by models which occur in reliability theory. An interesting feature of this subject is that the maximum likelihood estimators (they are not isotonic estimators) actually converge to the wrong distribution as sample size becomes infinite.

Chapter 6 is concerned with tests for exponentiality against monotone failure rate alternatives. The total time on test transformation is used to convert this problem to one of stochastic ordering. A test based on the cumulative total time on test statistic is shown to have isotonic power with respect to convex ordered alternative distributions in Section 6.2. Using the concept of contiguous alternatives, this test is shown to be asymptotically minimax over a class of generalized scores statistics and with respect to a certain extremal class of alternative distributions in Section 6.3. The asymptotic distribution of the cumulative total time on test statistic is computed for general alternative distributions. Section 6.5 is concerned with tests for trend in a series of events. An analogue of the cumulative total time on test statistic is shown to be isotonic with respect to star-ordering. The concept of total time on test is used in Section 6.5 to extend our unbiasedness results to tests for monotone failure rate based on incomplete data.

Chapter 7 develops the mathematical theory of isotonic regression in an abstract, measure-theoretic setting, where it is called "conditional expectation given a σ-lattice". The second section establishes the relationship between isotonic regression and conditional expectation given a σ-lattice. The third develops this generalization of the concept "conditional expectation given a σ-field" as projection in the Hilbert space of square-integrable random variables, while the fourth develops the appropriate Lebesgue–Radon–Nikodym approach.

Our bibliography serves the two purposes of giving details of all published work referred to in the text, and of providing a list of all the references known to us on matters connected with inference under order restrictions. Thus many of the papers listed in the bibliography are not referred to in the text, as our primary aim has been to give an account of the theory of isotonic regression and associated statistical methods.

Acknowledgements

We are grateful to the authors (where appropriate) and publishers concerned for permission to reproduce material from previously published work, as follows. Figures 1.9 and 1.10 are taken from Hartigan (1967b). Figure 2.1 is based on a figure appearing in Lombard and Brunk (1963). Table 3.2 incorporates values taken from Kudô and Fujisawa (1964). Tables 3.4–3.8, 3.10, 4.1, 4.2 and A.6 reproduce various tables of Bartholomew (1961a), with the incorporation of some newly computed and corrected values. Tables 4.3 and 4.4 are based on tables in Boswell and Brunk (1969), and Table 4.5 is from Hollander (1967). Table 6.1 is taken from Barlow and Proschan (1969), Table 6.2 is part of a table of Proschan (1963), and Figure 6.1 appears in Barlow and Doksum (1972). Tables A.1–A.3 incorporate various tables of Bartholomew (1959a, 1959b), and Table A.4 is from Chacko (1963).

The progress of this project was greatly accelerated by a happy coincidence which brought three of us together for the summer of 1969 in the Department of Industrial Engineering and Operations Research at the University of California at Berkeley. Without this period of close cooperation, the book might have been indefinitely delayed and we are most grateful to the University for making the facilities available. The work of R. E. Barlow during this summer was sponsored by the United States Navy, Office of Naval Research, under

Contract Number N00014-69-A-0200-1036, and that of H. D. Brunk by the Air Force Office of Scientific Research, Office of Aerospace Research, United States Air Force, under AFOSR Grant Number 69-1691; this office also sponsored much of the earlier research on this subject by him and his colleagues. H. D. Brunk also wishes to record here his debt of gratitude to Miriam Ayer, George Ewing, W. T. Reid, Edward Silverman, and Roy Utz for a most enjoyable collaboration in one of the many beginnings of this work.

We are grateful to S. Johansen for helpful suggestions resulting from a critical reading of Chapter 7; but of course, responsibility for such errors as remain is ours.

Last but by no means least, we wish to thank our wives, Barbara, Marian, Judy and Jean, for their patient understanding and loyal support.

Contents

Preface   v

1 Isotonic Regression   1
    1.1 Introduction   1
    1.2 Isotonic regression over a simply ordered finite set   5
    1.3 Isotonic regression over a quasi-ordered set   24
    1.4 Generalized isotonic regression problems   38
    1.5 Extremum problems associated with dual convex cones   46
    1.6 Complements   55

2 Estimation Under Order Restrictions   64
    2.1 Isotonization of estimates   64
    2.2 Isotonic estimation of regression   66
    2.3 Algorithms for calculation   72
    2.4 Maximum likelihood and Bayesian estimation of ordered parameters in exponential families   91
    2.5 Maximum likelihood estimation of stochastically ordered distributions   105
    2.6 Complements   110

3 Testing the Equality of Ordered Means: Likelihood Ratio Tests in the Normal Case   116
    3.1 Introduction   116
    3.2 The $\bar{\chi}^2$ and $\bar{E}^2$ tests   117
    3.3 The null hypothesis distributions of $\bar{\chi}^2$ and $\bar{E}^2$ 126
    3.4 The power function of $\bar{\chi}^2$ 153
    3.5 Complements   170

4 Testing the Equality of Ordered Means: Extensions and Generalizations 175
4.1 A one-sided test for the multivariate normal mean 175
4.2 Tests based on contrasts among the means 183
4.3 Ordered tests for non-normal distributions 191
4.4 Distribution-free tests for ordered alternatives 198
4.5 Complements 213

5 Estimation of Distributions 222
5.1 Introduction 222
5.2 Estimation of unimodal densities 223
5.3 Maximum likelihood estimation for distributions with monotone failure rate 231
5.4 Isotonic window estimators for the generalized failure rate function 242
5.5 Isotonic estimators for star-ordered families of distributions 254
5.6 Complements 261

6 Isotonic Tests for Goodness of Fit 263
6.1 Introduction 263
6.2 Tests for exponentiality against monotone failure rate alternatives 263
6.3 Asymptotic minimax property of the cumulative total time on test statistic 279
6.4 Generalizations 297
6.5 Complements 305

7 Conditional Expectation Given a σ-Lattice 307
7.1 Introduction 307
7.2 Formulation of the isotonic regression problem in terms of projection in L_2 307
7.3 Conditional expectation given a σ-lattice as projection in L_2 314
7.4 Conditional expectation given a σ-lattice as a Lebesgue-Radon-Nikodym derivative 326
7.5 Complements 354

Appendix: Tables 359

Bibliography and Author Index 365

Subject Index 379

CHAPTER 1

Isotonic Regression

1.1 INTRODUCTION

The theory of regression is bound up with conditional expectations and least squares. The *regression function* of a random variable $\tilde{y}$ on another, $\tilde{x}$, is the conditional expectation $\mu(x) = E(\tilde{y}|\tilde{x} = x)$, and furnishes the best fit to the distribution of $\tilde{y}$ by a function of $\tilde{x}$ in the sense of least squares. The *regression line* of $\tilde{y}$ on $\tilde{x}$ is the best fit by a linear function of $\tilde{x}$ in the sense of least squares. *Isotonic regression* is introduced in Section 1.2 by means of least squares, as a generalization of the regression function.

Regression analysis is directed toward making inferences about $\mu(x)$: estimating it, or testing hypotheses about it. In common statistical usage the term regression analysis is confined to the case where x is real or vector valued. However it is convenient for our purposes to use the term in a wider sense. Thus the symbol x may be used to represent a statement about the conditions under which $\tilde{y}$ is observed. For example, when concerned with the effects of two treatments, the means $E(\tilde{y}|\text{treatment 1})$ and $E(\tilde{y}|\text{treatment 2})$, which are written as $\mu(1)$ and $\mu(2)$ or μ_1 and μ_2, are compared. In effect the variable x is qualitative, though x may be formally allowed to range over the set X consisting of the two elements 1 and 2: $X = \{1, 2\}$. The one-way analysis of variance can be described similarly, as can the whole analysis stemming from the general linear hypothesis.

In discussing the regression of $\tilde{y}$ on $\tilde{x}$, then, $\tilde{y}$ is a random variable whose values are real numbers. But $\tilde{x}$ may or may not be a random variable. Its values may or may not be real numbers or vectors (ordered k-tuples) of real numbers. In general, $\tilde{x}$ ranges over an abstract set X. In defining regression of $\tilde{y}$ on $\tilde{x}$ via least squares, "weights" $w(x)$ associated with the values of $\tilde{x}$ must be used. If $\tilde{x}$ is a random variable, $w(x)$ is the probability that the random variable $\tilde{x}$ will be equal to x, or the density of $\tilde{x}$ at x. In sampling situations weights may be proportional to numbers of observations. In general, if the set X of values of $\tilde{x}$ is finite, a positive weight $w(x)$ is associated with each x in X; and a probability distribution, which is thought of as the distribution of a random variable $\tilde{y}$ given $\tilde{x} = x$, is associated with each x in X.

Attention is focused in this book on regression problems in which a knowledge of the xs used in an experiment determines an ordering, partial or total, of the corresponding numbers $\mu(x)$. A simple example is provided by the two sample problem referred to above in which the mean responses for two treatments are compared. If Treatment 1 is the application of a fertilizer to the soil in which a crop is grown and if Treatment 2 is simply to grow the crop without fertilizer then it may be possible to assert *a priori* that $\mu_1 \geq \mu_2$. This situation is of frequent occurrence and the method of dealing with it is well known. If the hypothesis $\mu_1 = \mu_2$ is being tested significance is attributed to an observed difference $\bar{y}_1 - \bar{y}_2$ only if it is positive. That is, a one-tail or one-sided test is made. This is, perhaps, the simplest example of isotonic regression analysis.

A similar situation arises when several means are being compared (Bartholomew (1959a)). Suppose, for example, that $\tilde{y}$ is a measure of performance or a test of ability carried out under varying degrees of stress, x. It may be expected that $\mu(x)$ is a decreasing function of x, but be impossible to quantify stress. However, if the experiment is carried out at levels of stress $x_1, x_2, \ldots, x_k$ which are known to satisfy the inequalities $x_1 \leq x_2 \leq x_3 \leq \ldots \leq x_k$ then it could be asserted that $\mu(x_1) \geq \mu(x_2) \geq \ldots \geq \mu(x_k)$. The usual one-way analysis of variance makes no use of order information of this kind. Hence it may be expected that the efficiency of our analysis is increased by taking such information into account. Isotonic regression analysis provides the means of doing this.

It may not be possible to arrive at a total, or simple, ordering as in the last example. For example, the conditions

$$\mu(x_1) \geq \mu(x_2),$$
$$\mu(x_1) \geq \mu(x_3)$$

may hold without it being possible to rank $\mu(x_2)$ and $\mu(x_3)$ with respect to one another. Partial orderings may also arise when x is vector valued. For example, it might be possible to assert only that $\mu(x_1', x_1'') \geq \mu(x_2', x_2'')$ if both $x_1' \leq x_2'$ and $x_1'' \leq x_2''$ (cf. Brunk, Ewing and Utz (1957)). The ranking of pairs of μs for which only one of these inequalities held would be unknown.

Estimation subject to order restrictions arises also when estimating variance components in random models of the analysis of variance. To take the simplest case, consider the one way layout with the same number of observations in each group. In the random model it is assumed that the observations y_{ti} have the following structure:

$$y_{ti} = \mu + a_t + e_{ti} \quad (t = 1, 2, \ldots, k, i = 1, 2, \ldots, n),$$

where $\{a_t\}$ and $\{e_{ti}\}$ are sets of mutually independent normal variates with zero expectations and μ is a constant. Denote the variance of the a_ts by σ_a^2

and that of the e_{ti}s by σ^2. The usual way of estimating σ_a^2 and σ^2 is by equalizing the mean squares of the analysis of variance table to their expectations. That is

$$\hat{\sigma}^2 = \text{residual mean square},$$

$$\hat{\sigma}^2 + n\hat{\sigma}_a^2 = \text{between groups mean square}.$$

Since variances cannot be negative, $\sigma^2 \leq \sigma^2 + n\sigma_a^2$. However, the sample estimates may turn out in the reverse order. The computation of the isotonic estimates ensures that all estimates of components of variance are nonnegative. More complicated designs give rise to various kinds of complete and partial ordering among the expected mean squares. (A discussion of the theory and some examples are given in Thompson (1962).)

The foregoing discussion has made use of examples which arise in the standard "normal" theory of analyses of data. That is, it is assumed that the conditional distribution of $\tilde{y}$ given $\tilde{x} = x$ is normal with constant variance. The general theory makes no distributional assumptions of this kind although many of the particular cases which will be considered relate to exponential families of distributions which, of course, include normal distributions. Another exponential family which often arises in practice is the family of binomial distributions. There are a number of contexts in which an experiment may give rise to several independent observed proportions with expectations p_i, say (Ayer and coworkers (1955); van Eeden (1956, 1957a and b)). Thus in a dosage–mortality experiment, groups of animals may be given different doses of a drug and it may be known that the greater the dose the greater the expected proportion that will die. An example, formally equivalent, is provided by the nonparametric estimation of a distribution function. Suppose that the lifetime of a certain article has distribution function $F(x)$. Then it is known that $F(x)$ must be a nondecreasing function. If n_i items are observed for length of time x_i, $x_1 < x_2 < \ldots < x_n$, and the proportion of failures, s_i, counted ($i = 1, 2, \ldots, n$) s_i/n_i could be used as an unbiased estimate of $F(x)$ at $x = x_i$. However, if this is done it is quite possible for it to happen for some i that

$$s_i/n_i > s_{i+1}/n_{i+1}$$

which, since $x_i < x_{i+1}$, makes the estimate of $F(x)$ decrease between x_i and x_{i+1}. The theory of isotonic regression enables use to be made of the fact that $F(x)$ is monotonic to improve our estimate.

In "debugging" (or running in) large scale systems it is common to plot a graph of the cumulative number of failures against elapsed operational time.

If $x_1 < x_2 < \ldots < x_n$ are the times at which the system has to be readjusted or repaired, then the slopes

$$\frac{i - (i-1)}{x_i - x_{i-1}} = \frac{1}{x_i - x_{i-1}}$$

of successive secants roughly estimate failure rates over successive time periods. As the "bugs" are removed from the system, the failure rate, $\mu(x)$, should tend to decrease. However, due to statistical fluctuations in the estimates, some reversals will occur. The theory of isotonic regression enables use to be made of the fact that $\mu(x)$ is monotonic to improve the estimate.

It is common practice, during the development of a system, to make engineering changes as the program develops. These changes are generally made in order to correct design deficiencies and, thereby, to increase reliability. Suppose that a test program is conducted in n stages. At each stage of experimentation, tests are run on similar items. The results of each stage of testing are used to improve the item for further testing in the next stage. For the ith stage the number, a_i, of inherent failures, the number, b_i, of assignable cause failures, and the number, c_i, of successes are recorded. (Inherent failures are defined as those which reflect the state-of-the-art and assignable cause failures as those which can be corrected by equipment or operational modifications.) The probability of an inherent failure, q_0, is assumed to be constant and not to change from stage to stage of testing. The probability of an assignable cause failure in the ith stage, $\mu(i)$, is assumed nonincreasing.

The unrestricted maximum likelihood estimates are

$$\hat{\mu}(i) = (1 - \hat{q}_0)b_i/(b_i + c_i), \quad i = 1, 2, \ldots, n,$$

where

$$\hat{q}_0 = \sum_1^n a_i / \sum_1^n (a_i + b_i + c_i).$$

Isotonic regression enables the estimates to be improved under the monotonicity restriction (see Barlow and Scheuer (1966)).

Analyses of many types of life test data indicate that the failure rate function $\mu(x) = f(x)/[1 - F(x)]$ (f is the density of the failure distribution F) often has a U shape. The actuarial method of estimating this failure rate requires first that one subdivide the time axis as follows: $0 < x_1 < x_2 < \ldots < x_n$. The failure rate estimate is a step function. The estimate over the interval $[x_i, x_{i+1})$ is

$$\mu_n(x_i) = \frac{N_i}{(x_{i+1} - x_i)\{1 - F_n(x_i)\}}$$

where N_i is the observed number of failures in $[x_i, x_{i+1})$ and F_n is the empirical distribution function. Since, by assumption, $\mu(x)$ is U shaped there must exist x_k such that

$$\mu(x_1) \geq \mu(x_2) \geq \ldots \geq \mu(x_k)$$

and

$$\mu(x_k) \leq \mu(x_{k+1}) \leq \ldots \leq \mu(x_n).$$

However, the estimate μ_n will in general be very ragged and certainly not U shaped. Again isotonic regression enables μ_n to be smoothed making use of the fact that μ is U shaped.

It sometimes happens that the x variable is time, in which case the basis of the ordering is the belief that the conditional expectation of $\tilde{y}$ is changing in a known way with time. Instead of having a finite set of xs at which $\tilde{y}$ has been observed, a continuous record of y over a period of time is now obtained, in principle at least. An example is provided by the Poisson process when the rate is a function of time (Boswell (1966)). In Chapter 5 the special case arising when the rate is a monotonic function of time will be considered and in Chapter 7 the basic theory of isotonic regression will be formulated so that it covers this case also.

The object of this book is to give a unified account of the statistical theory which has grown up to deal with problems in which conditional expectations are subject to order restrictions. Underlying the diverse applications there is a basic mathematical structure. In the remainder of this chapter the definitions and theorems required for the applications described later in the book will be set out.

1.2 ISOTONIC REGRESSION OVER A SIMPLY ORDERED FINITE SET

Isotonic regression is introduced by way of an example.

Example 1.1 *Maximum likelihood estimation of increasing means of normal distributions*

Bhattacharyya and Klotz (1966) use freezing dates and thawing dates for Lake Mendota for a period of 111 years to test against warming trend. Among models they consider is a bivariate normal model. For purposes of illustration a univariate normal model for freezing dates alone is considered here. Let y_i denote the number of days to freezing, measured from 23 November (the earliest freezing date) in the year $1854 + i$, $i = 1, 2, \ldots, 111$. According to a simple, useful (if not completely realistic) model, the y_i are independent observations on a normal distribution with unknown means μ_i, $i = 1, 2, \ldots, 111$, and common variance σ^2. In order to construct the likelihood ratio test of no warming trend,

$$H_0: \mu_1 = \mu_2 = \ldots = \mu_{111},$$

within the class of alternatives

$$H_1: \mu_1 \leq \mu_2 \leq \ldots \leq \mu_{111},$$

it is necessary to determine the maximum likelihood estimates of the μs under H_0 and under H_1.

The negative of the log-likelihood function is

$$(1/2\sigma^2) \sum_{i=1}^{111} (y_i - \mu_i)^2 - (111/2) \log(2\pi\sigma^2).$$

This is minimized under H_0 by

$$\mu_1 = \mu_2 = \ldots = \mu_{111} = \bar{y} = (1/111) \sum_{i=1}^{111} y_i.$$

The maximum likelihood estimates of $\mu_1, \ldots, \mu_{111}$ under H_1 minimize

$$\sum_{i=1}^{111} (y_i - \mu_i)^2 \tag{1.1}$$

subject to $\mu_1 \leq \mu_2 \leq \ldots \leq \mu_{111}$. If $y_1 \leq \ldots \leq y_{111}$, the minimum value, 0, is achieved by setting $\mu_i = y_i$, $i = 1, 2, \ldots, 111$. In the event, the observations y_i do not satisfy the inequalities. This quadratic programming problem is then obtained: minimize the quadratic objective function (1.1) subject to the linear inequalities $\mu_1 \leq \ldots \leq \mu_{111}$.

Suppose the following alternative hypothesis is of interest:

$$H_2: \mu_i = ai + b \qquad i = 1, 2, \ldots, 111$$

for some real constants a and b. The problem of maximum likelihood estimation of $\mu_1, \ldots, \mu_{111}$ subject to H_2 is the problem of minimizing (1.1) subject to H_2. This is a problem of *linear* regression, for H_2 requires μ_i to be a linear function of i. Similarly, since inequalities H_1 require the μ_i as a function of i be isotonic (that is, order preserving; in the present case, nondecreasing), this quadratic programming problem is referred to as an isotonic regression problem.

The problem of maximum likelihood estimation of partially ordered means of normal distributions is treated in Section 2.4; see Example 2.4.

For simplicity and because it suffices for most of the applications to be discussed, it is assumed throughout the early chapters of the book that the range X of the "independent" variable, x, is a finite set. (The reader will recognize that the results are applicable more generally; see Chapter 7.)

Example 1.2 Sample isotonic regression

Let $X = \{x_1, x_2, \ldots, x_k\}$, where $x_1 < x_2 < \ldots < x_k$. For $i = 1, 2, \ldots, k$, let $y_j(x_i)$, $j = 1, 2, \ldots, m(x_i)$, be a set of measurements of some quantity. That is, for $x \in X$, $y_1(x), \ldots, y_{m(x)}(x)$ are observations on a distribution. (In the special case of Example 1.1, $k = 111$, $X = \{1, 2, \ldots, 111\}$, $x_i = i$, $m(x_i) = 1$, $y_j(x_i) = y_i$, $j = 1$, $i = 1, 2, \ldots, 111$.) Let $\mu(x)$ denote the mean of the distribution. If μ is known or assumed to be linear in x, it may be desired to estimate μ by the sample linear regression. This is the solution of the problem of linear regression: to fit the data in the sense of least squares by a linear function of x, i.e. to minimize

$$\sum_{x \in X} \sum_{j=1}^{m(x)} [y_j(x) - f(x)]^2$$

in the class of linear functions f. Let

$$\bar{y}(x) = \frac{1}{m(x)} \sum_{j=1}^{m(x)} y_j(x), \qquad x \in X.$$

Since

$$\sum_{j=1}^{m(x)} [y_j(x) - f(x)]^2 = \sum_{j=1}^{m(x)} [y_j(x) - \bar{y}(x)]^2 + m(x)[\bar{y}(x) - f(x)]^2,$$

an equivalent problem is to minimize

$$\sum_{x \in X} [\bar{y}(x) - f(x)]^2 m(x) \tag{1.2}$$

in the class of linear functions f on X.

If no restriction were to be placed on μ, its least squares estimate would be obtained by minimizing (1.2) in the class of arbitrary functions f on X. The solution is clearly the function $\bar{y}: \{\bar{y}(x), x \in X\}$. In another situation, it might be known or assumed that μ is nondecreasing in x; that is, isotonic with respect to the simple order on X. A least squares estimate of μ would be obtained by minimizing the weighted sum of squares (1.2) in the class of nondecreasing functions f on X, the class of functions isotonic with respect to the simple order on X: functions f such that $x_i \leq x_j$ implies $f(x_i) \leq f(x_j)$. The solution may be called the *sample isotonic regression*.

Suppose for example that $X = \{1, 2\}$, i.e. $x_1 = 1$, $x_2 = 2$. Suppose one measurement, $\bar{y}_1 = \bar{y}(1) = 5$ is made on a first quantity, and one measurement $\bar{y}_2 = \bar{y}(2) = 3$ on a second (see Figure 1.1). Then $m(1) = m(2) = 1$. Set $f_i = f(i)$, $\mu_i = \mu(i)$, $i = 1, 2$. Suppose it is known that $\mu_1 \leq \mu_2$. Here $\bar{y}_1$ and $\bar{y}_2$ do not satisfy $\bar{y}_1 \leq \bar{y}_2$ and so will not serve as estimates of μ_1 and μ_2

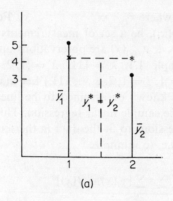

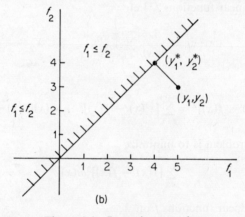

Figure 1.1 Isotonic regression

subject to $\mu_1 \leq \mu_2$. In Figure 1.1b, $(\bar{y}_1, \bar{y}_2)$ is plotted as a point in the Cartesian plane. It follows from the Pythagorean theorem that the foot, $(\bar{y}_1^*, \bar{y}_2^*)$ of the perpendicular onto the region $\{f_1 \leq f_2\}$ minimizes

$$\sum_{i=1}^{2} (\bar{y}_i - f_i)^2 = \sum_{x \in X} [\bar{y}(x) - f(x)]^2 m(x)$$

subject to $f_1 \leq f_2$.

Definition 1.1

Let X be the finite set $\{x_1, \ldots, x_k\}$ with the simple order $x_1 \prec x_2 \prec \ldots \prec x_k$. A real valued function f on X is *isotonic* if $x, y \in X$ and $x \prec y$ imply $f(x) \leq f(y)$. (The term "nondecreasing" would serve equally well here, but not in Section 1.3.) Let g be a given function on X and w a given positive function on X. An isotonic function g^* on X is an *isotonic regression of g with weights w with respect to the simple ordering* $x_1 \prec x_2 \prec \ldots \prec x_k$ if it minimizes in the class of isotonic functions f on X the sum

$$\sum_{x \in X} [g(x) - f(x)]^2 w(x). \tag{1.3}$$

When the weight function and the simple ordering are understood, we call g^* simply an isotonic regression of g.

Graphical interpretation—greatest convex minorant

A graphical interpretation of the isotonic regression is illuminating. Assuming still the simple ordering $x_1 \prec x_2 \prec \ldots \prec x_k$, plot the cumulative sums

$$G_j = \sum_{i=1}^{j} g(x_i) w(x_i)$$

against the cumulative sums

$$W_j = \sum_{i=1}^{j} w(x_i), \quad j = 1, 2, \ldots, k.$$

That is, plot the points $P_j = (W_j, G_j), j = 0, 1, 2, \ldots, k$ ($P_0 = (0, 0)$), in the Cartesian plane. These points constitute the *cumulative sum diagram* (CSD) of the given function g with weights w. The *slope* of the segment joining P_{j-1} to P_j is just $g(x_j), j = 1, 2, \ldots, k$. The slope of the chord joining P_{i-1} to P_j $(i \leq j)$ represents the weighted average

$$\text{Av}\{x_i, x_{i+1}, \ldots, x_j\} = \sum_{r=i}^{j} g(x_r) w(x_r) / \sum_{r=i}^{j} w(x_r).$$

It will be seen that the isotonic regression of g is given by the slope of the greatest convex minorant (GCM) of the CSD. This GCM is the graph of the supremum of all convex functions whose graphs lie below the CSD. Consider a line which intersects the CSD and is such that the entire CSD lies on or above it. If it intersects the CSD in more than one point then the segment joining its leftmost and rightmost intersections is a part of the graph of the GCM. The GCM (no confusion will result if we refer to both the function and its graph as the GCM) is made up of such segments. Graphically, *the*

GCM *is the path along which a taut string lies* if it joins P_0 and P_k and is constrained to lie below the CSD. (The graph of the GCM is called the Newton–Puiseux polygon, cf. Perron (1910).) The value of the isotonic regression g^* at a point x_j is just the slope of the GCM at the point P_j^* with abscissa

$$\sum_{i=1}^{j} w(x_i).$$

If P_j^* is a corner of the graph, $g^*(x_j)$ is the slope of the segment extending to the left. An illustrative example is shown in Table 1.1 and Figure 1.2.

It will be useful to express analytically some of the properties of the GCM. First, the CSD and GCM coincide at P_k; i.e.,

$$G_k^* = G_k. \tag{1.4}$$

Second, if for some index i the GCM at P_{i-1}^* lies strictly below the CSD at P_{i-1}, then the slopes of the GCM entering P_{i-1}^* from the left and leaving to the right are the same. That is,

$$G_{i-1}^* < G_{i-1} \Rightarrow g_i^* - g_{i-1}^* = 0, \quad i = 1, 2, \ldots, k. \tag{1.5}$$

Finally, if $P_r = P_r^*$, $P_s = P_s^*$, if $P_r P_s = P_r^* P_s^*$ is one side of the GCM and $P_s P_t = P_s^* P_t^*$ is the adjacent side to the right, and if $r < j < s$, then the slope of $P_j P_s$ is smaller (algebraically) than the slope of $P_j^* P_s^*$ (see Figure 1.3). If $s < j < t$ then the slope of $P_s P_j$ is larger (algebraically) than the slope of $P_s^* P_j^*$. That is,

$$\left.\begin{array}{l} \text{if } g^*(x) \text{ has the constant value } a \text{ for } x_r < x \leq x_s \text{ and the} \\ \text{constant value } b > a \text{ for } x_s < x \leq x_t, \text{ then} \\[6pt] \dfrac{\sum_{i=j+1}^{s} g(x_i) w(x_i)}{\sum_{i=j+1}^{s} w(x_i)} = \dfrac{G(x_s) - G(x_j)}{W(x_s) - W(x_j)} \leq a \quad \text{for } r \leq j < s, \\[6pt] \text{and} \\[6pt] \dfrac{\sum_{i=s+1}^{j} g(x_i) w(x_i)}{\sum_{i=s+1}^{j} w(x_i)} = \dfrac{G(x_j) - G(x_s)}{W(x_j) - W(x_s)} \geq b \quad \text{for } s < j \leq t. \end{array}\right\} \tag{1.6}$$

In particular,

$$g(x_s) \leq g^*(x_s) = a < b = g^*(x_{s+1}) < g(x_{s+1}).$$

These properties yield a simple proof that g^* is the isotonic regression of g with weights w.

Table 1.1 Example of CSD and GCM

j	$w(x_j)$	W_j	$g(x_j)$	G_j	G_j^*	g_j^*
1	1	1	−2	−2	−2	−2
2	2	3	5/2	3	−8/5	1/5
3	3	6	−4/3	−1	−1	1/5
4	2	8	1	1	1	1

$$W_j = \sum_{i=1}^{j} w(x_i), \quad G_j = \sum_{i=1}^{j} g(x_i)w(x_i), \quad G_j^* = \sum_{i=1}^{j} g^*(x_i)w(x_i),$$

$j = 1, 2, 3, 4.$

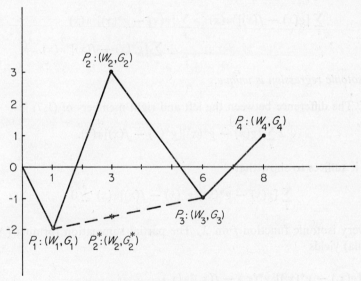

Figure 1.2 Example of CSD and GCM.

Slope at P of CSD: $\dfrac{G_j - G_{j-1}}{W_j - W_{j-1}} = g(x_j);$

Slope at P_j^* of GCM: $\dfrac{G_j^* - G_{j-1}^*}{W_j - W_{j-1}} = g^*(x_j), j = 1, 2, 3, 4$

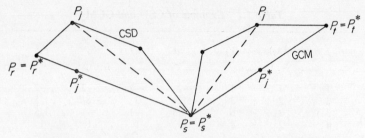

Figure 1.3 A portion of a Cumulative Sum Diagram and its Greatest Convex Minorant or Newton–Puiseux polygon

Theorem 1.1

If X is simply ordered, the slope g^* of the GCM furnishes the isotonic regression of g. Indeed, if f is isotonic on X then

$$\sum_x [g(x) - f(x)]^2 w(x) \geq \sum_x [g(x) - g^*(x)]^2 w(x)$$
$$+ \sum_x [g^*(x) - f(x)]^2 w(x). \quad (1.7)$$

The isotonic regression is unique.

Proof. The difference between the left and right members of (1.7) is

$$2 \sum_x [g(x) - g^*(x)][g^*(x) - f(x)] w(x).$$

Thus it suffices to show that

$$\sum_x [g(x) - g^*(x)][g^*(x) - f(x)] w(x) \geq 0 \quad (1.8)$$

for every isotonic function f on X. The partial summation formula (Abel's formula) yields

$$\sum_{i=1}^{k} [g(x_i) - g^*(x_i)][g^*(x_i) - f(x_i)] w(x_i)$$
$$= \sum_{i=1}^{k} \{[f(x_i) - f(x_{i-1})] - [g^*(x_i) - g^*(x_{i-1})]\}$$
$$\times [G_{i-1} - G^*_{i-1}] w(x_i) + [g^*(x_k) - f(x_k)][G_k - G^*_k],$$

where $x_0 = f(x_0) = g^*(x_0) = G_0 = G^*_0 = 0$. The last term on the right is 0 by (1.4). From (1.5), also $[g^*(x_i) - g^*(x_{i-1})][G_{i-1} - G^*_{i-1}] = 0$, $i = 1, 2, \ldots, k$. Since $G^*_{i-1} \leq G_{i-1}$ and $f(x_{i-1}) \leq f(x_i)$, $i = 1, 2, \ldots, k$, (1.8) follows. To verify uniqueness of the isotonic regression, suppose g_1 is another isotonic function on X such that

$$\sum_x [g(x) - g_1(x)]^2 w(x) \leq \sum_x [g(x) - f(x)]^2 w(x)$$

for each isotonic function f on X. Then since g_1 and g^* are both isotonic,

$$\sum_x [g(x) - g_1(x)]^2 w(x) = \sum_x [g(x) - g^*(x)]^2 w(x).$$

But also, from (1.7), since g_1 is isotonic,

$$\sum_x [g(x) - g_1(x)]^2 w(x) \geq \sum_x [g(x) - g^*(x)]^2 w(x) \\ + \sum_x [g^*(x) - g_1(x)]^2 w(x),$$

which now implies

$$\sum_x [g^*(x) - g_1(x)]^2 w(x) = 0,$$

hence $g^*(x) = g(x)$ for $x \in X$. ∎

The "Pool-Adjacent-Violators" Algorithm

It is clear that if, for some i, $g(x_{i-1}) > g(x_i)$, then the graph of the part of the GCM between points P_{i-2}^* and P_i^* is a straight line segment. Thus the CSD could be altered by connecting P_{i-2} with P_i by a straight line segment, without changing the GCM. This method of successive approximation to the GCM can be described algebraically as the *Pool-Adjacent-Violators algorithm*.

The isotonic regression g^* partitions X into sets on which it is constant, i.e. into *level sets* for g^*, called *solution blocks*. On each of these solution blocks the value of g^* is the weighted average of the values of g over the block, using weights w; i.e., the slope of the side of the GCM corresponding to this block. Thus it suffices to find the solution blocks, i.e. sets of consecutive elements of X on each of which g^* assumes a particular value. In describing the algorithm an arbitrary set of consecutive elements of X will be referred to as a *block*. The algorithm starts with the finest possible partition into blocks, namely the individual points of X, and joins the blocks together step by step until the final partition is reached.

If $g(x_1) \leq g(x_2) \leq \ldots \leq g(x_k)$, then this initial partition is also the final partition, and $g^*(x_i) = g(x_i)$, $i = 1, 2, \ldots, k$. If not, select any of the pairs of violators of the ordering; that is, select an i such that $g(x_i) > g(x_{i+1})$ (see Table 1.2, Figure 1.4). "Pool" ("amalgamate") these two values of g: i.e.,

Table 1.2 Pooling Adjacent Violators; Example

j	1	2	3	4	5
$w(x_j)$	1	2	1	3	1
$g(x_j)$	−1/1	2/2	4/1	−4/3	2/1
First pool			[4 + (−4)]/(1 + 3) = 0/4		
Second pool		(2 + 0)/(2 + 4) = 2/6			
$g^*(x_j)$	−1	1/3	1/3	1/3	2

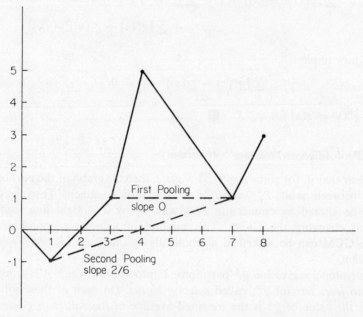

Figure 1.4 Graphical interpretation of pooling adjacent violators

join the two points x_i and x_{i+1} in a block $\{x_i, x_{i+1}\}$ ordered between $\{x_{i-1}\}$ and $\{x_{i+2}\}$, with associated average value

$$[w(x_i)g(x_i) + w(x_{i+1})g(x_{i+1})]/[w(x_i) + w(x_{i+1})]$$

and associated weight $w(x_i) + w(x_{i+1})$.

After each step in the algorithm, the average values associated with the blocks are examined to see whether or not they are in the required order. If so, the final partition has been reached, and the value of g^* at each point of a block is the "pooled" value associated with the block. If not, a pair of

REGRESSION OVER A FINITE SET

adjacent violating blocks is selected, and pooled to form a single block, with associated weight the sum of their weights and associated average value the weighted average of their average values, completing another step of the algorithm.

In order to program this algorithm for a computer, it would of course be necessary to prescribe a rule for selecting the pair of violators to be pooled at each step. One way would be to start at the "lowest" block and proceed to higher blocks until the first violating pair is reached. This is essentially the scheme of J. B. Kruskal who has written a program to carry it out as part of a large program in Kruskal (1964b). This algorithm is presented in Section 2.3.

Example 1.3

The computation is illustrated with a portion of the data from Bhattacharyya and Klotz (1966) (see Example 1.1). X is taken as the set of the first twelve positive integers: $X = \{1, 2, \ldots, 12\}$. For each $x \in X$, $g(x)$ denotes the number of days to freezing of Lake Mendota measured from 23 November in the year $1854 + x$. A warming trend is interpreted as a tendency for $g(x)$

x	1	2	3	4	5	6	7	8	9	10	11	12
$g(x)$	25	13	2	15	14	21	9	33	25	15	21	25

to increase with x. If g is thought of as the sum of a nondecreasing function μ plus a random error, the isotonic regression g^* of g is a least squares estimate of μ. To compute g^* by the Pool-Adjacent-Violators algorithm, it is noted that $g(1) > g(2)$, so $x = 1$ and $x = 2$ are pooled, obtaining the block $\{1, 2\}$ with weight $1 + 1 = 2$ and average value $(25 + 13)/2 = 38/2 = 19$. Now $19 > 2$, so blocks $\{1, 2\}$ and $\{3\}$ are pooled getting block $\{1, 2, 3\}$ with weight $2 + 1 = 3$ and average $(38 + 2)/(2 + 1) = 40/3 = 13.3$.

It could easily have been foreseen that the first three elements would be pooled, since g decreases as x increases through 1, 2, 3. After pooling in every monotonically nonincreasing run of values of g, the data shown in Table 1.3

Table 1.3

x	1	2	3	4	5	6	7	8	9	10	11	12
$g(x)$	25	13	2	15	14	21	9	33	25	15	21	25
First Pooling:		⎯⎯	⎯⎯		⎯⎯		⎯⎯		⎯⎯		⎯⎯			
Weight		3			2		2		3		1	1
Average		40/3			29/2		30/2		73/3		21/1	25/1
		13.3			14.5		15		24.3		21	25

are obtained as the result of a first pass through the data. In the second pass through the data, we note that Av {8, 9, 10} = 24.3 and Av {11} = 21 are in decreasing rather than increasing order; so we pool, getting total weight 3 + 1 = 4 and Av {8, . . ., 11} = 94/4 = 23.5 (see Table 1.4). After this

Table 1.4

Block	{1, 2, 3}	{4, 5}	{6, 7}	{8, 9, 10}	{11}	{12}
Weight	3	2	2	3	1	1
Average	40/3	29/2	30/2	73/3	21/1	25/1
	13.3	14.5	15	24.3	21	25
Second Pooling:	⌊___⌋	⌊__⌋	⌊__⌋	⌊_____	_____⌋	⌊__⌋
Weight	3	2	2	4		1
Average	13.3	14.5	15	94/4		25
				23.5		

pass through the data the averages are in increasing order and the isotonic regression is given by

$$g^*(1) = g^*(2) = g^*(3) = 13.3,$$
$$g^*(4) = g^*(5) = 14.5,$$
$$g^*(6) = g^*(7) = 15,$$
$$g^*(8) = \ldots = g^*(11) = 23.5,$$
$$g^*(12) = 25.$$

Figure 1.5 shows the cumulative sum diagram for Example 1.3. Note that since $w(x) = 1$ for $x \in X$,

$$\sum_{i=1}^{j} w(x_i) = j = x_j$$

in this example. The vertices of the GCM are at 0, 3, 5, 7, 11 and 12, corresponding to the partition into the solution blocks {1, 2, 3}, {4, 5}, {6, 7}, {8, . . ., 11}, {12} which are level sets of g^*. The slopes of the sides of the GCM, 40/3, 29/2, 30/2, 94/4 and 25/1 are the values of g^* at points in the solution blocks.

The reader may wish to verify Table 1.5 giving the isotonic regression of g for the 111 year period (Example 1.1). A comparison of the isotonic regression for 12 years with the isotonic regression for the 111 years illustrates the point that additional data may change the isotonic regression. Thus, for the partial data $g^*(12) = 25$ while for the complete set $g^*(12) = 19.3$.

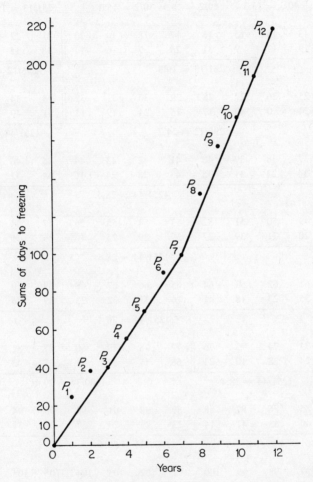

Figure 1.5 Cumulative Sum Diagram (CSD) and Greatest Convex Minorant (GCM) for first twelve freezing dates

Table 1.5

x	1	2	3	4	5	6	7	8	9	10	11	12
$g(x)$	25	13	2	15	14	21	9	33	25	15	21	25
$g^*(x)$		$40/3 = 13.3$		$29/2 = 14.5$		$30/2 = 15.0$			$251/13 = 19.3$			

x	13	14	15	16	17	18	19	20	21	22	23	24
$g(x)$	19	17	9	31	26	7	6	17	48	15	44	28
$g^*(x)$				$251/13 = 19.3$						$296/12 = 24.7$		

x	25	26	27	28	29	30	31	32	33	34	35	36
$g(x)$	24	0	40	17	25	24	19	12	31	40	52	33
$g^*(x)$				$296/12 = 24.7$						$1251/44 = 28.4$		

x	37	38	39	40	41	42	43	44	45	46	47	48
$g(x)$	34	23	11	35	43	28	24	16	34	32	22	32
$g^*(x)$						$1251/44 = 28.4$						

x	49	50	51	52	53	54	55	56	57	58	59	60
$g(x)$	20	21	39	27	39	29	25	16	35	31	50	23
$g^*(x)$						$1251/44 = 28.4$						

x	61	62	63	64	65	66	67	68	69	70	71	72
$g(x)$	35	23	18	41	16	32	32	23	39	26	23	13
$g^*(x)$						$1251/44 = 28.4$						

x	73	74	75	76	77	78	79	80	81	82	83	84
$g(x)$	24	28	10	23	68	17	32	31	27	43	14	35
$g^*(x)$		$1251/44 = 28.4$						$1000/34 = 29.4$				

x	85	86	87	88	89	90	91	92	93	94	95	96
$g(x)$	40	43	41	14	23	25	20	37	28	31	30	18
$g^*(x)$						$1000/34 = 29.4$						

x	97	98	99	100	101	102	103	104	105	106	107	108
$g(x)$	23	37	37	40	19	21	37	16	36	26	23	19
$g^*(x)$						$1000/34 = 29.4$						

x	109	110	111
$g(x)$	27	22	49
$g^*(x)$	$1000/34 = 29.4$		$49/1 = 49.0$

The max-min formulas

Closely related to the graphical representation of the isotonic regression as the slope of the GCM is the formula

$$g^*(x_i) = \max_{s \leq i} \min_{t \geq i} \text{Av}(s, t), \quad (1.9)$$

where

$$\text{Av}(s, t) = \sum_{r=s}^{t} g(x_r)w(x_r) / \sum_{r=s}^{t} w(x_r). \quad (1.10)$$

Equivalent formulas are:

$$g^*(x_i) = \min_{t \geq i} \max_{s \leq i} \text{Av}(s, t), \quad (1.11)$$

$$g^*(x_i) = \max_{s \leq i} \min_{t \geq s} \text{Av}(s, t), \quad (1.12)$$

and

$$g^*(x_i) = \min_{t \geq i} \max_{s \leq t} \text{Av}(s, t). \quad (1.13)$$

One may think of finding the slope, $g^*(x_i)$, of the GCM graphically as follows. Select $s \leq i$ and draw that ray proceeding to the right from P_s which intersects again the graph of partial sums but lies below all others which do so. The slope of this ray is

$$\min_{t \geq s} \sum_{r=s}^{t} g(x_r)w(x_r) / \sum_{r=s}^{t} w(x_r).$$

Repeat this for all other choices of $s \leq i$, selecting finally that for which the resulting minimum slope is as large as possible. A segment of this final ray is part of the graph of the GCM. On the other hand, this is a graphical way of calculating $g^*(x_i)$ by formula (1.12). Similarly each of the other three formulas may also be regarded as a representation of the slope of the GCM at P_s^*.

It is clear from any of the above formulas, and also from the graphical interpretation, that:

1. the addition of a constant to g results in the addition of the same constant to g^*:

$$(g + c)^* = g^* + c; \quad (1.14)$$

and

2. the multiplication of g by a *positive* constant results in the multiplication of g^* by the same constant:

$$(ag)^* = ag^* \quad \text{if} \quad a > 0. \quad (1.15)$$

In fact it will be seen in the corollary to Theorem 1.8 that the constant c in (1.14) may be replaced by any nondecreasing function of g^*, and the constant a in (1.15) by any positive, nondecreasing function of g^*.

The reverse order

Suppose it is required to minimize

$$\sum_x [g(x) - f(x)]^2 w(x)$$

in the class of functions f satisfying

$$f(x_1) \geq f(x_2) \geq \ldots \geq f(x_k).$$

These functions are *antitonic* with respect to the order $x_1 \prec x_2 \prec \ldots \prec x_k$, and *isotonic* with respect to the reverse order on X:

$$x_1 \succ x_2 \succ \ldots \succ x_k, \quad \text{or} \quad x_k \prec x_{k-1} \prec \ldots \prec x_1.$$

With appropriate relabelling, of course, the problem may be recast so that methods and descriptions above apply directly. Without relabelling, such a table as Table 1.1 is turned upside down; for one like Table 1.2, left and right are interchanged. In the graphical interpretation, the slopes g^* must now be *nonincreasing*, and a least concave majorant (LCM) replaces the GCM. In formulas (1.9), (1.11), (1.12) and (1.13), max and min are interchanged. Reversed data from Table 1.2 are used to illustrate (see Table 1.6).

A related problem

Various related problems are solved by isotonic regression. The following Lemma deals with a simple one which will be used to prove Theorem 1.2, which in turn will be used in Chapter 5.

Lemma

Let g be a given function on X and let w and c be given positive functions on X. Define $g_1(x) = g(x)/c(x)$, $x \in X$, and let g_1^* be the isotonic regression of g_1 with weights $c^2 w$. Then $f = cg_1^*$ minimizes

$$\sum_x [g(x) - f(x)]^2 w(x)$$

in the class of functions f on X such that f/c is isotonic.

Proof. This is immediate from the identity

$$\sum_x [g(x) - f(x)]^2 w(x) = \sum_x [g(x)/c(x) - f(x)/c(x)]^2 [c(x)]^2 w(x). \quad \blacksquare$$

REGRESSION OVER A FINITE SET

Table 1.6 Pooling Adjacent Violators—Reverse Order

j	1	2	3	4	5
$w(x_j)$	1	3	1	2	1
$g(x_j)$	2/1	−4/3	4/1	2/2	−1/1
First pool:		[(−4) + 4]/(3 + 1) = 0/4			
Second pool:		(0 + 2)/(4 + 2) = 2/6			
$g^*(x)$	2	1/3	1/3	1/3	−1

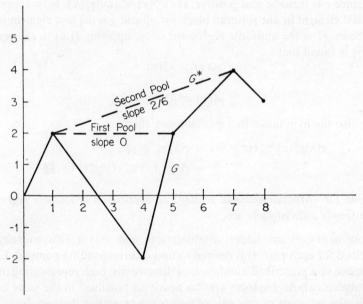

Figure 1.6 Graphical interpretation of Pool-Adjacent-Violators algorithm with reverse order. The antitonic regression, g^*, with respect to the natural order, or isotonic regression with respect to the reversed order, is the slope from the left of the least concave majorant, G^*, of the cumulative sum diagram

Theorem 1.2

Let X be simply ordered, and let g and c be given isotonic functions on X, with $c > 0$. Let w be a given positive function. Define $g_1(x) = g(x)/c(x)$, $x \in X$, and

let g_1^* be the antitonic regression of g_1 with weight function c^2w. Then $f = cg_1^*$ minimizes

$$\sum_x [g(x) - f(x)]^2 w(x)$$

in the class of functions f on X such that f/c is antitonic and f is isotonic.

Proof. From the above Lemma applied to the reverse order it is found that cg_1^* minimizes the weighted sum of squares in the class of functions f such that f/c is antitonic. Since this class includes all isotonic functions f such that f/c is antitonic, to complete the proof of the theorem it suffices to show that cg_1^* is isotonic. Since X is simply ordered, it is enough to show that $c(x)g_1^*(x) \leq c(y)g_1^*(y)$ whenever x immediately precedes y. Two cases are distinguished. Suppose first that x and y are in the same solution block for g_1^*: $g_1^*(x) = g_1^*(y)$. Then since c is isotonic and positive, $c(x)g_1^*(x) \leq c(y)g_1^*(y)$. Now suppose x is the last element in one solution block for g_1^* and y is the first element in the next. Since g_1^* is the antitonic regression of g_1, applying (1.6) to the reverse order it is found that

$$g_1(x) \geq g_1^*(x)$$

and

$$g_1(y) \leq g_1^*(y).$$

Using also the hypothesis that g is isotonic, then

$$c(x)g_1^*(x) \leq c(x)g_1(x) = g(x) \leq g(y)$$
$$= c(y)g_1(y) \leq c(y)g_1^*(y). \quad \blacksquare$$

Example 1.4 *Multidimensional scaling; an application of isotonic regression over a simply ordered finite set*

Suppose n objects are under consideration, and that a "dissimilarity" is prescribed for each pair. It is desired to find n corresponding points in Euclidean space of a prescribed number, t, of dimensions, each representing one of the n objects, whose distances are "as nearly as possible" in the same order. That is, the members of one pair of points are at greater distance from each other than are those of another pair if and only if the objects corresponding to the first pair are more dissimilar than the objects corresponding to the second pair. (Such problems arise, for example, in establishing classifications of artifacts uncovered in archaeological investigations; cf. Hodson, Sneath and Doran (1966). References to other applications of multidimensional scaling are found in Kruskal (1964a). The authors are indebted to D. G. Kendall for calling their attention to these applications of isotonic regression.) Let each of n points in t space be associated with one of the original objects. Let the distances between the ith and jth points be denoted by d_{ij}. The required isotonic relationship may be formalized as follows. Let X denote the set of all pairs $x = (i,j)$, $1 \leq i < j \leq n$. Define a simple order "$\precsim$" on X by the requirement $(i_1, j_1) \precsim (i_2, j_2)$ if $\delta_{i_1, j_1} \leq \delta_{i_2, j_2}$, where $\delta_{i,j}$ denotes the dissimilarity of the pair (i, j). It is hoped to find n points in t space such that $(i_1, j_1) \precsim (i_2, j_2)$ implies $d_{i_1, j_1} \leq d_{i_2, j_2}$; that is, such that distance, d, as a function on X, is isotonic. For example, Table 1.7 and Figure 1.7 represent

Table 1.7

Pair of objects $\{i, j\}$	Dissimilarity score
{1, 2}	1
{1, 3}	4
{2, 3}	2

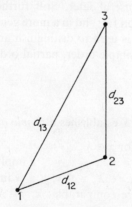

Figure 1.7 Dissimilarities represented by distances

a situation in which $n = 3$, and it is possible to find three points in Euclidean space of two dimensions ($t = 2$) whose distances are in the same order as the given dissimilarities: $d_{12} < d_{23} < d_{13}$, $\delta_{12} < \delta_{23} < \delta_{13}$. However, in some situations this may not be possible. Kruskal (1964a) proposes a measure of goodness of fit to the isotonic relationship by a prescribed set of points, Let $\hat{d}$ denote the isotonic regression of d on X. The "raw stress" of the n points is defined as

$$\sum_{j=1}^{n} \sum_{i=1}^{j-1} (d_{i,j} - \hat{d}_{i,j})^2.$$

A normalized stress is defined as the square root of the quotient of the raw stress by the sum of the squares of the distances. Programming techniques are used to determine n points so as to minimize the stress. Isotonic regression enters the problem only in determining the stress of a given set of points.

1.3 ISOTONIC REGRESSION OVER A QUASI-ORDERED SET

In Section 1.2 isotonic regression was considered with respect to a simple order on a finite set X, but several problems were mentioned in Section 1.1, and are studied in later chapters, in each of which a role is played by an ordering which is not simple: a partial ordering, or a quasi-ordering. In each of these more general situations an isotonic regression still exists. In this section, conditions are given which are necessary and sufficient in order that a function on X be the isotonic regression of a given object function with respect to a quasi-order on X. Also obtained are other properties of isotonic regression which will be useful later. Still further properties of isotonic regression are given in Section 1.4; and in a more general context, in Chapter 7.

The following definition is used to distinguish among the successively less restrictive kinds of order: simple order, partial order, and quasi-order.

Definition 1.2

A binary relation "$\lesssim$" on X establishes a *simple order* on X if

1. it is reflexive: $x \lesssim x$ for $x \in X$;
2. it is transitive: $x, y, z \in X$, $x \lesssim y$, $y \lesssim z$ imply $x \lesssim z$;
3. it is antisymmetric: $x, y \in X$, $x \lesssim y$, $y \lesssim x$ imply $x = y$;
4. every two elements are comparable: $x, y \in X$ implies either $x \lesssim y$ or $y \lesssim x$.

A *partial order* is reflexive, transitive and antisymmetric, but there may be noncomparable elements. A *quasi-order* is reflexive and transitive but not necessarily antisymmetric: it may admit non-identical elements x and y such that $x \lesssim y$ and $y \lesssim x$, as well as elements x and y such that neither $x \lesssim y$ nor $y \lesssim x$. Thus, every simple order is a partial order and every partial order is a quasi-order.

Definition 1.3

A real valued function f on X is *isotonic* with respect to a quasi-ordering "$\lesssim$" on X if $x, y \in X$, $x \lesssim y$ imply $f(x) \leq f(y)$.

Definition 1.4

Let g be a given function on X and w a given positive function on X. An isotonic function g^* on X is an *isotonic regression of g with weights w* if and only if it minimizes in the class of isotonic functions f on X the sum

$$\sum_{x \in X} [g(x) - f(x)]^2 w(x).$$

The existence of an isotonic regression is proved in the corollary to Theorem 1.6, and in a more general context in Chapter 7 (corollary to Theorem 7.2). Theorem 1.3 states that the isotonic regression is unique, and gives a necessary and sufficient condition, (1.18), that an isotonic function be the isotonic regression.

Theorem 1.3

An isotonic regression g^* of g with weights w is an isotonic function on X with respect to "$\lesssim$" and satisfies

$$\sum_{x \in X} [g(x) - g^*(x)][g^*(x) - f(x)] w(x) \geq 0 \qquad (1.16)$$

and

$$\sum_{x \in X} [g(x) - f(x)]^2 w(x) \geq \sum_{x \in X} [g(x) - g^*(x)]^2 w(x) \\ + \sum_{x \in X} [g^*(x) - f(x)]^2 w(x) \qquad (1.17)$$

for every isotonic function f on X.

Conversely, if an isotonic function u satisfies

$$\sum_{x} [g(x) - u(x)][u(x) - f(x)] w(x) \geq 0 \qquad (1.18)$$

for every isotonic function f on X then u is an isotonic regression of g with weights w. There is at most one such isotonic function.

Note that the only special property of the class of isotonic functions on X used in the proof of this theorem is that it is *convex*:

$$\left. \begin{array}{l} \text{if } f_1 \text{ and } f_2 \text{ are isotonic on } X \text{ and if } 0 \leq \alpha \leq 1 \\ \text{then } \alpha f_1 + (1 - \alpha) f_2 \text{ is isotonic.} \end{array} \right\} \qquad (1.19)$$

Proof. It is first shown that if g^* is an isotonic regression of g, i.e., if g^* is isotonic and minimizes

$$\sum_{x} [g(x) - f(x)]^2 w(x)$$

in the class of isotonic f on X, then (1.16) is satisfied. If $0 \leq \alpha \leq 1$ and if f is isotonic then $(1 - \alpha)g^* + \alpha f$ is also isotonic by (1.19). Therefore

$$\sum_x \{g(x) - [(1 - \alpha)g^*(x) + \alpha f(x)]\}^2 w(x)$$

assumes its minimum at $\alpha = 0$, the value of α for which $(1 - \alpha)g^* + \alpha f = g^*$. This sum is a quadratic function of α, whose derivative at $\alpha = 0$ must then be non-negative. Since its derivative at $\alpha = 0$ is just

$$2 \sum_x [g(x) - g^*(x)][g^*(x) - f(x)]w(x),$$

it is seen that (1.16) must hold if g^* is the isotonic regression of g. Inequality (1.17) follows from the identity

$$[g(x) - f(x)]^2 w(x) = [g(x) - g^*(x)]^2 w(x) + [g^*(x) - f(x)]^2 w(x)$$
$$+ 2[g(x) - g^*(x)][g^*(x) - f(x)]w(x).$$

Now suppose u is an isotonic function on X satisfying (1.18). Then as above

$$\sum_x [g(x) - f(x)]^2 w(x) \geq \sum_x [g(x) - u(x)]^2 w(x) + \sum_x [u(x) - f(x)]^2 w(x)$$

for every isotonic f. Since the last term is non-negative, u minimizes

$$\sum_x [g(x) - f(x)]^2 w(x)$$

in the class of isotonic functions f; i.e., u is an isotonic regression of g.

We complete the proof by showing that if two isotonic functions u_1 and u_2 both satisfy (1.18), i.e. if

$$\sum_x [g(x) - u_1(x)][u_1(x) - f(x)]w(x) \geq 0$$

and

$$\sum_x [g(x) - u_2(x)][u_2(x) - f(x)]w(x) \geq 0$$

for all isotonic f, then $u_1 = u_2$. To see this, we substitute u_2 for f in the first inequality, and u_1 for f in the second, then add. Simplifying, we have

$$- \sum_x [u_1(x) - u_2(x)]^2 w(x) \geq 0.$$

Since each term in the sum is obviously non-negative, each term must be 0. But $w(x) > 0$ for each x, so that $u_1(x) = u_2(x)$ for each x, i.e., $u_1 = u_2$. ∎

Theorem 1.4

An isotonic function u on X is the isotonic regression of g with weights w if and only if

$$\sum_{x} [g(x) - u(x)]u(x)w(x) = 0 \qquad (1.20)$$

and

$$\sum_{x} [g(x) - u(x)]f(x)w(x) \leq 0 \qquad (1.21)$$

for all isotonic f. The isotonic regression g^* satisfies also

$$\sum_{x} g(x)w(x) = \sum_{x} g^*(x)w(x). \qquad (1.22)$$

Before proceeding to the proof, this is illustrated with an example involving a simple order. Let $g_1 = 0, g_2 = 2, g_3 = 0$. Let it be required to find numbers g_1^*, g_2^*, g_3^* closest to the given gs in the sense of least squares with equal weights, and satisfying the inequalities $g_1^* \leq g_2^* \leq g_3^*$. These numbers furnish the isotonic regression of g, where $X = \{1, 2, 3\}$, the simple order on X specifies $1 \lesssim 2 \lesssim 3$, $w(x) = 1$, $x = 1, 2, 3$, and $g(1) = g_1 = 0$, $g(2) = g_2 = 2$, $g(3) = g_3 = 0$. The Pool-Adjacent-Violators algorithm is used; since $g_2 \geq g_3$, $\{2\}$ and $\{3\}$ are pooled, and it is found that $g^*(1) = 0, g^*(2) = g^*(3) = 1$. Properties (1.22) and (1.20) (with $u = g^*$) are immediately verified. Property (1.21) is readily verified for the special isotonic functions f_2, f_3 defined by $f_2(1) = 0$, $f_2(2) = f_2(3) = 1$ and $f_3(1) = f_3(2) = 0$, $f_3(3) = 1$. Since every isotonic f is of the form $f = a_1 + a_2 f_2 + a_3 f_3$ where a_2 and a_3 are non-negative, it then follows from (1.22) that (1.21) is satisfied for all isotonic f.

Properties (1.20) and (1.21) of isotonic regression require an additional property of the class of isotonic functions, that it is a *cone*:

$$\left. \begin{array}{l} \text{if } f \text{ is isotonic and } c \text{ is a non-negative} \\ \text{constant then } cf \text{ is isotonic.} \end{array} \right\} \qquad (1.23)$$

Many other classes of functions share properties (1.19) and (1.23). Theorems 1.3 and 1.4 (except for equation (1.22), which uses the additional property that constant functions are isotonic) remain valid if the class of isotonic functions is replaced by any other class of functions on X having these properties; that is, any other convex cone of functions on X. Such a class, for example, is the class of all linear functions. With this substitution, g^* becomes ordinary linear regression. Another example is furnished by the class of all linear combinations of a fixed finite set of functions; the coefficients of some functions in the linear combination may be restricted to be positive.

This remark can be stated formally as follows.

Theorem 1.5

Let X be a finite set and let g and w be given functions on X, $w > 0$. Let $\mathscr{C}$ be a convex cone of functions on X. Then a function u in $\mathscr{C}$ minimizes

$$\sum_x [g(x) - f(x)]^2 w(x)$$

for f in $\mathscr{C}$ if and only if

$$\sum_x [g(x) - u(x)]u(x)w(x) = 0 \qquad (1.20)$$

and

$$\sum_x [g(x) - u(x)]f(x)w(x) \leq 0 \qquad (1.21)$$

for all $f \in \mathscr{C}$. If u satisfies these conditions then also

$$\sum_x [g(x) - f(x)]^2 w(x) \geq \sum_x [g(x) - u(x)]^2 w(x)$$
$$+ \sum_x [u(x) - f(x)]^2 w(x) \qquad (1.24)$$

for all $f \in \mathscr{C}$.

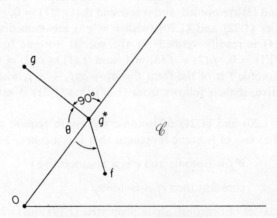

Figure 1.8 Isotonic regression

In Figure 1.8, the angle (two dimensional cone) with vertex O is intended to represent a class of admissible functions f. The orthogonality (1.20) of the vectors $g - g^*$ and g^* is represented by the right angle between segments g^*O and g^*g. Property (1.16), which in the presence of (1.20) is equivalent to (1.21), is represented by the obtuse angle between segments g^*g and g^*f. When the class of admissible functions is a linear space, there is equality in (1.17), which may then be interpreted as the Pythagorean theorem.

Proofs of Theorems 1.4 *and* 1.5. If u satisfies (1.20) and (1.21) then u satisfies also (1.18), so that u is an isotonic regression of g, and satisfies (1.24) by Theorem 1.3. To see conversely that $u = g^*$ satisfies (1.20), set $f(x) = cg^*(x)$ in (1.16), first with a constant $c > 1$ and then with a positive constant $c < 1$. Inequality (1.21) with $u = g^*$ then follows from (1.16) and (1.20) (with $u = g^*$). To prove (1.22) in Theorem 1.4, set $u = g^*$ in (1.21) and let f be first the constant function 1, then the constant function -1. ∎

A further property of isotonic regression to be used in subsequent chapters makes use of yet another property of the class of isotonic functions: it is a *lattice*. In order to state this property explicitly, note first that if a and b are two real numbers, $a \wedge b$ and $a \vee b$ denote respectively the smaller and the larger:

$$a \wedge b = \min(a, b), \qquad a \vee b = \max(a, b).$$

If f and g are two functions on X, the functions $f \wedge g$ and $f \vee g$ are defined by

$$(f \wedge g)(x) = f(x) \wedge g(x), \qquad x \in X,$$
$$(f \vee g)(x) = f(x) \vee g(x), \qquad x \in X.$$

Then to say that the class of isotonic functions on X is a *lattice* is simply to say that

$$\left.\begin{array}{l} \text{if } f \text{ and } g \text{ are isotonic functions} \\ \text{on } X, \text{ so are } f \wedge g \text{ and } f \vee g. \end{array}\right\} \quad (1.25)$$

Theorem 1.6 states that if the isotonic regression g^* of g exists, and if g is bounded above and below by isotonic functions, then g^* has the same bounds. This theorem is used in the following corollary to prove the existence of the isotonic regression, and in Section 2.1 in the proof of a consistency theorem on isotonic estimation. Corresponding theorems in a more general context are 7.2 and the corollaries to Theorems 7.13 and 7.15.

Theorem 1.6

If g_1 and g_2 are isotonic functions on X such that $g_1(x) \leq g(x) \leq g_2(x)$ for $x \in X$, and if g^* is an isotonic regression of g, then also $g_1(x) \leq g^*(x) \leq g_2(x)$ for $x \in X$. In particular, if a and b are constants such that $a \leq g(x) \leq b$ for $x \in X$, then also $a \leq g^*(x) \leq b$ for $x \in X$.

Proof Let f be isotonic and let $h = f \vee g_1$. By (1.25), h is also isotonic. Further, if $f(x) \geq g_1(x)$ for a particular $x \in X$, then $h(x) = f(x)$ so that $g(x) - f(x) = g(x) - h(x)$; while if $f(x) < g_1(x)$ then $0 \leq g(x) - h(x)$

$= g(x) - g_1(x) < g(x) - f(x)$. Thus for all $x \in X$, $[g(x) - h(x)]^2 \leq [g(x) - f(x)]^2$. Therefore

$$\sum_x [g(x) - h(x)]^2 w(x) \leq \sum_x [g(x) - f(x)]^2 w(x),$$

with strict inequality if $f(x) < g_1(x)$ for some $x \in X$. Thus a minimizing function g^* satisfies $g^*(x) \geq g_1(x)$ for all $x \in X$. The proof that $g^*(x) \leq g_2(x)$ is similar. ∎

Corollary

An isotonic regression of g exists.

Proof. Let a and b be bounds on $g: a \leq g(x) \leq b$ for $x \in X$. From the proof of Theorem 1.6 it follows that if f is any isotonic function on X, there is an isotonic function h bounded between a and b such that

$$\sum_x [g(x) - h(x)]^2 w(x) \leq \sum_x [g(x) - f(x)]^2 w(x).$$

Thus it suffices to show that there exists an isotonic function minimizing the sum

$$\sum_x [g(x) - f(x)]^2 w(x)$$

subject to the restrictions $a \leq f(x) \leq b$ for $x \in X$. This sum may be regarded as a continuous function of the k real arguments $f(x)$, $x \in X$, where k is the number of elements in X. As such, it achieves its minimum in the closed bounded subset of k-dimensional Euclidean space R^k determined by the requirements $a \leq f(x) \leq b$ for $x \in X$ and $x_1 \lesssim x_2$ implies $f(x_1) \leq f(x_2)$. This minimizing point in R^k is the required function g^* on X. ∎

Methods of calculating isotonic regression when X carries a nonsimple partial order are discussed in Section 2.3.

Example 1.5 Representation of similarity matrices by trees; an application of isotonic regression with respect to a partial order.

Hartigan (1967b) uses isotonic regression in representing a similarity matrix by a tree. Let there be given a set B of N objects. Let $S(i, j)$ denote a *similarity* between the ith and jth objects; it is the element in the ith row and jth column of a real symmetric matrix, the *similarity matrix*. Particular interest attaches to a similarity matrix associated in a special way with a *tree*. What is here called a *tree* is also called a directed tree or rooted tree in graph theory. A tree T consists of points or *nodes*, and directed segments between certain pairs of nodes, joining a node to its immediate ancestor. Each node has at most one immediate ancestor, and exactly one node (the root) has none. The set B is represented by the *barren nodes*, which are ancestors of no nodes. (Cf. Figure 1.9, borrowed from Hartigan (1967b).) A natural ordering is associated with a tree: if x and y are the same node or if x is an ancestor

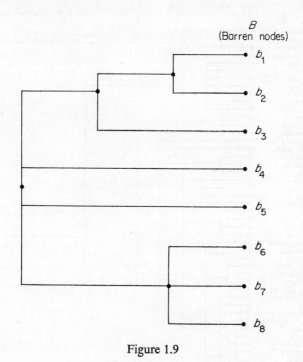

Figure 1.9

(not necessarily immediate ancestor) of y, it is written $x \lesssim y$; x is nearer the root than y. (This is the reverse of the order introduced by Hartigan; we adopt this convention so the function σ below will be isotonic rather than antitonic.) In the set X of nodes of the tree, as a partially ordered set, every pair x, y of nodes has an infimum or greatest lower bound, which is denoted by $x \wedge y$. Its defining properties are: $x \wedge y \lesssim x$, $x \wedge y \lesssim y$; and if also $z \lesssim x, z \lesssim y$ then $z \lesssim x \wedge y$. In other words, it is the farthest node from the root which is at least as close to the root as each of x and y; it is their closest common ancestor. A *node similarity*, σ, of a tree is an isotonic real function on X: $x \lesssim y$ implies $\sigma(x) \leq \sigma(y)$. A similarity matrix S has *exact tree structure* τ on B, denoted by $S \in \tau$, if there is a node similarity σ such that $S(i, j) = \sigma(b_i \wedge b_j)$ for all points b_i and b_j in B: the similarity of two barren nodes is the node similarity of their nearest common ancestor.

Figure 1.10 is also taken from Hartigan (1967b). Each barren node represents one of the fifty states of the United States. In the situation discussed by Hartigan, an element $S(i, j)$ in the similarity matrix represents the

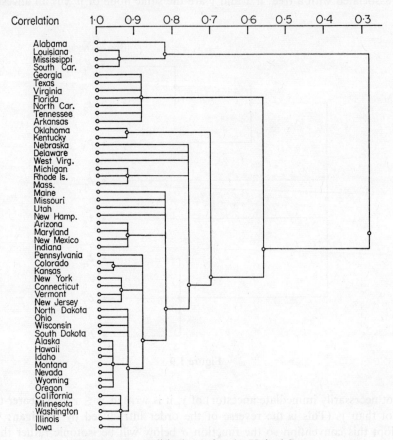

Figure 1.10 Political tree for the United States

correlation of the Democratic percentages of the presidential vote, 1916–1964, of states i and j. The node similarity σ which fits S best in the sense described below (for certain weights) is given by the horizontal scale labelled "correlation": if x is a node, $\sigma(x)$ is the scale marking directly above it on the "correlation" scale. Thus if b_i represents Oklahoma and b_j represents Missouri, $\sigma(b_i, b_j)$ is about 0.7.

Suppose a similarity matrix S is given. It is desired to find a tree τ such that S has exact tree structure τ; or, if no such tree exists, to find one such that S comes as close as possible, as measured by an appropriate criterion, to having exact tree structure τ. Let τ first be fixed. If S has exact tree structure τ, nothing more is required. But if not, Hartigan introduces a measure $\rho(S, \tau)$ of the amount by which S fails to have exact tree structure τ; it is a "squared distance" between S and the closest similarity matrix to S which does have exact tree structure τ:

$$\rho(S, \tau) = \inf \sum_{i,j} W(i,j)[S(i,j) - f(b_i \wedge b_j)]^2$$

where the infimum is taken over all isotonic functions f on X and where $W(i,j)$ are preassigned weights. The tree τ is then varied so as to minimize $\rho(S, \tau)$.

It is in finding $\rho(S, \tau)$ for fixed τ that isotonic regression plays a role. Let Ω denote the set of pairs (i,j) such that $1 \leq i \leq N$, $1 \leq j \leq N$ (i and j may be equal, or either may be the greater). Ω is thought of as a probability space, having the probability

$$W(i,j) / \sum_{(r,s) \in \Omega} W(r,s)$$

associated with the point (i,j) in Ω. Let $\tilde{y}$ denote the random variable on Ω whose value at (i,j) is $S(i,j)$. Let $\tilde{x}$ denote the random element on Ω whose value at (i,j) is $b_i \wedge b_j$, where b_i and b_j are the ith and jth barren nodes. Then the sum

$$\sum_{(i,j) \in \Omega} W(i,j)[S(i,j) - f(b_i \wedge b_j)]^2$$

is proportional to

$$E[\tilde{y} - f(\tilde{x})]^2.$$

For $x \in X$, let

$$\bar{y}(x) = E(\tilde{y} | \tilde{x} = x) = \frac{\sum_{(i,j): b_i \wedge b_j = x} W(i,j) S(i,j)}{\sum_{(i,j): b_i \wedge b_j = x} W(i,j)}.$$

Then, as in Example 1.2,

$$E[\tilde{y} - f(\tilde{x})]^2 = E[\tilde{y} - \bar{y}(\tilde{x})]^2 + E[\bar{y}(\tilde{x}) - f(\tilde{x})]^2.$$

The required node similarity f minimizes $E[\tilde{y} - f(\tilde{x})]^2$ in the class of isotonic functions f on X. It therefore also minimizes, in this class,

$$E[\bar{y}(\tilde{x}) - f(\tilde{x})]^2 = \sum_{x \in X} [\bar{y}(x) - f(x)]^2 w(x),$$

where
$$w(x) = \sum_{(i,j):b_i \wedge b_j = x} W(i,j).$$
That is,
$$\rho(S, \tau) = \sum_{i,j} W(i,j)[S(i,j) - \bar{y}^*(b_i \wedge b_j)]^2,$$
where $\bar{y}^*$ is the isotonic regression of $\bar{y}$.

This section is completed with a discussion of further properties of isotonic regression which will be used in Section 1.4 and in later sections and chapters. The first of these properties is that

$$\sum_x [g(x) - g^*(x)]\psi[g^*(x)]w(x) = 0 \qquad (1.26)$$

for every function ψ, generalizing (1.20) and (1.22). A definition and a lemma prepare the way for the proof.

Definition 1.5

The *average* of g over A with weights w is defined by:
$$\text{Av } A = \sum_{x \in A} g(x)w(x) / \sum_{x \in A} w(x)$$
for $A \neq \phi$, $A \subset X$.

Lemma

If c is a real number and if the subset $[g^* = c]$ of X on which g^* takes the value c is not empty, then $c = \text{Av }([g^* = c])$.

Proof. We have
$$\sum_x [g(x) - g^*(x)]^2 w(x) = \sum_{[g^* \neq c]} [g(x) - g^*(x)]^2 w(x)$$
$$+ \sum_{[g^* = c]} [g(x) - c]^2 w(x).$$

Now consider
$$\sum_{[g^* = c]} [g(x) - t]^2 w(x)$$

as a function of the real argument t; it is quadratic, achieving its minimum at $t = \text{Av }([g^* = c])$. Thus if c were not equal to $\text{Av }([g^* = c])$ an isotonic function could be found coinciding with g^* on the set $[g^* \neq c]$ and differing slightly from g^* on $[g^* = c]$ which would yield a smaller weighted sum of squares. This contradiction establishes the lemma. ∎

Theorem 1.7

For an arbitrary real valued function ψ on the reals,

$$\sum_x [g(x) - g^*(x)]\psi[g^*(x)]w(x) = 0.$$

Proof. This theorem follows from the identity

$$\sum_x [g(x) - g^*(x)]\psi[g^*(x)]w(x) = \sum_c \psi(c) \sum_{[g^*=c]} [g(x) - c]w(x),$$

for by the above Lemma

$$\sum_{[g^*=c]} [g(x) - c]w(x) = 0$$

for each real number c. ∎

Theorem 1.7 leads directly to Theorem 1.8 which provides another characterization of isotonic regression. For that reason it is included here, although it will first be used in Section 2.5. The proof of Theorem 1.8 depends, in part, on a remark concerning a way of rewriting the product of an isotonic function and a non-negative function of another isotonic function.

Lemma

Let f and h be isotonic functions on X and let ρ be a non-negative function of a real argument. Then the product $f\rho(h)$ is the sum of a function of h and an isotonic function on X.

Proof. We shall use the following notation for the indicator function of a set. If $A \subset X$,

$$1_A(x) = 1 \quad \text{for} \quad x \in A,$$
$$1_A(x) = 0 \quad \text{for} \quad x \notin A.$$

For example, if h is a real valued function defined on X and c is a real number,

$$1_{[h \leq c]}(x) = 1 \quad \text{if} \quad h(x) \leq c,$$
$$1_{[h \leq c]}(x) = 0 \quad \text{if} \quad h(x) > c.$$

If B is a set of real numbers then

$$1_B(t) = 1 \quad \text{if} \quad t \in B,$$
$$1_B(t) = 0 \quad \text{if} \quad t \notin B.$$

For example, $1_{(-\infty,c]}(t) = 1$ if $t \leq c$, $1_{(-\infty,c]}(t) = 0$ if $t > c$, and

$$1_{(-\infty,c]}[h(x)] = 1_{[h \leq c]}(x).$$

Set
$$f_m = \min_{x \in X} f(x), \qquad f^m = \max_{x \in X} f(x).$$
Let C denote the set of reals assumed as values by h:
$$C = \{h(x) : x \in X\}.$$
Consider the function ψ of a real argument defined by
$$\psi(t) = (f^m - f_m) \sum_{c \in C} \rho(c) 1_{(-\infty, c]}(t).$$
It will be shown that the function
$$f_1 = f\rho(h) - [f_m \rho(h) + \psi(h)]$$
is isotonic on X. Since the function in brackets is a function of h, this will complete the proof of the lemma. Thus it is necessary to show that $x \lesssim y$ implies $f_1(x) \leq f_1(y)$. We have
$$f_1(y) = [f(y) - f_m]\rho[h(y)] - (f^m - f_m) \sum_{c : c \geq h(y)} \rho(c),$$
and a similar expression is obtained for $f_1(x)$. Since h is isotonic,
$$f_1(y) - f_1(x) = [f(y) - f_m]\rho[h(y)] - [f(x) - f_m]\rho[h(x)]$$
$$+ (f^m - f_m) \sum_{c : h(x) \leq c < h(y)} \rho(c)$$
$$= [f(y) - f_m]\rho[h(y)] + [f^m - f(x)]\rho[h(x)]$$
$$+ (f^m - f_m) \sum_{c : h(x) < c < h(y)} \rho(c) \geq 0. \qquad \blacksquare$$

Theorem 1.8

A necessary and sufficient condition that an isotonic function u on X be the isotonic regression g^* of g is that for every isotonic function f on X and every non-negative function ρ of a real argument,
$$\sum_x [g(x) - u(x)] \rho[u(x)] f(x) w(x) \leq 0.$$

This theorem is related to Corollary A to Theorem 7.18 and Theorem 7.19.

Proof of sufficiency. Suppose the inequality holds for all such ρ and f. Set $\rho \equiv 1$; it is found that (1.21) is satisfied. Now set $f \equiv 1$. On setting first $\rho \equiv 1$, it is found that (1.22) is satisfied. Now set $f \equiv 1$ and $\rho(t) = t$ for real t. It is found that
$$\sum_x [g(x) - u(x)] u(x) w(x) \leq 0.$$

Let M be an upper bound on u, and set $\rho(t) = (M - t) \vee 0$ for real t. Using (1.22) it is found that

$$\sum_x [g(x) - u(x)]u(x)w(x) \geq 0,$$

so that (1.20) is satisfied. An application of Theorem 1.4 completes the proof.

Proof of necessity. By the preceding lemma

$$f\rho(g^*) = \psi(g^*) + f_1,$$

where f_1 is isotonic. By Theorem 1.8

$$\sum_x [g(x) - g^*(x)]\psi[g^*(x)]w(x) = 0.$$

By Theorem 1.4,

$$\sum_x [g(x) - g^*(x)]f_1(x)w(x) \leq 0.$$

Hence

$$\sum_x [g(x) - g^*(x)]\rho[g^*(x)]f(x)w(x) \leq 0. \quad \blacksquare$$

An immediate corollary of Theorem 1.8 generalizes a remark in Section 1.2 (see also Corollary C of Theorem 7.19).

Corollary

If ρ is a non-negative function of a real argument, if ψ is an arbitrary function of a real argument, and if g^* is the isotonic regression of g, then the isotonic regression of $\rho(g^*)g + \psi(g^*)$ is $\rho(g^*)g^* + \psi(g^*)$, provided the latter is isotonic.

Theorem 1.8 yields a theorem of Robertson (1965) which will also be used in Section 2.5.

Theorem 1.9

Let a and b be positive functions on X, and let g^* be the isotonic regression of a/b with weights b with respect to a quasi-order on X. Then $1/g^*$ is the antitonic regression of b/a with weights a, i.e., the isotonic regression of b/a with weights a with respect to the reverse quasi-order on X.

Proof. Since $a/b > 0$, by Theorem 1.6 $g^* > 0$. Let $u = 1/g^*$. By Theorem 1.8 it suffices to show that for every antitonic function f (isotonic with respect to the reverse quasi-order) on X and every non-negative function ρ on the reals,

$$\sum_x \left[\frac{b(x)}{a(x)} - \frac{1}{g^*(x)}\right] \rho\left[\frac{1}{g^*(x)}\right] f(x)a(x) \leq 0.$$

But the left member can be rewritten as

$$\sum_x \left[\frac{a(x)}{b(x)} - g^*(x)\right] \left[\frac{1}{g^*(x)}\right] \rho\left[\frac{1}{g^*(x)}\right] [-f(x)]b(x).$$

Since $-f$ is isotonic, this sum is non-positive by Theorem 1.8. ∎

1.4 GENERALIZED ISOTONIC REGRESSION PROBLEMS

The problem of maximum likelihood estimation of ordered binomial parameters is closely related to the problem of isotonic regression discussed in Sections 1.2 and 1.3. It is also an instance of a class of problems studied in Section 2.4, maximum likelihood estimation of ordered parameters in sampling from distributions belonging to an exponential family. These are all generalized isotonic regression problems in the sense in which that term is used in this section.

Example 1.6 Maximum likelihood estimation of ordered binomial parameters

A problem met in bioassay can be interpreted as the problem of estimating ordered binomial parameters, or alternatively as the problem of estimating the distribution of minimum stimulus which will produce a specified response. Suppose that for each of various levels x of a stimulus (e.g., dose of an insecticide) the probability of a response (e.g., death of the insect) is $\mu(x)$, known to be nondecreasing in x. (Alternatively, $\mu(x)$ is the distribution function of the random variable: "minimum stimulus which will produce the response.") If X is the finite set of stimulus levels, it is simply ordered by the natural order. Suppose for $x \in X$ there are $m(x)$ independent trials at stimulus level x, $a(x)$ responses occur, and $\bar{y}(x) = a(x)/m(x)$ is the average number of responses per trial. If $m(x)$ is large for each x, the ratios $\bar{y}(x)$ can be expected to be in increasing order and are natural estimates of the probabilities $\mu(x)$. But if some of the ratios $\bar{y}(x)$ are in the wrong order, another estimate is required, and the isotonic regression of $\bar{y}$ with weights m is an obvious candidate. It will be seen that this restricted least squares estimate coincides with the maximum likelihood estimate.

Let $b(x) = m(x) - a(x)$ denote the number of nonresponses among the

$m(x)$ trials at stimulus level x. If f denotes an arbitrary function bounded between 0 and 1 on X, the likelihood at f of the sample is

$$\prod_{x \in X} [f(x)]^{a(x)}[1 - f(x)]^{b(x)} \qquad (1.27)$$

and the negative log-likelihood can be written as

$$-\sum_{x} \{\bar{y}(x) \log f(x) + [1 - \bar{y}(x)] \log [1 - f(x)]\} m(x). \qquad (1.28)$$

Thus the solution of the problem of maximum likelihood estimation of μ is the function minimizing (1.28) in the class of isotonic functions on X.

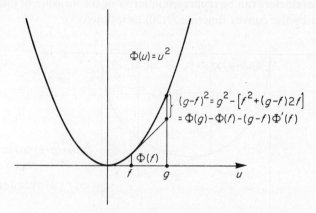

Figure 1.11 Graphical interpretation of the squared error measure of discrepancy

The class of admissible functions f is the same here as for the isotonic regression problem in Example 1.2 (except that here $0 \leq f(x) \leq 1$); but (1.2) and (1.28) look very different. Nevertheless, they are closely related. To see this clearly, consider first a graphical interpretation of (1.2). In Figure 1.11, $(g - f)^2$ is interpreted as the excess of the rise in the graph of the function $\Phi(u) = u^2$ from f to g over the rise of its tangent line at f. The corresponding excess is shown in Figure 1.12 for the function $\Phi(u) = u \log u + (1 - u) \log (1 - u)$. This excess is non-negative for *every convex* function Φ, whether $f < g$ or $g < f$; it is strictly positive if Φ is strictly convex and if $f \neq g$. This excess is denoted by $\Delta_\Phi(g, f)$, or by $\Delta(g, f)$ when Φ is understood:

$$\Delta(g, f) = \Phi(g) - \Phi(f) - (g - f)\varphi(f), \qquad (1.29)$$

where φ is the derivative of Φ at f; or if Φ does not have a derivative at f

(its graph has a "corner" at f), $\varphi(f)$ denotes any number between the left and right derivatives at f. When

$$\begin{aligned}\Phi(u) &= u \log u + (1-u) \log(1-u), \quad 0 < u < 1,\\ \Phi(0) &= 0,\\ \Phi(1) &= 0,\end{aligned}\right\} \qquad (1.30)$$

$$\begin{aligned}\Delta(g,f) = {} & g \log g + (1-g) \log(1-g)\\ & - [g \log f + (1-g) \log(1-f)]\end{aligned}$$

is obtained. Noting that the first two terms on the right of this equation do not involve f, the problem of maximum likelihood estimation of ordered binomial parameters can be rephrased in terms of the measure of discrepancy determined by the convex function (1.30) as follows.

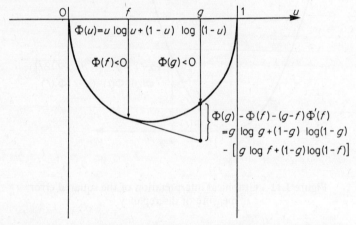

Figure 1.12 Graphical interpretation of the measure of discrepancy determined by the convex function
$$\Phi(u) = u \log u + (1-u) \log(1-u)$$

Binomial extremum problem

To minimize

$$\sum_x \{\bar{y}(x) \log \bar{y}(x) + [1 - \bar{y}(x)] \log[1 - \bar{y}(x)]$$
$$\qquad - \bar{y}(x) \log f(x) - [1 - \bar{y}(x)] \log[1 - f(x)]\} m(x),$$

or

$$\sum_x \Delta[y(x), f(x)] m(x) \qquad (1.31)$$

in the class of isotonic functions f on X.

Since $0 \leq y(x) \leq 1$ for $x \in X$, Theorem 1.6 implies that the isotonic regression $\bar{y}^*$ of $\bar{y}$ also satisfies $0 \leq \bar{y}^*(x) \leq 1$ for $x \in X$. It will be seen as a consequence of Theorem 1.10 that in fact $\bar{y}^*$ minimizes (1.31) (and thus (1.28)) as well as the sum of squares (1.2) in the class of isotonic functions on X, and is thus the maximum likelihood estimate of μ.

We now generalize the Binomial Extremum Problem replacing $u \log u + (1 - u) \log (1 - u)$ by an arbitrary convex function. Let Φ be a convex function which is finite on an interval I containing the range of the function g and infinite elsewhere, and let φ be an arbitrary determination of its derivative (i.e., having at a "corner" any value between the left and right derivatives), defined and finite on I. Then φ is nondecreasing. For numbers u and v, set

$$\left.\begin{aligned}\Delta(u, v) = \Delta_\Phi(u, v) = \Phi(u) - \Phi(v) - (u - v)\varphi(v), \\ \text{if } u, v \in I, \\ \Delta(u, v) = \infty \end{aligned}\right\} \quad (1.32)$$

if $v \in I$, $u \notin I$. The number Δ is non-negative, and may be interpreted as in Figures 1.11 and 1.12 as the excess of the rise of the graph of Φ between v and u over the rise of the line tangent to the graph at v. The fact that Δ is non-negative may be expressed as follows:

$$\Phi(u) \geq \Phi(v) + (u - v)\varphi(v) \quad (1.33)$$

for $u, v \in I$, with strict inequality if $u \neq v$ and if Φ is strictly convex. Clearly the addition of a linear function to Φ does not affect Δ. Also, directly from the definition (1.29),

$$\Delta(r, t) = \Delta(r, s) + \Delta(s, t) + (r - s)[\varphi(s) - \varphi(t)], \quad (1.34)$$

if $s, t \in I$.

Theorem 1.10

If f is isotonic on X and if the range of f is in I then

$$\sum_x \Delta[g(x), f(x)]w(x) \geq \sum_x \Delta[g(x), g^*(x)]w(x)$$
$$+ \sum_x \Delta[g^*(x), f(x)]w(x). \quad (1.35)$$

Consequently g^* minimizes

$$\sum_x \Delta[g(x), f(x)]w(x) \quad (1.36)$$

in the class of isotonic f with range in I, and maximizes

$$\sum_x \{\Phi[f(x)] + [g(x) - f(x)]\varphi[f(x)]\}w(x).$$

The minimizing (maximizing) function is unique if Φ is strictly convex.

Proof. Using (1.34) with $r = g(x)$, $t = f(x)$, $s = g^*(x)$, the last term on the right becomes $[g(x) - g^*(x)]\{\varphi[g^*(x)] - \varphi[f(x)]\}$. Since φ is nondecreasing, $\varphi[f(x)]$ is isotonic if f is, and then

$$\sum_x [g(x) - g^*(x)]\varphi[f(x)]w(x) \le 0$$

by Theorem 1.4. Also

$$\sum_x [g(x) - g^*(x)]\varphi[g^*(x)]w(x) = 0$$

by Theorem 1.7. The first conclusion of the theorem now follows from (1.34). The next conclusion then follows from the fact that $\Delta \ge 0$, and the last from the observation that the first term of $\Delta[g(x), f(x)]$, which is $\Phi[g(x)]$, does not depend on f. The uniqueness conclusion follows from the fact that the second term on the right of (1.35) is strictly positive unless $f = g^*$. ∎

Corollary

Let $\psi_1, \psi_2, \ldots, \psi_p$ be arbitrary real valued functions and let $h_1, h_2, \ldots, h_m$ be isotonic functions on X. Then g^* minimizes

$$\sum_x \Delta[g(x), f(x)]w(x)$$

in the class of isotonic functions f with range in I satisfying any or all of the side conditions

$$\sum_x [g(x) - f(x)]\psi_j[f(x)]w(x) = 0, \quad j = 1, 2, \ldots, p,$$

$$\sum_x f(x)h_j(x)w(x) \ge \sum_x g(x)h_j(x)w(x), \quad j = 1, 2, \ldots, m.$$

Proof. By Theorems 1.4 and 1.7, g^* satisfies all of these side conditions. ∎

Example 1.7 The "geometric" extremum problem

Let $\tilde{y}_i$ ($i = 1, 2, \ldots, k$) have the geometric density

$$f(y_i; p_i) = q_i p_i^{y_i} \quad \text{for} \quad y_i = 0, 1, 2, \ldots,$$

where $p_i + q_i = 1$. Given independent observations $y_1, y_2, \ldots, y_k$, the likelihood is

$$L(y_1, y_2, \ldots, y_k) = \prod_{i=1}^{k} q_i p_i^{y_i}. \qquad (1.37)$$

Let $\mu_i = p_i/q_i$ so that μ_i is the mean of $f(y_i; p_i)$. Making this change of parameter in (1.37),

$$L(y_1, y_2, \ldots, y_k) = \prod_{i=1}^{k} (1 + \mu_i)^{-(1+y_i)} \mu_i^{y_i}.$$

Consider the problem of restricted maximum likelihood estimation:

$$\text{Maximize} \prod_{i=1}^{k} (1 + \mu_i)^{-(1+y_i)} \mu_i^{y_i}$$

subject to $\mu_i \leq \mu_j$ when $i \leq j$. This is equivalent to:

$$\text{Maximize} \sum_{i=1}^{k} \{y_i \log \mu_i - [1 + y_i] \log [1 + \mu_i]\}$$

subject to $\mu_i \leq \mu_j$ when $i \leq j$. If

$$\Phi(u) = u \log u - (1 + u) \log (1 + u)$$

for $u > 0$, we see that

$$\Delta_\Phi(g, f) = g \log g - (1 + g) \log (1 + g) - g \log f$$
$$+ (1 + g) \log (1 + f).$$

Let $X = \{1, 2, \ldots, k\}$, $g(i) = y_i$, $f(i) = \mu_i$, and $w(i) = 1$, $i = 1, 2, \ldots, k$. Noting that the first two terms of the formula above for $\Delta_\Phi(g, f)$ do not involve f, we see from Theorem 1.10 that the isotonic regression $g_i^* = y_i^*$ solves the problem.

Example 1.8 The "Poisson" extremum problem

Suppose that a system (i.e., computer, etc.) is put into operation and n_1 failures or errors occur during the first time period $[0, T_1]$. The device may then be modified so as to improve performance. Subsequently n_2 failures may occur in the second time interval of length T_2 and so on for k time periods. It is assumed that the number of failures, say N_i, observed during the ith time period, has a Poisson distribution with mean $\lambda_i T_i$; i.e.,

$$\Pr(N_i = j) = (\lambda_i T_i)^j e^{-\lambda_i T_i}/j!, \qquad j = 0, 1, \ldots.$$

Let $(n_1, n_2, \ldots, n_k)$ be the sample failure outcome in k time periods. The likelihood of this event is

$$\prod_{i=1}^{k} (\lambda_i T_i)^{n_i} e^{-\lambda_i T_i}/n_i!. \tag{1.38}$$

Since the system is believed to be improving, it is desired to maximize the likelihood (1.38) subject to $\lambda_1 \geq \lambda_2 \geq \ldots \geq \lambda_k$. An equivalent problem is to:

$$\text{Maximize} \sum_{i=1}^{k} \left[\frac{n_i}{T_i} \log \lambda_i - \lambda_i \right] T_i \tag{1.39}$$

subject to $\lambda_1 \geq \lambda_2 \geq \ldots \geq \lambda_k$. It can be seen as follows that the solution $\hat{\lambda}_1, \ldots, \hat{\lambda}_k$ of this extremum problem satisfies

$$\sum_{i=1}^{k} \left(\frac{n_i}{T_i} - \hat{\lambda}_i\right) T_i = 0. \tag{1.40}$$

If $\hat{\lambda}_1 \geq \ldots \geq \hat{\lambda}_k$ and if $\rho > 0$ then also $\rho\hat{\lambda}_1 \geq \ldots \geq \rho\hat{\lambda}_k$. Thus

$$\sum_{i=1}^{k} \left[\frac{n_i}{T_i} \log \rho\hat{\lambda}_i - \rho\hat{\lambda}_i\right] T_i$$

achieves its maximum as a function of ρ at $\rho = 1$. On setting its derivative at $\rho = 1$ equal to zero (1.40) is obtained.

Let $\Phi(u) = u \log u$ so that

$$\Delta_\Phi(g, f) = g \log g - g \log f - (g - f).$$

Since the first term does not depend on f, it follows from Theorem 1.10 that the isotonic regression g^* of g maximizes

$$\sum_{x} [g(x) \log f(x) + g(x) - f(x)] w(x)$$

in the class of isotonic positive functions f. By the Corollary to Theorem 1.10 (with $\psi = 1$) it also maximizes

$$\sum_{x} [g(x) \log f(x) - f(x)] w(x) \tag{1.41}$$

in the class of isotonic positive functions f satisfying

$$\sum_{x} [g(x) - f(x)] w(x) = 0. \tag{1.42}$$

Let $X = \{1, 2, \ldots, k\}$, $g(i) = n_i/T_i$, $f(i) = \lambda_i$ and $w(i) = T_i$, $i = 1, 2, \ldots, k$. It is found then that the isotonic regression g^* maximizes (1.41) subject to (1.42) and f positive, isotonic, and hence also (1.39) subject to (1.40) and

$\lambda_1 \geq \ldots \geq \lambda_k$. Here the isotonicity is with respect to the reverse of the natural order on X. Computational methods discussed in Section 1.2 are available; a formula is

$$g_i^* = \min_{s \leq i} \max_{t \geq i} \sum_{j=s}^{t} n_j / \sum_{j=s}^{t} T_j, \quad i = 1, 2, \ldots, k.$$

Example 1.9 The "gamma" extremum problem

Let $\alpha_1, \alpha_2, \ldots, \alpha_k$ be given positive numbers, and let x_i have a gamma density

$$f(x_i; \lambda_i, \alpha_i) = (1/\lambda_i)^{\alpha_i} \frac{x_i^{\alpha_i - 1} e^{-x/\lambda_i}}{\Gamma(\alpha_i)} \tag{1.43}$$

for $\lambda_i > 0$, $i = 1, 2, \ldots, k$. Let $x_1, x_2, \ldots, x_k$ be independent observations from $f(x_1; \lambda_1, \alpha_1), \ldots, f(x_k; \lambda_k, \alpha_k)$ respectively (each may be the sum of independent observations from a gamma distribution). The problem of maximum likelihood estimation of $\lambda_1, \lambda_2, \ldots, \lambda_k$ under order restrictions which require $\lambda_{(\cdot)}$ to be isotonic, as a function of its index, with respect to some partial order is considered:

$$\text{Maximize} \sum_{i=1}^{k} [-\log \lambda_i - x_i/\alpha_i \lambda_i] \alpha_i \tag{1.44}$$

subject to $\lambda_{(\cdot)}$ isotonic.

Let $\Phi(u) = -\log u$ so that

$$\Delta_\Phi(g, f) = -\log g + \log f + (g - f)(1/f). \tag{1.45}$$

The problem:

$$\text{Minimize} \sum_x \Delta_\Phi[g(x), f(x)] w(x) \tag{1.46}$$

subject to f isotonic, is equivalent to:

$$\text{Minimize} \sum_x [\log f(x) + g(x)/f(x)] w(x) \tag{1.47}$$

subject to f isotonic. Let $X = \{1, 2, \ldots, k\}$, $g(i) = x_i/\alpha_i$, $f(i) = \lambda_i$, and $w(i) = \alpha_i$, $i = 1, 2, \ldots, k$. By Theorem 1.10, the isotonic regression g^* of g solves (1.47) and hence (1.44).

Example 1.10 Maximizing a product

Maximum likelihood estimation of ordered multinomial parameters (Example 2.1) leads to the Poisson extremum problem. More generally, suppose it is required to maximize the product of several factors, given relations of order and equality among them and a linear side condition. After grouping the factors required to be equal, the problem takes the form: given numbers $c(x)$ and positive numbers $r(x)$, $x \in X$, a number s, and a partial order on X, to maximize

$$\prod_{x \in X} [f(x)]^{r(x)} \tag{1.48}$$

in the class of isotonic functions f on X satisfying

$$\sum_x c(x)f(x) = s. \tag{1.49}$$

Set

$$g(x) = sr(x)/[c(x) \sum_{t \in X} r(t)],$$

and

$$w(x) = r(x)/g(x) = c(x) \sum_{t \in X} r(t)/s, \quad x \in X. \tag{1.50}$$

Then the problem is that of maximizing

$$\sum_x g(x)w(x) \log f(x) \tag{1.51}$$

in the class of isotonic functions f on X satisfying

$$\sum_x [g(x) - f(x)]w(x) = 0. \tag{1.52}$$

The solution is the isotonic regression g^* of g with weights w. In computing g^*, the weighted average

$$\sum_{x \in A} g(x)w(x) / \sum_{x \in A} w(x)$$

of g over a subset A of X takes the form

$$s \sum_{x \in A} r(x)/R \sum_{x \in A} c(x),$$

where

$$R = \sum_{x \in X} r(x).$$

1.5 EXTREMUM PROBLEMS ASSOCIATED WITH DUAL CONVEX CONES

There is another class of extremum problems solved by isotonic regression, in which convex functions appear in the sum to be minimized, but the inequalities restricting the solution have a different form. An instance of such a problem has in fact already been met in the remark in Section 1.2 that the

EXTREMUM PROBLEMS WITH CONVEX CONES   47

isotonic regression is the slope of the path along which a taut string lies, which joins the endpoints of the CSD and is nowhere above the CSD. This section is begun by describing formally the taut string problem. In Proposition 1.1 and its Corollary it will be verified that the solution is given directly by an isotonic regression. Applications of Proposition 1.1 to a production planning problem and an inventory problem appear in Examples 1.12 and 1.13, and an application to estimation of stochastically ordered distributions in Section 2.5 (see also Section 1.6).

Example 1.11   The taut string

As in Section 1.2, suppose $X = \{x_1, \ldots, x_k\}$, $x_1 \prec x_2 \prec \ldots \prec x_k$, and let g and w be given functions on X, $w > 0$. The CSD of g is the set of points in the Cartesian plane consisting of the origin, $P_0:(0, 0)$ and the points P_i with coordinates (W_i, G_i), where

$$W_i = \sum_{j=1}^{i} w(x_j), \qquad G_i = \sum_{j=1}^{i} g(x_j)w(x_j),$$

$i = 1, 2, \ldots, k$. Imagine a taut string which joins P_0 and P_k and passes nowhere above the CSD. For each $x_i \in X$, consider the slope $f(x_i)$ of the segment of the string arriving at P_i from the left. Let F_i be the ordinate of the string at P_i:

$$F_i = \sum_{j=1}^{i} f(x_j)w(x_j).$$

In order for the string to be taut, the slope f, as a function on X, must minimize the total length of the string,

$$\sum_{x} [1 + f^2(x)]^{\frac{1}{2}} w(x), \tag{1.53}$$

in the class of functions f on X satisfying

$$F_i \leq G_i, \qquad i = 1, 2, \ldots, k - 1 \tag{1.54}$$

(the string starts at P_0 and passes through or below each point P_i) and

$$F_k = G_k \tag{1.55}$$

(the string passes through the endpoint P_k of the CSD; see Figure 1.2).

Conditions (1.54) and (1.55) can be given a geometric interpretation in the space of functions on X. To do this, let f' denote the function $g - f$:

$$f'(x) = g(x) - f(x), \qquad x \in X.$$

Then (1.54) and (1.55) are equivalent to

$$\sum_{j=i}^{k} f'(x_j)w(x_j) \leq 0, \qquad i = 2, 3, \ldots, k \tag{1.56}$$

and

$$\sum_{j=1}^{k} f'(x_j)w(x_j) = 0. \tag{1.57}$$

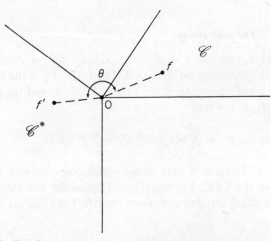

Figure 1.13 Dual convex cones—a schematic diagram. The inner product $\Sigma_x f(x)f'(x)w(x)$ is represented by $|f||f'|\cos\theta$, which is nonpositive when f and f' are in $\mathscr{C}$ and $\mathscr{C}^*$ respectively

The class $\mathscr{C}_0^*$ of all functions on X satisfying (1.56) and (1.57) is a convex cone in the class of all functions on X: it is *convex*, that is

$$\left.\begin{array}{l} \text{if } f_1' \text{ and } f_2' \text{ are in } \mathscr{C}_0^* \text{ (satisfy (1.56)} \\ \text{and (1.57)) and if } 0 \leq \alpha \leq 1 \text{ then} \\ \alpha f_1' + (1-\alpha)f_2' \text{ is in } \mathscr{C}_0^*; \end{array}\right\} \tag{1.58}$$

and it is a *cone*:

$$\left.\begin{array}{l} \text{if } f' \text{ is in } \mathscr{C}_0^* \text{ and } c \text{ is a non-negative} \\ \text{constant then } cf' \text{ is in } \mathscr{C}_0^*. \end{array}\right\} \tag{1.59}$$

We saw in Section 1.3 that properties (1.19) and (1.23) are expressed by the statement that the class of all functions on X isotonic with respect to a partial order on X is also a convex cone. It will be observed in the Remark below that if $\mathscr{C}_0$ is the class of all functions on $X = \{x_1, \ldots, x_k\}$ isotonic with respect to the simple order $x_1 \prec x_2 \prec \ldots \prec x_k$ (i.e., satisfying $f(x_1) \leq \ldots \leq f(x_k)$) then the convex cone $\mathscr{C}_0^*$ defined above is the dual cone of $\mathscr{C}_0$ according to the following definition.

Definition 1.6

Let X be a finite set and let w be a given positive weight function on X. Let $\mathscr{C}$ be a convex cone of functions on X, i.e., a class of functions on X satisfying (1.58) and (1.59) with $\mathscr{C}_0^*$ replaced by $\mathscr{C}$. The class $\mathscr{C}^*$ of all functions f' on X satisfying

$$\sum_x f(x)f'(x)w(x) \leq 0 \qquad (1.60)$$

for all $f \in \mathscr{C}$ is a convex cone, the *dual cone* (also called *polar cone* or *negative conjugate cone*) of $\mathscr{C}$. It is also clear that $\mathscr{C}$ is the dual cone of $\mathscr{C}^*$.

Remark

If $X = \{x_1, \ldots, x_k\}$ is simply ordered, $x_1 \prec x_2 \prec \ldots \prec x_k$, and if $\mathscr{C}_0$ is the convex cone of isotonic functions on X, then the class $\mathscr{C}_0^*$ of functions on X satisfying (1.56) and (1.57) is the dual cone of $\mathscr{C}_0$. To see this, fix i_0, and consider the function f_0 defined on X by:

$$f_0(x_i) = 0 \quad \text{for} \quad i = 1, 2, \ldots, i_0 - 1,$$

$$f_0(x_i) = 1 \quad \text{for} \quad i = i_0, i_0 + 1, \ldots, k.$$

This isotonic function is the *indicator function* of the set $U_0 = \{x_{i_0}, x_{i_0+1}, \ldots, x_k\}$; the set U_0 is called an *upper set* in Chapter 2. If

$$\sum_x f(x)f'(x)w(x) \leq 0$$

for all f in $\mathscr{C}_0$ then in particular

$$\sum_x f_0(x)f'(x)w(x) \leq 0.$$

On setting $i_0 = 2, 3, \ldots, k$, (1.56) is obtained. The constant function identically equal to 1 on X, the indicator function of X, is isotonic, as is its negative, -1. Setting $f = 1$ and $f = -1$ in the inequality

$$\sum_x f(x)f'(x)w(x) \leq 0$$

yields (1.57). Since *every* isotonic function on X is a positive combination of these indicators and -1, (1.56) and (1.57) imply also

$$\sum_x f(x) f'(x) w(x) \leq 0$$

for all $f \in \mathscr{C}_0$. Thus $\mathscr{C}_0^*$ is the dual cone of $\mathscr{C}_0$. ∎

In the terminology of dual cones, the taut string problem may be restated as follows.

Taut string problem

To minimize

$$\sum_x [1 + f^2(x)]^{\frac{1}{2}} w(x)$$

in the class of functions f such that $g - f \in \mathscr{C}_0^*$.

Having described g^* in Section 1.2 in terms of the taut string, one would expect g^* to be the solution of this problem. Indeed, it is now easy to verify that g^* solves each of the wider class of problems in which the convex function $(1 + u^2)^{\frac{1}{2}}$ is replaced by an arbitrary convex function. A more general proposition is proved first which will have other applications as well (see Example 1.13 and Section 2.5).

Proposition 1.1

Let $\mathscr{K}$ be an arbitrary class of functions on X. Let Φ be convex and φ a determination of its derivative. Let g_1 be a given function on X, and w a positive weight function. Then a function $u \in \mathscr{K}$ minimizes

$$\sum_x \{\Phi[f(x)] - f(x) g_1(x)\} w(x), \quad f \in \mathscr{K},$$

provided there is a function g_2 on X such that

$$\sum_x [u(x) - g_2(x)]\{\varphi[u(x)] - g_1(x)\} w(x) = 0 \quad (1.61)$$

and

$$\sum_x [f(x) - g_2(x)]\{\varphi[u(x)] - g_1(x)\} w(x) \geq 0 \quad (1.62)$$

for all $f \in \mathscr{K}$. The minimizing function is unique if Φ is strictly convex.

Proof. If $f \in \mathscr{K}$ then

$$\Phi(f) - f g_1 - [\Phi(u) - u g_1] = \Delta(f, u) + (f - u)\varphi(u) - (f - u) g_1,$$

by equation (1.29). Since $\Delta \geq 0$, we have

$$\Phi(f) - fg_1 - [\Phi(u) - ug_1] \geq (f - u)[\varphi(u) - g_1]$$
$$= [(f - g_2) + (g_2 - u)][\varphi(u) - g_1],$$

with strict inequality if Φ is strictly convex and $f \neq u$. It now follows from (1.61) and (1.62) that

$$\sum_x \{\Phi[f(x)] - f(x)g_1(x)\}w(x) \geq \sum_x \{\Phi[u(x)] - u(x)g_1(x)\}w(x),$$

with strict inequality if Φ is strictly convex and if there is an $x \in X$ such that $f(x) \neq u(x)$. ∎

Corollary

Let g and w be given functions on X, $w(x) > 0$ for $x \in X$. Let Φ be a convex function and φ a determination of its derivative. Let $\mathscr{C}$ be the class of functions on X isotonic with respect to a quasi-order on X, and let $\mathscr{C}^*$ be its dual cone. Then g^* minimizes

$$\sum_x \Phi[f(x)]w(x) \qquad (1.63)$$

in the class of functions f such that $g - f \in \mathscr{C}^*$. The minimizing function is unique if Φ is strictly convex.

Proof. We apply Proposition 1.1 with $g_1 \equiv 0$, $g_2 \equiv g$, and $\mathscr{K}$ the class of functions f such that $g - f \in \mathscr{C}^*$. It suffices to verify that $u \equiv g^*$ satisfies (1.61) and (1.62). The former becomes

$$\sum_x [g^*(x) - g(x)]\varphi[g^*(x)]w(x) = 0,$$

a conclusion of Theorem 1.7. The latter is

$$\sum_x [f(x) - g(x)]\varphi[g^*(x)]w(x) \geq 0.$$

Since g^* is isotonic and φ is nondecreasing, also $\varphi(g^*) \in \mathscr{C}$. Then if $f \in \mathscr{K}$ we have $g - f \in \mathscr{C}^*$ and the above inequality follows from the definition of the dual cone, completing the proof. ∎

As has been seen above, the fact that g^* solves the Taut String problem is the special case of the Corollary to Proposition 1.1 in which $\Phi(u) = (1 + u^2)^{\frac{1}{2}}$.

Example 1.12 A production planning problem

The dual cone of the cone of isotonic functions on X enters naturally in the following problem. Suppose an amount y_i of goods is to be manufactured during the ith period, $i = 1, 2, \ldots, k$. Suppose the cost of manufacturing an amount y is $\Phi(y)$, where Φ is a convex function. Suppose furthermore there is a demand for r_i goods in the ith period, $i = 1, 2, \ldots, k$. The problem is to choose the production vector $(y_1, \ldots, y_k)$ so as to minimize total production cost,

$$\sum_{i=1}^{k} \Phi(y_i) \tag{1.64}$$

subject to the requirement that demand never exceeds supply:

$$\sum_{j=1}^{i} (y_j - r_j) \geq 0, \qquad i = 1, 2, \ldots, k - 1, \tag{1.65}$$

and that total demand is equal to total production:

$$\sum_{j=1}^{k} (y_j - r_j) = 0. \tag{1.66}$$

To restate the problem in the terminology used in this chapter, set $X = \{1, 2, \ldots, k\}$, $x_i = i$, $i = 1, 2, \ldots, k$, $f(x) = y_x$, $x \in X$, $g(x) = r_x$, $x \in X$, $f'(x) = g(x) - f(x)$, $x \in X$. In these terms, (1.65) and (1.66) can be rewritten:

$$\left.\begin{array}{l} \sum_{j=i}^{k} f'(x_j) \geq 0, \\[1em] \sum_{j=1}^{k} f'(x_j) = 0. \end{array}\right\} \tag{1.67}$$

A function f' satisfies (1.67) if and only if $-f'$ satisfies (1.56) and (1.57) with $w(x) = 1$ for $x \in X$. Thus (1.67) describes a convex cone which is just the negative of the cone $\mathscr{C}_1^*$ discussed in the above Remark: $\mathscr{C}_1^* = -\mathscr{C}_0^*$. The cone $\mathscr{C}_1^*$ is the dual cone of $\mathscr{C}_1 = -\mathscr{C}_0$, where $\mathscr{C}_1$ consists of the functions f on X satisfying

$$f(x_1) \geq f(x_2) \geq \ldots \geq f(x_k).$$

These are the isotonic functions with respect to the simple order on X:

$$x_1 \succ x_2 \succ \ldots \succ x_k. \tag{1.68}$$

The objective function is

$$\sum_{i=1}^{k} \Phi(y_i) = \sum_{x} \Phi[f(x)] w(x).$$

EXTREMUM PROBLEMS WITH CONVEX CONES

By the corollary to Proposition 1.1, the isotonic regression g^* of $g = r$ with weights $w = 1$ with respect to the simple order (1.68) solves the problem.

The GCM and Pool-Adjacent-Violators techniques are available for computing g^* in this problem, after relabelling. Alternatively, one may pool for the reverse order, and interpret the solution via the LCM, as in Section 1.2 under the heading: "The Reverse Order".

Example 1.13 An inventory problem

Assume that there are k facilities in series. Random demand, $\tilde{D}$, occurs at Facility k only. Shortages are passed up the line to $k - 1$, then to $k - 2$ if $k - 1$ is out of stock, etc., until a given demand is either met or backlogged at Facility 1. Note that Facility k may be a retail store, whereas Facility 1 may be a distant warehouse where storage costs are much lower.

At the beginning of a typical time period, each facility places an order with immediate delivery at unit cost $c \geq 0$. Set

$$X = K = \{1, 2, \ldots, k\}.$$

For $x \in K$, let $f(x)$ denote the *total* stock on hand at facilities $x, x + 1, \ldots, k$, *after* ordering at the beginning of a period. Note that

$$f(1) \geq f(2) \geq \ldots \geq f(k),$$

i.e., f is isotonic with respect to the reverse of the natural order on K. An ordering policy consists of choosing a function f isotonic with respect to this simple order,

$$1 \succ 2 \succ \ldots \succ k.$$

Let $p(x) > 0$ denote the unit shortage cost at Facility x. Let $H(x)$ denote the unit storage cost at Facility x, and set

$$h(x) = H(x) - H(x - 1), \quad x \in K,$$

where $H(0) = 0$. Note that the restriction

$$h(x) \geq 0, \quad x \in K,$$

is reasonable, since distant warehouses may have smaller holding costs than retail stores within a city. The following costs are incurred:

ordering cost, $cf(1)$,

expected shortage or *penalty costs,* $\sum_x p(x) E[\tilde{D} - f(x)]^+$,

where a^+ denotes the larger of a and 0:
$$a^+ = \max(a, 0) = a \vee 0;$$
and *expected holding cost*,
$$\sum_x h(x) E[f(x) - \tilde{D}]^+.$$
To verify this formula for expected holding cost, note that if
$$f(j) - D > 0 \quad \text{and} \quad f(j+1) - D \leq 0,$$
then the holding cost is
$$[f(j) - D]H(j) + [f(j-1) - f(j)]H(j-1)$$
$$+ \ldots + [f(1) - f(2)]H(1)$$
$$= \sum_{x=1}^{j} [f(x) - D]h(x) = \sum_x h(x)[f(x) - D]^+.$$
It is desired to minimize total costs
$$cf(1) + \sum_x \{p(x) E[\tilde{D} - f(x)]^+ + h(x) E[f(x) - \tilde{D}]^+\} \quad (1.69)$$
in the class of functions f isotonic with respect to the reverse order. Set
$$F(t) = \Pr\{\tilde{D} \leq t\};$$
for simplicity, F is assumed strictly increasing, and its inverse is denoted by F^{-1}. Set
$$\mu = E(\tilde{D}).$$
For arbitrary real a,
$$a^+ = a + (-a)^+,$$
so that
$$E[\tilde{D} - f(x)]^+ = \mu - f(x) + E[f(x) - \tilde{D}]^+.$$
Set
$$\Phi(u) = E[u - \tilde{D}]^+$$
$$= \int_0^u F(t)\,dt, \quad u \geq 0,$$
$$= 0, \quad u < 0.$$
Then Φ is convex, and
$$\varphi(u) = \Phi'(u) = F(u), \quad u \geq 0;$$

(1.69) becomes

$$\sum_x \{[p(x) + h(x)]\Phi[f(x)] - p(x)f(x)\} + cf(1) + \mu \sum_x p(x).$$

Thus the problem is reduced to that of minimizing

$$\sum_x \{\Phi[f(x)] - f(x)g_1(x)\}w(x)$$

in the class of functions f on K isotonic with respect to the reverse order, where

$$w(x) = p(x) + h(x), \quad x \in K,$$

$$g_1(1) = [p(1) - c]/[p(1) + h(1)], \quad g_1(x) = p(x)/[p(x) + h(x)],$$
$$x = 2, 3, \ldots, k.$$

Let g_1^* be the isotonic regression of g_1 with respect to the reverse order, and set

$$u(x) = F^{-1}[g_1^*(x)].$$

By Theorem 1.4,

$$\sum_x [g_1(x) - g_1^*(x)]f(x)w(x) \leq 0$$

for every function f isotonic with respect to the reverse order on K. On setting $g_2 = 0$ in Proposition 1.1, it is found that (1.62) is verified. Also, Theorem 1.7 yields (1.61) in the present context. Thus it follows from Proposition 1.1 that $u(x)$ is the desired solution. ∎

1.6 COMPLEMENTS

The use of the GCM for representing isotonic regression is due to W. T. Reid (cf. Brunk (1956)). It was used in a related context (cf. Section 5.6) by Grenander (1956). The Pool-Adjacent-Violators algorithm and the max-min formulas were given by Ayer and coworkers (1955) in a study of maximum likelihood estimation of completely ordered parameters (bioassay). The least squares property of the solution, in effect identifying it with the isotonic regression, is noted also in that paper. The problem was studied independently by van Eeden (1956), who admitted also partial orders and gave methods of solution. The Pool-Adjacent-Violators algorithm for complete order is implicit in van Eeden's results. It is also Miles's method $\mathscr{A}_1$ (1959); cf. also Bartholomew (1959a).

If X is simply ordered and if the slope g of the CSD is antitonic (non-increasing) then the CSD is concave. Thus the graph of the GCM is a line segment, and its slope, g^*, is constant:

$$g^*(x) = G(x)/W(x) = \sum_{t \in X} g(t)w(t) / \sum_{t \in X} w(t), \qquad x \in X.$$

In this case (1.21) becomes

$$\sum_x g(x)f(x)w(x) \leq [\sum_x g(x)w(x)][\sum_x f(x)w(x)] / \sum_x w(x),$$

referred to as Tchebycheff's Inequality in Hardy, Littlewood and Pólya (1952) page 43. Integral generalizations given by Ciesielski (1958) are related in the same way to the corollary to Theorem 7.14 and Corollary A to Theorem 7.8.

Among interesting properties of the isotonic regression g^* not yet mentioned is that of being the isotonic function most highly correlated with g. A precise statement of this property in a more general setting is given in Section 7.3.

Leik and Gove (1969) consider the isotonic regression problem in the setting of Example 1.2 in attempting an analogue for this situation of the correlation ratio.

The term "monotone regression" appears in Lombard and Brunk (1963) and in Kruskal (1965; page 253). "Isotonic", i.e. "order preserving", is preferred to "monotonic" which may be interpreted as "order preserving" *or* "order reversing".

Some recent work on multidimensional scaling, the subject of Example 1.4, is to be found in Horan (1969), in Klahr (1969), and in Stenson and Knoll (1969). Another reference is Shepard (1966).

The class of extremum problems solved by the isotonic regression, of which those for finite X are described in Theorem 1.10, was discovered independently by W. T. Reid and by Brunk, Ewing and Utz (1957). Application of the results of this paper to maximum likelihood estimation of ordered parameters was made in Brunk (1955). (In spite of the apparent discrepancy in dates, the work of Brunk, Ewing and Utz (1957) preceded that of Brunk (1955).) Further statistical applications are noted in Brunk (1965).

An interesting special case of Theorem 1.10 has $\Phi(u) = -u^\alpha$: if $0 < \alpha < 1$, $g(x) > 0$ for $x \in X$, then g^* minimizes

$$\sum_x \{[(1 - \alpha)f(x) + \alpha g(x)]w(x) / [f(x)]^{1-\alpha}\}$$

in the class of isotonic f.

More general problems of maximizing strictly unimodal functions were also studied by van Eeden (1957a, 1957b and 1958). (Robertson and Waltman (1968) remove the restriction to "strictly" unimodal functions; see also Brunk and Johansen (1970).) In addition to generalizing the objective function to be minimized, van Eeden considers also more general restrictions on the minimizing functions: there may be given functions a and b on X, and the function f required not only to be isotonic but also to satisfy

$$a(x) \leq f(x) \leq b(x), \quad x \in X.$$

Let this be called the *bounded isotonic regression* problem. It may be supposed without loss of generality that a and b are isotonic. Three numbers are associated with each subset A of X:

$$\text{Av } A = \sum_{x \in A} g(x)w(x) / \sum_{x \in A} w(x);$$

$$\bar{a}(A) = \sup_{x \in A} a(x);$$

and

$$\underline{b}(A) = \inf_{x \in A} b(x).$$

If $\bar{a}(A) \leq \underline{b}(A)$, define

$$M(A) = \text{mid } (\text{Av } A, \bar{a}(A), \underline{b}(A))$$
$$= [\bar{a}(A) \vee \text{Av } A] \wedge \underline{b}(A)$$
$$= [\text{Av } A \wedge \underline{b}(A)] \vee \bar{a}(A).$$

Then it can be shown that appropriately modified versions of the Pool-Adjacent-Violators algorithm and the max-min formulas, in which "Av" is replaced by M, give the solution to the bounded isotonic regression problem. This same function also minimizes, in the class of isotonic functions bounded by a and b, the sum

$$\sum_{x \in X} \Delta[g(x), f(x)]w(x)$$

where

$$\Delta(u, v) = \Phi(u) - \Phi(v) - (u - v)\Phi'(v)$$

and where Φ is strictly convex. For partial orders, M may replace Av also in modifications of the Minimum Violator algorithm, the Minimum Lower Sets algorithm, and the max-min formulas, all in Section 2.3. Proofs can be based on appropriate generalizations of Theorem 2.4. These results may be considered as specializing results of van Eeden (1958) to the class of objective functions considered here.

For the case of simple order, it may be supposed without loss of generality that X is the set K of the first k positive integers, $K = \{1, 2, \ldots, k\}$, endowed with the natural order. Given a function g on K, i.e., given numbers $g_1, \ldots, g_k$ and given also positive weights $w_1, \ldots, w_k$, the isotonic regression problem is to minimize

$$\sum_{i=1}^{k}(g_i - f_i)^2 w$$

in the class of functions f satisfying

$$f_{i+1} \geq f_i, \quad i = 1, 2, \ldots, k-1.$$

Now suppose $a_1, a_2, \ldots, a_{k-1}$ are given real numbers, and it is required to minimize

$$\sum_{i=1}^{k}(g_i - f_i)^2 w_i$$

in the class $\mathscr{D}(a_1, \ldots, a_{k-1})$ of functions f on K such that

$$f_{i+1} \geq f_i + a_i, \quad i = 1, 2, \ldots, k-1.$$

This problem also is solved by isotonic regression as follows. Set

$$g_1' = g_1, \quad g_i' = g_i - \sum_{j=1}^{i-1} a_j, \quad i = 2, 3, \ldots, k,$$

and associate with each $f \in \mathscr{D}(a_1, \ldots, a_{k-1})$ the function f' on K defined by

$$f_1' = f_1, \quad f_i' = f_i - \sum_{j=1}^{i-1} a_j, \quad i = 2, 3, \ldots, k.$$

Then

$$\sum_{i=1}^{k}(g_i - f_i)^2 w_i = \sum_{i=1}^{k}(g_i' - f_i')^2 w_i,$$

and $f \in \mathscr{D}(a_1, \ldots, a_{k-1})$ if and only if f' is isotonic. Thus the required minimizing function in $\mathscr{D}(a_1, \ldots, a_{k-1})$ takes values given by

$$(g'^*)_1, \quad (g'^*)_i + \sum_{j=1}^{i-1} a_j, \quad i = 2, 3, \ldots, k,$$

where $(g'^*)_i$ is the value at $x = i$ of the isotonic regression of g'.

W. T. Reid (1968) has studied the situation in which X is an interval of the real line, endowed with the natural order. A function f on X is isotonic if and only if its difference quotients $[f(x_2) - f(x_1)]/(x_2 - x_1)$ are non-negative, and Reid has shown how to treat a wide class of minimization problems under the additional condition that the difference quotients be bounded above by a given positive number. The situation is of course much simpler when X is finite and the objective function is a weighted sum of squares; this case is discussed briefly here. $X = K = \{1, 2, \ldots, k\}$ is taken again with the natural order. Let g be a given function on K, $g(i) = g_i$, $i = 1, 2, \ldots, k$, and w_i, $i = 1, 2, \ldots, k$, given positive weights. Suppose given also extended reals $a_1, \ldots, a_{k-1}$ and $b_1, \ldots, b_{k-1}$ with $a_i \leq b_i$, $i = 1, 2, \ldots, k-1$. Consider the problem of minimizing

$$\sum_{i=1}^{k} (g_i - f_i)^2 w_i$$

in the class $\mathscr{D}(\mathbf{a}, \mathbf{b})$ ($\mathbf{a}' = (a_1, \ldots, a_{k-1})$, $\mathbf{b}' = (b_1, \ldots, b_{k-1})$) of functions f on K such that

$$a_i \leq f_{i+1} - f_i \leq b_i, \quad i = 1, 2, \ldots, k-1.$$

(The isotonic regression problem is obtained on setting $a_i = 0$, $b_i = \infty$ for $i = 1, 2, \ldots, k-1$.) Associate with each f on K its cumulative sum function

$$F_i = \sum_{j=1}^{i} f_j w_j, \quad i = 1, 2, \ldots, k.$$

Then

$$G_i = \sum_{j=1}^{i} g_j w_j, \quad i = 1, 2, \ldots, k,$$

and if g^* is the minimizing function in $\mathscr{D}(\mathbf{a}, \mathbf{b})$,

$$G_i^* = \sum_{j=1}^{i} g_j^* w_j.$$

Theorem 1.11

If $g^* \in \mathscr{D}(\mathbf{a}, \mathbf{b})$ then

$$\sum_{i=1}^{k} (g_i - g_i^*)^2 w_i \leq \sum_{i=1}^{k} (g_i - f_i)^2 w_i \quad \text{for all} \quad f \in \mathscr{D}(\mathbf{a}, \mathbf{b}) \quad (1.70)$$

if and only if

$$\sum_{i=1}^{k} (g_i - g_i^*) w_i = 0, \quad i.e., \quad G_k^* = G_k; \quad (1.71)$$

$$G_{i-1}^* < G_{i-1} \quad \text{implies} \quad g_i^* - g_{i-1}^* = a_{i-1}, \quad (1.72)$$

$$i = 2, 3, \ldots, k,$$

and

$$G_{i-1}^* > G_{i-1} \quad \text{implies} \quad g_i^* - g_{i-1}^* = b_{i-1}, \quad (1.73)$$

$$i = 2, 3, \ldots, k.$$

Note that if, for some i, $a_i = -\infty$ then (1.72) requires $G_{i-1}^* \geq G_{i-1}$; and if $b_i = \infty$ then (1.73) implies $G_{i-1}^* \leq G_{i-1}$.

ISOTONIC REGRESSION

Proof. Suppose g^* is a minimizing function in $\mathcal{D}(\mathbf{a}, \mathbf{b})$, so that (1.70) holds. Then

$$0 \leq \sum_{i=1}^{k} (g_i - f_i)^2 w_i - \sum_{i=1}^{k} (g_i - g_i^*)^2 w_i$$

$$= \sum_{i=1}^{k} (g_i^* - f_i)^2 w_i + 2 \sum_{i=1}^{k} (g_i - g_i^*)(g_i^* - f_i) w$$

for every $f \in \mathcal{D}(\mathbf{a}, \mathbf{b})$.

Now if ε is a real number then $g^* + \varepsilon \in \mathcal{D}(\mathbf{a}, \mathbf{b})$. Set $f = g^* + \varepsilon$ in the above inequality, obtaining

$$0 \leq \varepsilon^2 \sum_{i=1}^{k} w_i + 2\varepsilon \sum_{i=1}^{k} (g_i - g_i^*) w_i$$

for every real ε. This implies (1.71). It was noted in connection with Theorem 1.3 that only the convexity of the class of isotonic functions was used in the proof of that theorem. The class $\mathcal{D}(\mathbf{a}, \mathbf{b})$ is convex, so that here also

$$\sum_{i=1}^{k} (g_i - g_i^*)(g_i^* - f_i) w_i \geq 0 \quad \text{for every} \quad f \in \mathcal{D}(\mathbf{a}, \mathbf{b}). \quad (1.74)$$

The key identity in this argument is

$$\left.\begin{array}{l}\sum_{i=1}^{k} (g_i - g_i^*)(g_i^* - f_i) w \\ = \sum_{i=1}^{k} [(f_i - f_{i-1}) - (g_i^* - g_{i-1}^*)](G_{i-1} - G_{i-1}^*) w_i \\ + (g_k^* - f_k)(G_k - G_k^*) \\ (G_0 = G_0^* = 0). \end{array}\right\} \quad (1.75)$$

By (1.71), $G_k = G_k^*$. Now suppose there is an $i > 1$ such that $G_{i-1}^* < G_{i-1}$, and $g_i^* - g_{i-1}^* > a_{i-1}$. For this fixed i, choose c real such that $a_{i-1} \leq c < g_i^* - g_{i-1}^*$ and define f by $f_j = g_j^*$, $j = 1, 2, \ldots, i-1$, $f_i = f_{i-1} + c$, $f_j = f_{j-1} + g_j^* - g_{j-1}^*$, $j = i+1, \ldots, k$. Then from (1.75),

$$\sum_{i=1}^{k} (g_i - g_i^*)(g_i^* - f_i) w_i = [c - (g_i^* - g_{i-1}^*)](G_{i-1} - G_{i-1}^*) w_i < 0,$$

contradicting (1.74) and thus establishing (1.72). The proof of (1.73) is similar.

Now suppose conversely that (1.71), (1.72) and (1.73) are all satisfied. Then from (1.75) it follows that

$$\sum_{i=1}^{k}(g_i - g_i^*)(g_i^* - f_i)w_i \geq 0$$

for every $f \in \mathscr{D}(\mathbf{a}, \mathbf{b})$, so that by Theorem 1.3 (for the class $\mathscr{D}(\mathbf{a}, \mathbf{b})$ rather than the class of isotonic functions), (1.70) holds. ∎

This theorem can be made the basis of an algorithm for computing g^*, which takes a particularly simple form and admits a graphical interpretation when $b_i = \infty, i = 1, 2, \ldots, k-1$. When also $a_i = 0, i = 1, 2, \ldots, k-1$, this is the greatest convex minorant interpretation described in Section 1.2.

Theorem 1.9 is a special case of a theorem of Robertson (1965), of which a somewhat more general version appears in Chapter 7 as Theorem 7.19.

Professor Arthur Veinott kindly pointed out the production planning problem (Example 1.12) to the authors; a generalized form of it is discussed by Arrow, Karlin and Scarf (1958; pages 9–11 and Chapter 4). Example 1.13 appears in Veinott (1971); this paper also relates the estimation of ordered parameters to the solution of a separable quadratic network flow problem. Dantzig (1971) treats the case $\Phi(u) = u^2/2$ of the corollary to Proposition 1.1, but with a more general class of restrictions.

The case $g_2 = 0, \mathscr{K} = \mathscr{C}$ (convex cone of isotonic functions) of Proposition 1.1 can be obtained via a change of variable in Theorem 1.10. Let Φ, Ψ be conjugate, strictly convex functions:

$$\Phi(u) = \sup_t [ut - \Psi(t)],$$

$$\Psi(t) = \sup_u [ut - \Phi(u)].$$

Let φ and ψ respectively denote their derivatives. Then

$$\Phi(u) - u\varphi(u) = -\Psi[\varphi(u)].$$

Thus (1.29) can be rewritten

$$\Delta_\Phi(g_1, f_1) = \Phi(g_1) - g_1\varphi(f_1) + \Psi[\varphi(f_1)].$$

Since φ is strictly increasing, $f_1 \in \mathscr{C}$ if and only if $f \equiv \varphi(f_1) \in \mathscr{C}$. Then the problem of Theorem 1.10 is to minimize

$$\sum_x \{\Psi[f(x)] - f(x)g_1(x)\}w(x), \qquad f \in \mathscr{C}.$$

The solution is $u = \varphi(g_1^*)$, which satisfies (1.61) and (1.62) with φ replaced by ψ when $g_2 = 0$.

This problem may be regarded as a dual of the problem of the corollary to

Proposition 1.1. Theorem 31.4 of Rockafellar (1970) can be applied (which in turn applies Fenchel's Duality theorem), or saddle-point theory used, to obtain the following result.

Let w be a positive weight function on X. Let $\mathscr{C}$ be a convex cone of real valued functions on X, and let $\mathscr{C}'$ be its dual cone. For each $x \in X$, let Φ_x, Ψ_x be conjugate, strictly convex functions, with derivatives φ_x and ψ_x respectively. Let u and u' be functions on X satisfying

$$u'(x) = \psi_x[u(x)], \quad u = \varphi_x[u'(x)], \quad x \in X, \quad u \in \mathscr{C}, \quad u' \in -\mathscr{C}',$$

and

$$\sum_x u(x)u'(x)w(x) = 0.$$

Then u minimizes

$$\sum_x \Psi_x[f(x)]w(x)$$

in $\mathscr{C}$, and u' minimizes

$$\sum_x \Phi_x[f'(x)]w(x)$$

in $-\mathscr{C}'$.

Alternatively, the proof is very quick when, as is assumed here for convenience, Φ and Ψ are finite and strictly convex. For

$$\Psi_x(f) \geq \Psi_x(u) + (f - u)\psi_x(u).$$

so that

$$\sum_x \Psi_x[f(x)]w(x) \geq \sum_x \Psi_x[u(x)]w(x) + \sum_x f(x)u'(x)w(x)$$
$$- \sum_x u(x)u'(x)w(x).$$

The last term on the right is zero. Suppose $f \in \mathscr{C}$; since $u' \in -\mathscr{C}'$, the second term on the right is non-negative. Thus u minimizes

$$\sum_x \Psi_x[f(x)]w(x)$$

in $\mathscr{C}$. The proof of the second statement is similar.

Now consider the case in which Φ and Ψ are conjugate strictly convex functions, g is a given function on X, and

$$\Psi_x(t) = \Psi(t) - tg(x),$$
$$\Phi_x(u) = \Phi[u + g(x)], \quad x \in X.$$

Let $\mathscr{C}$ be the class of functions isotonic with respect to a quasi-order on X, and $\mathscr{C}^*$ its dual. Let g^* be the isotonic regression of g with weights w. Set $u = \varphi(g^*)$. Applying the above theorem it is found that u minimizes

$$\sum_x \{\Psi[h(x)] - h(x)g(x)\}w(x)$$

for $h \in \mathscr{C}$; and g^* minimizes

$$\sum_x \Phi[f(x)]w(x)$$

in the class of functions f such that $g - f \in \mathscr{C}^*$.

CHAPTER 2

Estimation Under Order Restrictions

2.1 ISOTONIZATION OF ESTIMATES

Several problems were mentioned in Section 1.1 in which parameters were to be estimated subject to order restrictions. In each case an estimate was at hand which would have been appropriate in the absence of the order conditions. Let such an estimate be called a "basic estimate". When this basic estimate does not satisfy the required order restrictions, one may consider replacing it by its isotonic regression.

An instance occurs in estimating a stimulus-response curve. For each of various levels x of a stimulus the probability of a response is $\mu(x)$, known to be nondecreasing in x. If X is the finite set of stimulus levels, then X is completely ordered and the unknown function μ on X is isotonic. Suppose that for x in X, there are $m(x)$ independent trials at the stimulus level, and that

$$\bar{y}(x) = (\text{number of responses})/m(x)$$

is the average number of responses per trial. The function $\bar{y}(x)$, $x \in X$, is the natural estimate of $\mu(x)$, but it may not be nondecreasing. Its isotonic regression $\bar{y}^*$ with weights $m(x)$, $x \in X$, can then be used as an estimate of μ, and will be better than $\bar{y}$ itself in the sense of least squares (cf. Theorem 2.1 below). Similar remarks apply to other examples mentioned in Section 1.1. In most of those examples the isotonic regression of the basic estimate turns out to coincide with the maximum likelihood estimate under the order restrictions. An example is given here (Example 2.2) in which there is no question of a maximum likelihood estimate, for no stochastic model is even specified. But even in such a case the isotonic regression of the basic estimate is better than the basic estimate itself, in the least squares sense.

Theorem 2.1

Let μ be an unknown function on a finite set X, known to be isotonic with respect to a quasi-order on X. Let $w(x)$, $x \in X$, be a set of positive weights. Let g be an estimate of μ. Let g^* be the isotonic regression of g with weights w. Then

$$\sum_x [\mu(x) - g^*(x)]^2 w(x) \leq \sum_x [\mu(x) - g(x)]^2 w(x).$$

Proof. This is an immediate consequence of Inequality (1.17), valid for all isotonic f:

$$\sum_x [g(x) - f(x)]^2 w(x) \geq \sum_x [g(x) - g^*(x)]^2 w(x)$$
$$+ \sum_x [g^*(x) - f(x)]^2 w(x),$$

since μ is isotonic and

$$\sum_x [g(x) - g^*(x)]^2 w(x) \geq 0;$$

f is simply replaced in (1.17) by μ. ∎

In some of the most interesting applications to be discussed in later chapters (see Chapter 5) the basic estimators are not consistent. Nevertheless, it is of interest to note here that if the basic estimator *is* consistent, so is its isotonic regression. Consider then a set X, not necessarily finite, endowed with a partial order. Let μ be an isotonic function on X. Let $\{X_n\}$ be an expanding class of finite subsets of X: $X_n \subset X_{n+1} \subset X$; set $X' = \cup_n X_n$. (In particular applications, X_n may coincide with X for each n.) Let μ_n denote the restriction of μ to X_n: $\mu_n(x) = \mu(x)$ for $x \in X_n$. Let $\tilde{g}_n$ be an estimator of μ, $n = 1, 2, \ldots$

Theorem 2.2

For $n = 1, 2, \ldots,$ let $w_n(x)$, $x \in X_n$ (or $\tilde{w}_n(x)$, $x \in X_n$) be positive reals (or positive random variables). Let $\{\tilde{g}_n, X_n\}$ be a consistent (strongly consistent) sequence of estimators of μ at each $x \in X' = \cup_n X_n$. Let μ be isotonic on X. Let $\tilde{g}_n^*$ denote the isotonic regression of $\tilde{g}_n$ on X_n with weights w_n ($\tilde{w}_n$), $n = 1, 2, \ldots$ Then $\{\tilde{g}_n^*, X_n\}$ also is a consistent (strongly consistent) sequence of estimators of μ at each $x \in X'$.

Proof. The conclusion is immediate from Theorem 1.6, for $\mu(x) - \varepsilon < \tilde{g}_n(x) < \mu(x) + \varepsilon$, $x \in X_n$ implies $\mu(x) - \varepsilon < \tilde{g}_n^*(x) < \mu(x) + \varepsilon$, $x \in X_n$. A similar theorem can be stated for uniform consistency. ∎

Example 2.1 *Maximum likelihood estimation of ordered multinomial parameters*

Let random variables $\tilde{r}_1, \tilde{r}_2, \ldots, \tilde{r}_k$ have the multinomial distribution with parameters $p_1, p_2, \ldots, p_k$ and n:

$$\Pr(\tilde{r}_1 = r_1, \ldots, \tilde{r}_k = r_k) = n! \prod_{i=1}^{k} p_i^{r_i} / \prod_{i=1}^{k} r_i! \quad \text{if} \quad \sum_{i=1}^{k} r_i = n.$$

For convenience, set $X = \{1, 2, \ldots, k\}$, and write $p_x = p(x)$, $r_x = r(x)$ for $x \in X$. Suppose that in addition to

$$\sum_{i=1}^{k} p_i = 1$$

the parameters are known to satisfy certain order conditions, which are assumed expressed by requiring p as a function on X to be isotonic with respect to a given quasi-order on X. Now suppose observed values $r_1, \ldots, r_k$ of the random variables are available. In the absence of order restrictions, natural estimates of $p_1, \ldots, p_k$ would be $r_1/n, \ldots, r_k/n$. Thus r/n, as a function on X, can be thought of as a basic estimate of p. Under the order restrictions then, the isotonic regression of r/n with prescribed weights might be considered as an estimate. It will be seen that the isotonic regression of r/n with *equal* weights is, in fact, the maximum likelihood estimate of p.

The problem of maximum likelihood estimation of $p_1, \ldots, p_k$ subject to the stated side conditions is precisely the problem of Example 1.10, with $c(x) = 1$ for $x \in X$, and $s = 1$: to maximize

$$\prod_{x} [f(x)]^{r(x)}$$

in the class of isotonic f satisfying

$$\sum_{x} f(x) = 1.$$

The solution is the isotonic regression of $\{g(x) = r(x)/n, x \in X\}$ with equal weights. (While (1.50) gives

$$w(x) = r(x)/g(x) = \sum_{t \in X} r(t),$$

any equal weights can be used, for example $w(x) = 1$, $x \in X$.) Algorithms for calculating the isotonic regression are given in Sections 1.2 and 2.3 for simple order, and in Section 2.3 for partial order.

2.2 ISOTONIC ESTIMATION OF REGRESSION

In the stimulus-response situation (Sections 1.1, 2.1) and in the problem of ordered means of normal distributions (Sections 1.1, 3.2, Example 2.4), as in Examples 1.1 and 1.2, it is desired to estimate a regression function in the presence of order restrictions. In general, suppose there is a probability distribution associated with each x in a finite set X; let $\mu(x)$ denote its mean. Suppose also that for each x a random sample $\{y_j(x), j = 1, 2, \ldots, m(x)\}$, is available from the distribution. Suppose further that a quasi-order on X is prescribed, and that μ is known to be isotonic with respect to this quasi-order. The sample average is a natural estimate of $\mu(x)$ for each x, but the sample averages may not satisfy the required order conditions. In this case the isotonic regression of the sample regression function is proposed as an estimate of μ.

Definition 2.1

The *sample regression function* $\bar{y}$ is defined by

$$\bar{y}(x) = \frac{1}{m(x)} \sum_{j=1}^{m(x)} y_j(x), \qquad x \in X. \tag{2.1}$$

It may be regarded as a *basic estimate* of the regression function μ.

Definition 2.2

Let w be a positive function on X. The *isotonized sample regression function with weights* w is the isotonic regression $\bar{y}^*$ of $\bar{y}$. That is, it minimizes

$$\sum_x [\bar{y}(x) - f(x)]^2 w(x)$$

in the class of isotonic functions f. Algorithms for its computation in the case of a simple order on X are discussed in Section 1.2, and for partial order in Section 2.3.

It follows from Theorem 2.1 that if the unknown regression function μ is known to be isotonic, then $\bar{y}^*$ is certainly closer to μ in the sense of least squares than is $\bar{y}$.

If X is held fixed but $m(x) \to \infty$ for each $x \in X$, the sample regression function $\bar{y}$ is a strongly consistent estimator of the regression function μ. It follows then from Theorem 2.2 that the isotonic regression $\bar{y}^*$ of $\bar{y}$ is also strongly consistent.

Example 2.2

Results and analysis of an experiment relating juice composition of mandarin oranges to unacceptance by tasters were reported in Lombard and Brunk (1963). Samples of the fruit picked at various stages of maturity were available. Juice composition was characterized by two figures: x_1, "soluble solids", and x_2, "ratio of soluble solids to acid". For each sample a determination was made of x_1 and x_2. One of the measures of acceptability of the fruit considered was probability of acceptance. (Actually, "unacceptance" was used in the study. Required modifications are obvious.) In illustrating the foregoing definitions and terminology in this example, let $x = (x_1, x_2)$ be an ordered pair of positive numbers, representing juice composition of a particular fruit sample. X is the set of all such pairs x for all the juice samples tasted. The associated random variable $\tilde{y}$ can be described verbally as follows. Let a taster score $y = 1$ if the juice is acceptable, $y = 0$ if it is not. Then $\tilde{y}$ is the score given a sample with juice composition $x = (x_1, x_2)$ by a taster selected at random from a hypothetical population of tasters. For each $x \in X$, $\tilde{y}$ has the Bernoulli distribution: $\Pr[\tilde{y} = 1 | \tilde{x} = x] = \mu(x)$, the probability that the taster will find the sample acceptable; and $\Pr[\tilde{y} = 0 | \tilde{x} = x] = 1 - \mu(x)$.

A useful partial ordering of the set X is the ordering by components: if $a = (a_1, a_2)$ and $b = (b_1, b_2)$ are in X, then $a \lesssim b$ if and only if $a_1 \leq b_1$ and $a_2 \leq b_2$. It is reasonable to assume $\mu(x)$ is isotonic; here this means simply that probability of acceptance is nondecreasing as a function of x_1, soluble solids, and also as a function of x_2, ratio of soluble solids to acid.

If for each x in X thirty tasters are chosen independently and at random from a population of tasters, then $y_j(x)$ is the score given to the fruit sample of composition x by the jth taster, $j = 1, 2, \ldots, m(x) = 30$. The sample regression function is $\bar{y}(x)$, the proportion of the tasters who found the fruit with juice composition x acceptable. If $\bar{y}(x)$ is found to be isotonic, it furnishes the natural estimate of $\mu(x)$. If not, the isotonic sample regression function $\bar{y}^*$ may be computed and used as an estimate of μ.

If the same thirty tasters are used for each fruit sample, then $\tilde{y}$ can be allowed to represent the average of the scores given by the thirty tasters in randomly selected circumstances to a fruit sample of composition x. Each average

$$\bar{y}(x) = \sum_{j=1}^{30} y_j(x)/30$$

can be thought of as a sample of size 1 from the distribution of $\tilde{y}$ given $\tilde{x} = x$. The conditional mean $\mu(x)$ represents the expected score of the panel on a fruit sample of composition x. If $\bar{y}(x)$ had been found to be isotonic it would have furnished the natural estimate of $\mu(x)$. It was not, and the isotonic sample regression (called monotone regression in the study; the terms are equivalent in this context) was computed and used as an estimate of μ. Since the partial order here is not a simple order, the algorithm discussed in Section 1.2 does not apply. Algorithms for partial orders form the subject matter of Section 2.3.

Approximate level curves for the estimate $\bar{y}^*$ are shown in Figure 2.1.

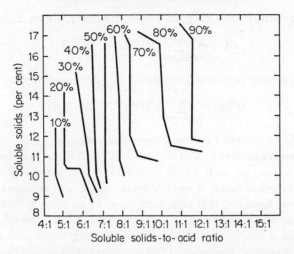

Figure 2.1 Approximate level curves of the isotonic regression $\bar{y}^*$ as an estimate of probability of acceptance for any reason, as rated by an untrained taste panel, of Dancy, Kinnow and Kara mandarin fruits, based on soluble solids and soluble solids/acid ratio

Estimation of cumulative regression

In the case of a simple order, estimation of an isotonic regression function is closely related to estimation of a cumulative regression function. Indeed, in the graphical interpretation discussed in Section 1.2, the ordinates of the cumulative sum diagram may be thought of as values of a cumulative sample regression function, a basic estimate of the cumulative regression function as defined below. The ordinates of the greatest convex minorant (GCM) provide an estimate of the cumulative regression function under the isotonicity condition. Let $X = \{x_1, \ldots, x_k\}$ where x_i is real, $i = 1, 2, \ldots, k$, and let the order on X be specified by $x_1 \prec x_2 \prec \ldots \prec x_k$. If μ is the regression function, the *cumulative regression function with weight function w* may be defined by

$$M(x) = \sum_{t \leq x, t \in X} \mu(t) w(t), \quad x \in X. \tag{2.2}$$

Similarly

$$\bar{Y}(x) = \sum_{t \leq x, t \in X} \bar{y}(t) m(t), \quad x \in X, \tag{2.3}$$

is the *cumulative sample regression function with weights $m(x)$*. For a given weight function w, the *cumulative weight function* is defined by

$$W(x) = \sum_{\leq x, t \in X} w(t), \quad x \in X, \tag{2.4}$$

and for an arbitrary function g, its *cumulative sum function*, G, is defined by

$$G(x) = \sum_{t \leq x, t \in X} g(t)w(t), \qquad x \in X. \tag{2.5}$$

If g^* is the isotonic (antitonic) regression of g with weights w, then

$$G^*(x) = \sum_{t \leq x, t \in X} g^*(t)w(t), \qquad x \in X, \tag{2.6}$$

is the *cumulative isotonic (antitonic) regression function*.

As was remarked above, if X is held fixed and if the number of observations $m(x) \to \infty$ for each $x \in X$, then $\bar{y}(x) \to \mu(x)$ almost surely for each $x \in X$. On the other hand, if $m(x)$ is small, $\bar{y}(x)$ is not expected to be near $\mu(x)$. Suppose, however, that near each point of an interval there are many points in X at which observations are made, and that μ is continuous on the interval. Then it can be shown (Brunk (1970)) that, under mild hypotheses, the cumulative sample regression $\bar{Y}$ is a consistent estimator of the cumulative regression M (see also Section 5.2). Other situations are discussed in Chapter 5 (see in particular Section 5.3) in which a function g, regarded as a basic estimator of an unknown function, is not consistent, but (perhaps after a change of variable) the cumulative sum function *is* a consistent estimator of the corresponding unknown cumulative sum function. In such circumstances, the following lemma is useful in proving consistency of the isotonic estimator.

Lemma

Let X be endowed with a simple order. Let h be a given isotonic (antitonic) function on X and H the corresponding cumulative sum function, with given positive weights w:

$$H(x) = \sum_{t \leq x, t \in X} h(t)w(t).$$

Let g be an arbitrary function on X, G its cumulative sum function, g^* the isotonic (antitonic) regression of g with weights w, and G^* the cumulative isotonic (antitonic) regression function. Then

$$\max_{x \in X} |H(x) - G^*(x)| \leq \max_{x \in X} |H(x) - G(x)|.$$

Proof. Let $x_1 \prec x_2 \prec \ldots \prec x_k$ be the simple ordering of X. Consider first the cumulative sum diagram of h (Section 1.2). For $i = 1, 2, \ldots, k$, set

$$W_i = \sum_{j=1}^{i} w(x_j), \qquad W_0 = 0,$$

and

$$H_i = \sum_{j=1}^{i} h(x_j)w(x_j), \qquad H_0 = 0.$$

Plot the points (W_i, H_i), $i = 0, 1, 2, \ldots, k$ in the Cartesian plane. Join successive points by straight line segments to obtain the cumulative sum diagram of h with weights w, which are considered also as the graph of a function $\mathscr{H}$ defined for $0 \leq W \leq W_k$ by

$$\mathscr{H}(W_i) = H_i, \qquad i = 0, 1, 2, \ldots, k,$$

$\mathscr{H}$ is linear between W_{i-1} and W_i, $i = 1, \ldots, k$. Define similarly a function $\mathscr{G}$ whose graph is the cumulative sum diagram of g with weights w, and a function $\mathscr{G}^*$ whose graph is the cumulative sum diagram of g^* with weights w. We note that $\mathscr{H}$ and $\mathscr{G}$ are convex if h and g are isotonic (increasing) and concave if h and g are antitonic (decreasing). The proof of the lemma for the isotonic case is given here.

Let

$$\max_{x \in X} |H(x) - G^*(x)| = \max_{W} |\mathscr{H}(W) - \mathscr{G}^*(W)|$$

be achieved at

$$W_r = \sum_{j=1}^{r} w(x_j).$$

Since $\mathscr{G}^*$ is the greatest convex minorant (GCM) of $\mathscr{G}$, either $\mathscr{G}^*(W_r) = \mathscr{G}(W_r)$ or W_r is interior to an interval over which $\mathscr{G}^*$ is linear, between points where $\mathscr{G}^* = \mathscr{G}$. If $\mathscr{G}^*(W_r) = \mathscr{G}(W_r)$ then clearly

$$\max_{W} |\mathscr{H}(W) - \mathscr{G}(W)| \geq |\mathscr{H}(W_r) - \mathscr{G}(W_r)|$$

$$= |\mathscr{H}(W_r) - \mathscr{G}^*(W_r)|$$

$$= \max_{W} |\mathscr{H}(W) - \mathscr{G}^*(W)|$$

$$= \max_{x \in X} |H(x) - G^*(x)|.$$

Suppose now that W_r is interior to an interval over which $\mathscr{G}^*$ is linear, between points where $\mathscr{G} = \mathscr{G}^*$. Since $\mathscr{H}$ is convex, if $\mathscr{H}(W_r) \geq \mathscr{G}^*(W_r)$, the absolute difference $|\mathscr{H} - \mathscr{G}|$ would be at least as large at an endpoint of the interval as at W_r, so that the maximum is also achieved at a point where

$\mathscr{G} = \mathscr{G}^*$, the case considered above. It remains to consider the case $\mathscr{H}(W_r) < \mathscr{G}^*(W_r)$. Since also $\mathscr{G}^*(W_r) \leq \mathscr{G}(W_r)$,

$$\max_W |\mathscr{H}(W) - \mathscr{G}(W)| \geq \mathscr{G}(W_r) - \mathscr{H}(W_r)$$
$$\geq \mathscr{G}^*(W_r) - \mathscr{H}(W_r)$$
$$= \max_W |\mathscr{H}(W_r) - \mathscr{G}^*(W_r)|$$
$$= \max_{x \in X} |H(x) - G^*(x)|. \blacksquare$$

2.3 ALGORITHMS FOR CALCULATION

The problem of calculating g^* given g, the weights w, and the partial order on X, is an instance of a quadratic programming problem. An extensive literature on methods of computing solutions of such problems exists. This section, however, will be restricted to methods which have been advanced for computing isotonic regression specifically. The "Pool-Adjacent-Violators" algorithm was presented in Section 1.2. Descriptions and justifications are given here for a more specific version of this algorithm and several algorithms for partial orders which are not necessarily simple. The algorithms all involve averaging g over suitably selected subsets of X; the term *amalgamation of means* has frequently been used in connection with such computations.

The "Up-and-Down Blocks" algorithm developed by J. B. Kruskal is essentially a version of the "Pool-Adjacent-Violators" algorithm and applies in the case of a simple order.

The "Up-and-Down Blocks" algorithm

The elements of X, arranged in order, are denoted by $x_1, x_2, \ldots, x_k$. Again the term *block* is used for a set of consecutive elements of X. If B_-, B, B_+ are three consecutive blocks in order, it is said that B is *up-satisfied* if Av B < Av B_+, where Av denotes the weighted average of g over the indicated subset of X; and it is said that B is *down-satisfied* if Av B_- < Av B. Extending these definitions, it is supposed that any block containing x_1 is down-satisfied, and that any block containing x_k is up-satisfied. The algorithm begins with a partition of X into blocks which are just its individual elements, and proceeds to amalgamate blocks as necessary until the partition of X into solution blocks (see Chapter 1) is obtained. The values of g^* are then given by averaging g over the solution blocks. At each stage of the algorithm one block is specified as *active*; this may be amalgamated with an adjacent block or, if it is up-satisfied and down-satisfied, the next block becomes the active block.

The operation of the algorithm is explained by the flow-chart which appears as Figure 2.2, in which the following abbreviations have been used:

US: the active block is up-satisfied;
DS: the active block is down-satisfied;
PU: pool the active block with the next higher block: the new block becomes active;
PD: pool the active block with the next lower block: the new block becomes active;
NH: the next higher block becomes active.

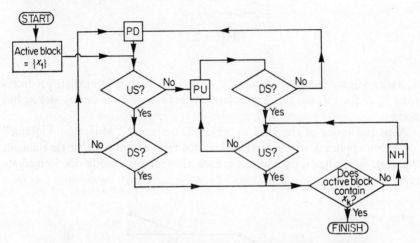

Figure 2.2 Flow-chart for the "Up-and-Down Blocks" algorithm

The Pool-Adjacent-Violators and Up-and-Down Blocks algorithms are applicable only in the case of a simple order, in which case each element of X (except first and last) has exactly one immediate predecessor and one immediate successor. W. A. Thompson, Jr., has proposed an algorithm for the somewhat more general case of a partial order in which each element of X with one exception (the "first" element) has exactly one immediate predecessor.

Definition 2.3

An element x in the partially ordered set X is an *immediate predecessor* of an element y if $x \lesssim y$ but there is no z in X distinct from x and y such that $x \lesssim z \lesssim y$.

It is helpful to visualize the order relations by means of a graph. One way is to represent by dots in a column at the left points of X having no predecessor:

i.e., points x such that there is no y for which $y \lesssim x$. Dots in a column to the right of these represent points which have as predecessor only points represented in the first column, etc. Arcs are then drawn connecting pairs of points ordered by "$\lesssim$". For example, the partial ordering $x_1 \lesssim x_3 \lesssim x_5$, $x_1 \lesssim x_4$, $x_2 \lesssim x_4$, $x_2 \lesssim x_5$ is represented by Figure 2.3. In this partial order,

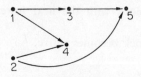

Figure 2.3

x_4 has two immediate predecessors, x_1 and x_2; x_5 has two immediate predecessors, x_2 and x_3; x_3 has one immediate predecessor and each of x_1 and x_2 has none.

A partial order of the kind to which Thompson's "Minimum Violator" algorithm applies is represented by a rooted tree: except for a single element, the root, which has no predecessor, each element has exactly one immediate predecessor (see Figure 2.4).

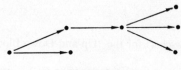

Figure 2.4

If each element has at most one immediate predecessor but there are several with none, then the graph consists of completely separated rooted trees: X can be partitioned into subsets such that no element of any is order-related to any element of any other. Then the Minimum Violator algorithm can be applied separately to the various subsets. In general *if the partial order is such that X can be partitioned into subsets which are independent in the sense that no element of any subset is order-related to any element of any other, then the isotonic regression can be calculated separately for each subset.*

The Minimum Violator algorithm, for a partial order represented by a rooted tree, again gives a procedure for arriving at the partition of X into sets of constancy of g^*.

The Minimum Violator algorithm

This algorithm applies when each element except the root has exactly one immediate predecessor. Again, the algorithm starts with the finest possible partition into blocks, the individual points of X. We look for *violators*: y is a *violator* if $g(y) < g(x)$ where x is the immediate predecessor of y. A *minimum violator* is a violator y for which $g(y)$ attains its minimum value among the values of g at violators. The algorithm begins by selecting a minimum violator and pooling it with its immediate predecessor to form a block. At an arbitrary stage of the algorithm we have a partition into blocks. Each block has a *weight*, the sum of the weights of its individual elements; a *value*, the weighted average of the values of g at its individual points; and a *root*, that one of its points whose immediate predecessor is not in the block. The *immediate predecessor* of any block is the block containing the immediate predecessor of its root. When a block and its immediate predecessor are pooled, the root of the new block is the root of the immediate predecessor block.

A block is a *violator* if its value is smaller than that of its immediate predecessor block. A *minimum violator* block is a violator whose value is at least as small as that of any other violator. Each step of the algorithm consists in pooling a minimum violator with its immediate predecessor block. This is continued until there are no violators. At this point the blocks are sets of constancy for g^* and the value of g^* at each point of any block is just the value of the block.

If the partial order is such that each element has exactly one immediate successor, except for one with no successor, an analogous maximum violator algorithm can be used. Of course, these algorithms apply in particular to the special case of a complete order.

Algorithms discussed thus far apply to special partial orders only. For discussions of algorithms for general partial orders, it is convenient to have a concept generalizing that of "block"; this is "level set", defined below, in Definition 2.5.

Definition 2.4

A subset L of X is a *lower set* with respect to the partial order "$\lesssim$" if $y \in L$, $x \in X$, $x \lesssim y$ imply $x \in L$. A subset U of X is an *upper set* if $y \in U$, $x \in X$, $x \gtrsim y$ imply $x \in U$. The class of all lower sets is denoted by $\mathscr{L}$, and the class of all upper sets by $\mathscr{U}$.

Each of the classes $\mathscr{L}$ and $\mathscr{U}$ is closed under arbitrary union and intersection. A set U is an upper set if and only if its complement U^c is a lower set.

These classes of sets and the family of isotonic functions on X can be characterized in terms of each other. A function f on X is isotonic if and only if any *one* of the following four conditions is satisfied (and then the others are also):

$$[f < a] = \{x : f(x) < a\} \in \mathscr{L} \text{ for every real } a;$$

$$[f \leq a] \in \mathscr{L} \text{ for every real } a;$$

$$[f > a] \in \mathscr{U} \text{ for every real } a;$$

$$[f \geq a] \in \mathscr{U} \text{ for every real } a.$$

Also, a set U is upper if and only if its indicator 1_U, defined by $1_U(x) = 1$ if $x \in U$, $1_U(x) = 0$ if $x \in U^c$, is isotonic. A set L is lower if and only if the negative of its indicator, -1_L, is isotonic.

Definition 2.5

A subset B of X is a *level set* if there are a lower set L and an upper set U such that $B = L \cap U$. We denote the class of level sets by $\mathscr{LS}$.

Since X is both upper and lower, each upper set is a level set, as is also each lower set.

The terminology is justified by the remark that level sets are sets of constancy of isotonic functions.

Theorem 2.3

A subset B of X is a level set if and only if there are an isotonic function f and a real c such that $B = [f = c]$.

Proof. If f is isotonic and c is a real number then $[f = c] = [f \leq c] \cap [f \geq c]$. But $[f \leq c] \in \mathscr{L}$ and $[f \geq c] \in \mathscr{U}$, so that $[f = c] \in \mathscr{LS}$. Suppose now that $B \in \mathscr{LS}$. Then there are $L \in \mathscr{L}$ and $U \in \mathscr{U}$ such that $B = L \cap U$. Set $f = 1_U - 1_L$. Then f is isotonic and $[f = 0] = B$. ∎

Minimum Lower Sets algorithm

Recall that Av B denotes the weighted average of g over a subset B of X. Select a lower set of minimum average; if more than one have the same lowest average, take the largest (their union). Let L_1, and also B_1, denote this "minimum lower set". This level set is the set on which g^* assumes its smallest value:

$$g^*(x) = \text{Av } B_1 = \min \{\text{Av } L : L \in \mathscr{L}\} \quad \text{for} \quad x \in B_1.$$

Now consider the averages of level sets of the form $L \cap L_1^c$; level sets consisting of lower sets with L_1 subtracted. Select again the largest level set of minimum average among these: $B_2 = L_2 \cap L_1^c$, where

$$\mathrm{Av}\, B_2 = \min\, \{\mathrm{Av}\, (L \cap L_1^c) : L \in \mathscr{L}\}.$$

The level set B_2 is the set on which g^* assumes its next smallest value:

$$g^*(x) = \mathrm{Av}\, B_2 \quad \text{for} \quad x \in B_2.$$

This process is continued until X is exhausted.

There is of course an analogous *Maximum Upper Sets algorithm*, in which the largest upper set of maximum average is selected first.

The fundamental basis for the justification of these algorithms lies in a property of the isotonic regression in relationship to the averaging function $\mathrm{Av}\,(\cdot)$, expressed in Theorem 2.4.

Theorem 2.4

For real a,

$$\mathrm{Av}\, (L \cap [g^* \geq a]) \geq a,$$
$$\mathrm{Av}\, (L \cap [g^* > a]) > a,$$
$$\mathrm{Av}\, (U \cap [g^* \leq a]) \leq a,$$

and

$$\mathrm{Av}\, (U \cap [g^* < a]) < a,$$

provided in each case that $L \in \mathscr{L}$, $U \in \mathscr{U}$, *and that the set over which the weighted average is taken is not empty.*

Proof. We observe first that, for arbitrary a, b such that $-\infty \leq a < b \leq \infty$,

$$\sum_{x \in [a < g^* < b]} [g(x) - g^*(x)]w(x) = 0.$$

This is an immediate consequence of the lemma before Theorem 1.7, which states that for each real c,

$$\sum_{x \in [g^* = c]} [g(x) - g^*(x)]w(x) = 0.$$

Further, if $L \in \mathscr{L}$,

$$-\sum_{x \in L} [g(x) - g^*(x)]w(x) = \sum_{x \in X} [g(x) - g^*(x)][-1_L(x)]w(x) \leq 0,$$

since $-1_L(\cdot)$ is isotonic. Now suppose a is real and $U \cap [g^* < a] \neq \phi$. Then

$$\sum_{x \in U \cap [g^* < a]} [g(x) - a]w(x) < \sum_{x \in U \cap [g^* < a]} [g(x) - g^*(x)]w(x)$$
$$= \sum_{x \in [g^* < a]} [g(x) - g^*(x)]w(x) - \sum_{x \in U^c \cap [g^* < a]} [g(x) - g^*(x)]w(x) \leq 0.$$

Then

$$\sum_{x \in U \cap [g^* < a]} g(x)w(x) < a \sum_{x \in U \cap [g^* < a]} w(x);$$

but this is equivalent to the last conclusion of the theorem. The proofs of the others are similar. ∎

Another essential ingredient in justifying the algorithms is a remark about reduction of a problem of isotonic regression. In the following definition $x \prec y$ is used to mean $x \lesssim y$ and $x \neq y$, $x \succ y$ to mean $x \gtrsim y$ and $x \neq y$.

Definition 2.6

Elements x_1 and x_2 are *poolable* with respect to the partial order "$\lesssim$" on X if $x \prec x_1$, $x_2 \prec y$ imply $x \prec y$ and if $x \prec x_2$, $x_1 \prec y$ imply $x \prec y$. The *reduced problem obtained by pooling poolable elements x_1 and x_2* is the isotonic regression problem in which a new space X' has one fewer elements than X, the single element $x' = \{x_1, x_2\}$ replacing the two elements x_1 and x_2. The partial order "$\lesssim'$" on X' is defined by: $x \lesssim' x'$ provided either $x \lesssim x_1$ or $x \lesssim x_2$; $x' \lesssim' x$ provided either $x_1 \lesssim x$ or $x_2 \lesssim x$; otherwise "$\lesssim'$" agrees with "$\lesssim$". It is easy to verify that "$\lesssim'$" is also a partial order. The weight function w' and the object function g' are defined by $w'(x) = w(x)$, $g'(x) = g(x)$ if $x \in X'$, $x \neq x'$, $w'(x') = w(x_1) + w(x_2)$,

$$g'(x') = \mathrm{Av}(\{x_1, x_2\})$$
$$= [g(x_1)w(x_1) + g(x_2)w(x_2)]/[w(x_1) + w(x_2)].$$

It will be said that solutions g^* and $g^{*\prime}$ of the original problem and the reduced problem *agree* if $g^*(x) = g^{*\prime}(x)$ for $x \in X$, $x \notin \{x_1, x_2\}$, and $g^*(x_1) = g^*(x_2) = g^{*\prime}(x')$.

Theorem 2.5

If there is a pair x_1, x_2 of poolable elements in an isotonic regression problem such that $g^*(x_1) = g^*(x_2)$, then the solution of the reduced problem obtained by pooling them agrees with the solution of the original problem.

Proof. If $f(x_1) = f(x_2) = f$ and if $f'(\cdot)$ is defined on X' by $f'(x) = f(x)$ if $x \in X'$, $x \neq x'$, and $f'(x') = f$ ($=f(x_1) = f(x_2)$), then $f'(\cdot)$ is isotonic on X' with respect to "$\lesssim'$" if and only if $f(\cdot)$ is isotonic on X with respect to "$\lesssim$". The sum of the terms in

$$\sum_{x \in X} [g(x) - f(x)]^2 w(x)$$

corresponding to the elements x_1 and x_2 is

$$[g(x_1) - f]^2 w(x_1) + [g(x_2) - f]^2 w(x_2)$$
$$= [g'(x') - f]^2 w'(x') + [g(x_1) - g'(x')]^2 w(x_1)$$
$$+ [g(x_2) - g'(x')]^2 w(x_2).$$

Since the last two terms of the right member are independent of f, the problem of finding the isotonic function $f(\cdot)$ on X minimizing

$$\sum_{x} [g(x) - f(x)]^2 w(x)$$

given $f(x_1) = f(x_2)$ is equivalent to the problem of finding the isotonic function f' on X' minimizing

$$\sum_{x \in X'} [g'(x) - f'(x)]^2 w'(x);$$

that is, the solutions agree. ∎

The justification of the Pool-Adjacent-Violators, the Up-and-Down Blocks, and the Minimum Violator algorithms is completed by Theorem 2.6.

Theorem 2.6

If x_1 is the only immediate predecessor of x_2 and if

(i) $g(x_2) = \min_{x \in X} g(x)$

or if

(ii) $g(x_1) \geq g(x_2)$ and x_2 is also the only immediate successor of x_1,

then $g^*(x_1) = g^*(x_2)$.

Proof. Suppose the contrary, that $g^*(x_1) = a < b = g^*(x_2)$. Set $L = [g^* \leq a] \cup \{x_2\}$; L is a lower set since if $y \in [g^* \leq a]$ and $x \lesssim y$ then $g^*(x) \leq a$ so that $x \in L$, while if $x \prec x_2$ then $x \lesssim x_1$ so that $g^*(x) \leq g^*(x_1) = a$, and again $x \in L$. By Theorem 2.4, $g(x_2) = \text{Av}(L \cap [g^* \geq b]) \geq b$. On the other hand, under hypothesis (i),

$$g^*(x_1) = a = \text{Av}[g^* = a] \geq g(x_2) \geq b,$$

a contradiction. To arrive at a contradiction under hypothesis (ii), set $U = [g^* \geq b] \cup \{x_1\}$. Again U is an upper set. By Theorem 2.4, $g(x_1)$ = Av $(U \cap [g^* \leq a]) \leq a$, contradicting $g(x_1) \geq g(x_2) \geq b$. ∎

Theorems 2.5 and 2.6 provide the justification for the Pool-Adjacent-Violators, Up-and-Down Blocks, and Minimum Violator algorithms. The Minimum Lower Sets algorithm is justified in Theorem 2.7.

Theorem 2.7

Let a, b be adjacent values of g^*: $a < b$, $[g^* = a] \neq \phi$, $[g^* = b] \neq \phi$, $[g^* = c] = \phi$ for $a < c < b$. Then Av $[g^* = b] =$ Av $([g^* \leq b] \cap [g^* > a])$ $\leq$ Av $(L \cap [g^* > a])$ *for all $L \in \mathscr{L}$ such that $L \cap [g^* > a] \neq \phi$, and* Av $[g^* = b] <$ Av $(L \cap [g^* > a])$ *for all $L \in \mathscr{L}$ such that L contains $[g^* \leq b]$ properly.*

Thus $[g^* = b]$ is the largest level set of form $L \cap [g^* > a]$, $L \in \mathscr{L}$, which has minimum average among such level sets, as described in the Minimum Lower Sets algorithm.

Proof. Av $([g^* \leq b] \cap [g^* > a]) =$ Av $[g^* = b] = b$ by the lemma before Theorem 1.7. Also, if $L \in \mathscr{L}$, Av $(L \cap [g^* > a]) =$ Av $(L \cap [g^* \geq b]) \geq b$ by Theorem 2.4, so that $[g^* = b]$ has minimum average among level sets of form $L \cap [g^* > a]$, $L \in \mathscr{L}$. If $L \in \mathscr{L}$ and L contains $[g^* \leq b]$ properly, then $L \cap [g^* > b] \neq \phi$ and Av $(L \cap [g^* > b]) > b$. Since Av $(L \cap [g^* \leq b]$ $\cap [g^* > a]) \geq b$ by the above (note $L \cap [g^* \leq b] \in \mathscr{L}$), we have also Av $(L \cap [g^* > a]) =$ Av $\{(L \cap [g^* \leq b] \cap [g^* > a]) \cup (L \cap [g^* > b])\} > b$ = Av $[g^* = b]$, which is the last conclusion of the theorem. ∎

Theorem 2.4 also yields proofs of max-min formulas for g^* which, for simple order, specialize to those given in Section 1.2.

Theorem 2.8

For $x \in X$,

$$g^*(x) = \max_{U: x \in U \in \mathscr{U}} \min_{L: x \in L \in \mathscr{L}} \text{Av}(L \cap U)$$

$$= \max_{U: x \in U \in \mathscr{U}} \min_{L \in \mathscr{L}} \text{Av}(L \cap U)$$

$$= \min_{L: x \in L \in \mathscr{L}} \max_{U: x \in U \in \mathscr{U}} \text{Av}(L \cap U)$$

$$= \min_{L: x \in L \in \mathscr{L}} \max_{U \in \mathscr{U}} \text{Av}(L \cap U).$$

Proof. Set $a = g^*(x)$. By Theorem 2.4, for $L \in \mathscr{L}$, $\text{Av}(L \cap [g^* \geq a]) \geq a$ if $L \cap [g^* \geq a] \neq \phi$. Also, by the lemma before Theorem 1.7,

$$\text{Av}([g^* \leq a] \cap [g^* \geq a]) = \text{Av}[g^* = a] = a.$$

Thus if $U_a = [g^* \geq a]$,

$$\min_{L \in \mathscr{L}} \text{Av}(L \cap U_a) = \min_{L : x \in L \in \mathscr{L}} \text{Av}(L \cap U_a) = a.$$

On the other hand, if $x \in U \in \mathscr{U}$, then $U \cap [g^* \leq a] \neq \phi$, and $\text{Av}(U \cap [g^* \leq a]) \leq a$ by Theorem 2.4. Hence

$$\min_{L \in \mathscr{L}} \text{Av}(L \cap U) \leq \min_{L : x \in L \in \mathscr{L}} \text{Av}(L \cap U) \leq a,$$

so that

$$g^*(x) = a = \min_{L \in \mathscr{L}} \text{Av}(L \cap U_a) = \max_{U : x \in U \in \mathscr{U}} \min_{L \in \mathscr{L}} \text{Av}(L \cap U),$$

verifying the first two formulas in the statement of the theorem. The proof of the last two is similar. ∎

An interesting algorithm for an arbitrary partial order has been given by Alexander (1970). In effect it reduces the problem of finding the isotonic regression for a partial order to that for a simple order. Algorithms for simple order are discussed in the early part of this section. Let g^* denote the isotonic regression for a given partial order. There is some simple order (often more than one) with respect to which g^* is nondecreasing; call it a g^*-*simple order*. If a g^*-simple order could be found, g^* could be calculated as the isotonic regression of g with respect to it. The Minimax Order algorithm leads to such a simple order.

To provide motivation, consider the g^*-simple order. The isotonic regression g^* for this simple order is given by the slopes of the GCM of the cumulative sum diagram (CSD) (Section 1.2) for the simple order. The corresponding slopes of the CSD itself are the corresponding values of g, and a g^*-simple order is a simple order for which the slopes of the CSD are as close as possible (least squares) to the slopes of the GCM. They would be equal, and g^* would coincide with g, if the CSD were itself convex and thus coincided with the GCM. The aim of the Minimax Order algorithm is to find a simple order for which the CSD is as nearly convex as possible. Since the endpoints of the CSD are fixed this is achieved by pushing the GCM down, by finding the simple order for which the GCM ("maximum" convex minorant) is as low as possible; i.e., by finding the simple order with the "minimax" convex minorant. It is proved in the corollary to Theorem 2.10 that the GCM for a g^*-simple order is the GCM of the union of the CSDs of all simple orders consistent with the given partial order.

Definition 2.7

A simple order is *consistent* with a partial order if every pair ordered by the partial order is ordered in the same way by the simple order. Two adjacent blocks in the simple order are *independent* if no pair of elements, one from each block, is ordered by the partial order. The *solution* associated with a simple order is the isotonic regression of the object function g with respect to the simple order, and the *solution partition* of X into *solution blocks* is the partition of X into sets on which the solution is constant.

A simple order consistent with the partial order can be found by the following scheme. Order first those elements of X which have no predecessors. Any order will do, but it might be desired to order them in the order of increasing g. Then order those elements each of which has exactly one predecessor, etc. The resulting simple order is consistent with the partial order, for if $x_1 \prec x_2$ then x_2 has more predecessors than x_1 and appears later in the simple order.

In the algorithm to be described, a simple order will be modified by interchanging independent adjacent sub-blocks of a solution block, depending on their excesses over the solution block average, described in the following definition.

Definition 2.8

If a is a real number and F is a subset of X then the *excess* of F over a is

$$v_a(F) = \sum_{x \in F} [g(x) - a]w(x).$$

(This is a key concept in Chapter 7.)

Suppose a simple order is at hand, and F_1 and F_2 are two adjacent blocks in it, within a solution block B whose average is a. Suppose $v_a(F_1) > v_a(F_2)$. If F_1 and F_2 are interchanged (Figure 2.5) the effect is to replace two adjacent segments of the CSD by the opposite sides of a parallelogram of which they were sides, in such a way as to reduce the height of the middle vertex above the chord of the GCM for the solution block B. This seems intuitively a step in the right direction (toward lowering the GCM), if perhaps a small one.

Minimax Order Algorithm

Begin with an arbitrary simple order consistent with the given partial order. It may often save time to interchange adjacent *independent* blocks in which the second has the higher average: i.e., if B_2 follows B_1 immediately in the simple order and if Av B_1 > Av B_2, then interchange B_1 and B_2.

ALGORITHMS FOR CALCULATION

The object of the algorithm is to produce a simple order in which each solution block has the following property: *the excesses of any two adjacent independent sub-blocks are in nondecreasing order.* According to Theorem 2.9 which follows, the isotonic regression for such a simple order is also the isotonic

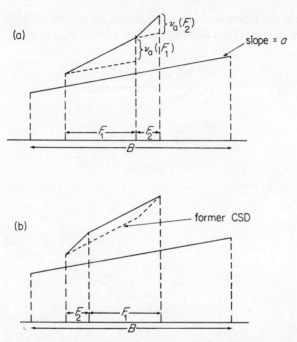

Figure 2.5 The effect of interchanging two blocks whose excesses are in decreasing order. (a) Before interchange; (b) after interchange

regression for the partial order. To this end, the algorithm interchanges independent adjacent subblocks whose excesses are in the wrong order.

Having an initial simple order, fix attention on one of its solution blocks, B. Let its average be

$$a = \mathrm{Av}\, B = \sum_{x \in B} g(x) w(x) / \sum_{x \in B} w(x).$$

If there are independent adjacent subblocks of B, B_2 following B_1, and if the excess of B_1 over a is greater than the excess of B_2 over a, interchange B_1 and B_2. After all possible such interchanges have been made for every solution block, obtain the solution partition for the new simple order, and repeat the process. When a stage is reached for which the new simple order has the same solution partition as the previous one, the solution for this simple order is the solution for the original partial order.

Example 2.3

The problem described in Figure 2.6 serves as an illustration. The circled numbers are indices which identify the four points, x_1, x_2, x_3, x_4 respectively,

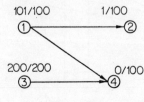

Figure 2.6

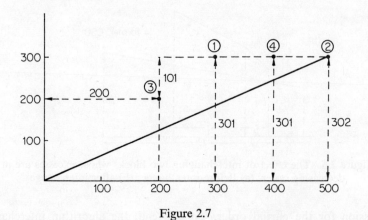

Figure 2.7

of X. Thus the partial order is $x_1 \lesssim x_2$, $x_1 \lesssim x_4$, $x_3 \lesssim x_4$. Each ratio is the corresponding value $g(x)$ of the object function g, and the denominator of the ratio is the weight $w(x)$. We start with the simple order

$$③ \to ① \to ④ \to ② \quad \text{or} \quad x_3 \lesssim x_1 \lesssim x_4 \lesssim x_2,$$

in which x_1 precedes x_2, x_1 precedes x_4, and x_3 precedes x_4, so that this simple order is consistent with the partial order. The cumulative sum diagram (see Section 1.2) for this simple order is shown in Figure 2.7, together with the greatest convex minorant (Section 1.2).

There is just one solution block in the solution partition. Its average is $a = 302/500$. The excesses over a of the various elements of X are:

$$\nu_a\{x_3\} = 200(200/200 - 302/500) = 396/5,$$
$$\nu_a\{x_1\} = 100(101/100 - 302/500) = 203/5,$$
$$\nu_a\{x_4\} = 100(0/100 - 302/500) = -302/5,$$

and

$$\nu_a\{x_2\} = 100(1/100 - 302/500) = -297/5.$$

Adjacent blocks $\{x_4\}$ and $\{x_2\}$ are independent, but their excesses are already in increasing order. The only other adjacent independent blocks are $\{x_3\}$ and

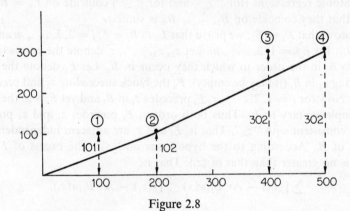

Figure 2.8

$\{x_1\}$, and their excesses are in decreasing order, so we interchange these blocks to obtain the new simple order $x_1 \lesssim x_3 \lesssim x_4 \lesssim x_2$. Now adjacent blocks $\{x_3, x_4\}$ and $\{x_2\}$ are independent, and

$$\nu_a\{x_3, x_4\} = \nu_a\{x_3\} + \nu_a\{x_4\} = 94/5$$

which is greater than $\nu_a\{x_2\}$. So these blocks are interchanged, getting the new simple order $x_1 \lesssim x_2 \lesssim x_3 \lesssim x_4$. The cumulative sum diagram for this simple order is illustrated in Figure 2.8. The solution partition has two solution blocks. There are no independent adjacent sub-blocks in either, and the solution for this simple order is the solution for the original partial order:

$$g^*(x_1) = g^*(x_2) = 102/200, \qquad g^*(x_3) = g^*(x_4) = 200/300.$$

The justification of the algorithm depends in part on the following theorems.

4

Theorem 2.9

Let "$\lesssim'$" be a simple order consistent with a given partial order "$\lesssim$" on X. Let every solution block of the solution partition of "$\lesssim'$" have the property that the excesses over its average of each two independent adjacent sub-blocks are in nondecreasing order. Then the solution for "$\lesssim'$" coincides with that for "$\lesssim$".

Proof. Let the solution blocks of "$\lesssim'$" be $B_1, B_2, \ldots, B_m$. Let L_1 be the set on which the solution g^* for "$\lesssim$" assumes its smallest value; according to the Minimum Lower Sets algorithm, L_1 is the *largest lower set* (according to "$\lesssim$") of *minimum average*. The plan of the proof is to show that $L_1 = B_1$. Since B_1 is the set on which the solution for "$\lesssim'$" takes its smallest value, the two isotonic regressions (for "$\lesssim'$" and for "$\lesssim$") coincide on $L_1 = B_1$. The proof that they coincide on $B_2, \ldots, B_m$ is similar.

To prove that $L_1 = B_1$, we prove that $L_1 \cap B_j = \phi, j = 2, 3, \ldots, m$ and that $B_1 \subset L_1$. Fix $j, j = 1, 2, \ldots, m$. Let $z_1, z_2, \ldots, z_r$ denote the points (if any) of $L_1 \cap B_j$ in the order in which they occur in B_j. Let F_1 denote the block preceding z_1 in B_j (it may be empty), F_2 the block succeeding z_1 and preceding z_2, etc. Now for $i = 1, 2, \ldots, r$, F_i precedes z_i in B_j and yet F_i is in the upper set complementary to L_i. Thus both orders, F_i precedes z_i and z_i precedes F_i, are consistent with "$\lesssim$". That is, F_i and z_i are adjacent independent sub-blocks of B_j. According to the hypotheses on "$\lesssim$", the excess of F_i over Av B_j is no greater than that of $\{z_i\}$. That is,

$$\sum_{x \in F_i} [g(x) - \text{Av } B_j] w(x) \le [g(z_i) - \text{Av } B_j] w(z_i).$$

Adding the right member to each side,

$$\sum_{x \in F_i + \{z_i\}} [g(x) - \text{Av } B_j] w(x) \le 2[g(z_i) - \text{Av } B_j] w(z_i).$$

Summing over i,

$$\nu_{\text{Av} B_j} (\sum_{i=1}^{r} [F_i + \{z_i\}]) = \sum_{x \in \sum_{i=1}^{r} [F_i + \{z_i\}]} [g(x) - \text{Av } B_j] w(x)$$

$$\le 2 \sum_{i=1}^{r} [g(z_i) - \text{Av } B_j] w(z_i).$$

But

$$\sum_{i=1}^{r} [F_i + \{z_i\}]$$

is a leading (left) sub-block of B_j, and according to the greatest convex minorant construction, its excess over Av B_j is non-negative. Thus the left member of the last string of inequalities is non-negative, and so therefore is the right:

$$\sum_{i=1}^{r} [g(z_i) - \text{Av } B_j]w(z_i) \geq 0,$$

whence

$$\sum_{x \in L_1 \cap B_j} g(x)w(x) = \sum_{i=1}^{r} g(z_i)w(z_i) \geq [\sum_{x \in L_1 \cap B_j} w(x)] \text{ Av } B_j.$$

But again by the GCM construction, $\text{Av } B_j > \text{Av } B_1$ for $j = 2, 3, \ldots, m$. Summing over $j = 1, 2, \ldots, m$,

$$\sum_{x \in L_1} g(x)w(x) \geq [\sum_{x \in L_1} w(x)] \text{ Av } B_1,$$

with strict inequality unless $L_1 \cap B_j = \phi$ for $j = 2, 3, \ldots, m$. Thus

$$\text{Av } L_1 \geq \text{Av } B_1$$

with strict inequality unless $L_1 \cap B_j = \phi$ for $j = 2, 3, \ldots, m$. Now B_1 is a lower set for "$\lesssim$". Since L_1 is the *largest* lower set of *minimum* average, both

$$\text{Av } L_1 \leq \text{Av } B_1,$$

which with the preceding inequality implies

$$\text{Av } L_1 = \text{Av } B_1$$

(hence $B_1 \subset L_1$), and $L_1 \cap B_j = \phi$ for $j = 2, 3, \ldots, m$. Thus $B_1 = L_1$. The proof is completed as indicated at the beginning. ∎

Theorem 2.10

There exists a simple order satisfying the hypotheses of Theorem 2.9.

Proof. Let g^* be the isotonic regression of g with respect to the partial order "$\lesssim$". Let a denote one of the values assumed by g^*. Arrange the elements of the subset $[g^* = a]$ of X so that the excesses over a of independent adjacent sub-blocks are in nondecreasing order. Having done this for each a, order the blocks $[g^* = a]$ according to increasing a. The resulting simple order satisfies the hypotheses of Theorem 2.9. ∎

Justification of Minimax Order Algorithm

By Theorem 2.10 there exists a simple order satisfying the hypotheses of Theorem 2.9, and by the latter theorem the solution for such a simple order is the isotonic regression g^* for the given partial order. It remains to verify that the algorithm will inevitably lead to such a simple order. The basic idea of the proof is simply that the completion of a stage of the algorithm either leaves the GCM unchanged, in which case the solution has been reached, or lowers the GCM. Since there are only a finite number of consistent simple orders, it is not possible to continue indefinitely to lower the GCM. In order to see that a stage in the algorithm either leaves the GCM unchanged or lowers it, note that points on the cumulative sum graph corresponding to endpoints of solution blocks suffice to determine the GCM. Since interchanges are made only *within* blocks, these points will also be on the cumulative sum graph for the new simple order. Hence its GCM must be on or below the GCM for the preceding simple order.

An interesting consequence of the foregoing discussion is that a simple order whose solution is g^* has a GCM lying on or below that for any other consistent simple order. In this sense it is a *minimax simple order*. This is stated formally as a corollary.

Corollary

The GCM for a consistent simple order "$\lesssim$" whose solution is g^* is the GCM of the union of cumulative sum graphs for all consistent simple orders.

Proof. This follows from the fact that the algorithm starts with an *arbitrary* consistent simple order, and at each stage leaves the GCM unchanged or lowers it. ∎

A rather different approach to the problem of computing isotonic regression is due to van Eeden (1957a, 1957b, 1958). The essence of her method is to begin by finding the isotonic regression of g for the trivial partial order which imposes no order restrictions on the element of X (this, of course, is equal to g), and then to generate the required partial order by adding order restrictions one at a time, recalculating the isotonic regression at each stage. The important result needed for this method is contained in the following theorem, before stating which a definition must be given of what is meant by the addition of an order restriction to a partial ordering.

Definition 2.9

Let "$\lesssim$" be a partial order on X, and let x_i, x_j be elements of X such that $x_i \not\lesssim x_j$. Then it is said that the partial order "$\lesssim'$" on X is obtained by the addition of the restriction $x_i \lesssim x_j$ to "$\lesssim$" if it satisfies the following condition: for any $x, y \in X$, $x \lesssim' y$ if, and only if, $x \lesssim y$ or $x \lesssim x_i$ and $x_j \lesssim y$.

In other words, the partial order "$\lesssim'$" imposes on the elements of X all the restrictions of "$\lesssim$" and, in addition, the restriction $x_i \lesssim' x_j$ and all other restrictions implied by transitivity. The condition $x_i \not\gtrsim x_j$ is required to ensure that "$\lesssim'$" is a partial order. However, all the results presented here may be generalized to deal with quasi-orders, in which case that condition is not required.

Theorem 2.11

Let "$\lesssim$" be a partial order on X, and "$\lesssim'$" the partial order obtained by adding the restriction $x_i \lesssim x_j$ to "$\lesssim$". Let g^* and $g^{*'}$ be the isotonic regressions of g, with weights w, with respect to "$\lesssim$" and "$\lesssim'$" respectively. Then

(i) *if* $g^*(x_i) \leq g^*(x_j)$,

$$g^{*'} = g^*$$

and

(ii) *if* $g^*(x_i) > g^*(x_j)$,

$$g^{*'}(x_i) = g^{*'}(x_j).$$

Proof. Define $\mathscr{F}$ to be the class of all functions on X which are isotonic with respect to "$\lesssim$", and $\mathscr{F}'$ to be the class of all functions on X which are isotonic with respect to "$\lesssim'$". Then $\mathscr{F}' \subset \mathscr{F}$.

(i) Let
$$S(f, g) = \sum_{x \in X} w(x)[f(x) - g(x)]^2,$$
where f is a function on X. Then $S(f, g)$ attains its minimum for $f \in \mathscr{F}$ when $f = g^*$. But if $g^*(x_i) \leq g^*(x_j)$, then $g^* \in \mathscr{F}' \subset \mathscr{F}$. Thus $S(f, g)$ must attain its minimum for $f \in \mathscr{F}'$ when $f = g^*$, i.e. $g^{*'} = g^*$.

(ii) Suppose $g^{*'}(x_i) < g^{*'}(x_j)$: a contradiction will be established. A function h_θ on X is defined by $h_\theta = \theta g^{*'} + (1 - \theta)g^*$, and θ_0 is defined to be such that $h_{\theta_0}(x_i) = h_{\theta_0}(x_j)$. Then, as $g^*(x_i) > g^*(x_j)$ and $g^{*'}(x_i) < g^{*'}(x_j)$, it follows that $0 < \theta_0 < 1$. Now the convexity of $\mathscr{F}$ implies that $h_\theta \in \mathscr{F}$, $0 \leq \theta \leq 1$; it is deduced that $h_\theta \in \mathscr{F} - \mathscr{F}'$, $0 \leq \theta < \theta_0$, and $h_\theta \in \mathscr{F}'$, $\theta_0 \leq \theta \leq 1$. Defining

$$S(\theta) = \sum_{x \in X} w(x)[h_\theta(x) - g(x)]^2,$$

it is known from the fact that g^* is the isotonic regression of g with respect to "$\lesssim$" that $S(\theta)$ achieves its minimum in $0 \leq \theta \leq 1$ at $\theta = 0$. Thus its derivative with respect to θ must be non-negative at that point, and as it is a positive quadratic function $S(\theta)$ must be a strictly monotonic increasing function of θ for $\theta > 0$. Hence $S(\theta)$ achieves its minimum in $\theta_0 \leq \theta \leq 1$ at $\theta = \theta_0$. But, as $g^{*\prime}$ is the isotonic regression of g with respect to "$\lesssim'$", the minimum of $S(\theta)$ in $\theta_0 \leq \theta \leq 1$ must be achieved at $\theta = 1$. The original assumption must, then, be false, and it follows that $g^{*\prime}(x_i) = g^{*\prime}(x_j)$. ∎

Van Eeden's Algorithm

In order to use the above theorem as the basis of an algorithm for calculating isotonic regression with respect to a partial order "$\lesssim$", $O_1, O_2, \ldots, O_p$ are defined to be a set of order restrictions on X which together define "$\lesssim$". Corresponding to each O_i, R_i is defined to be the corresponding constraint satisfied by a function which is isotonic with respect to "$\lesssim$", i.e. if O_i is $x_i \lesssim y_i$, and if f is a function defined on X, it is said that f satisfies R_i if $f(x_i) \leq f(y_i)$. The method is to start with no elements of X ordered and to add $O_1, O_2, \ldots, O_p$ one at a time, at each stage finding the isotonic regression with respect to the partial order defined by the restrictions added so far. Theorem 2.11 states that if, at the stage when O_j is added, the isotonic regression found at the previous stage satisfies R_j, then it is also the isotonic regression at this stage; but if R_j is not satisfied, then the new isotonic regression is the solution of the reduced problem obtained by pooling the elements ordered by O_j. The reduced problem may be solved by the same methods as the main problem, introducing its restrictions one by one and recalculating the solution at each stage. Further reductions may be necessary, and a complicated sequence of operations results in the calculation of the isotonic regression with respect to the partial order defined by $O_1, O_2, \ldots, O_j$. Then, if $j < p$, O_{j+1} is introduced and the process continues.

In Figure 2.9 a flow-chart is given for an algorithm, implementing the above methods, for finding the isotonic regression of a function g with respect to the partial order on X defined by order restrictions $O_1, O_2, \ldots, O_p$. $R_1, R_2, \ldots, R_p$ are as defined above and f is a function on X which is initially set equal to g and subsequently has its value modified until at the termination of the procedure it is equal to g^*, the isotonic regression of g. The value taken by f at any stage is defined by specifying some of $\{R_1, R_2, \ldots, R_p\}$ to be "boundary-satisfied": it is said that R_i is boundary-satisfied if $f(x_i) = f(y_i)$, and we denote by BS the set of constraints which are required to be boundary-satisfied. Initially BS $= \phi$ and subsequently f is recalculated every time BS is modified. The way this is done is as follows. Given the (nonempty) composition of BS, f is constrained to be constant on certain subsets of X containing two or more elements. On each of these subsets the value of f is calculated as the weighted average of the values of g on that subset; elsewhere $f = g$.

The algorithm of Figure 2.9 is rather complicated. Its use, however, may often be simplified by choosing a suitable order in which to introduce the Rs. In addition, the choice of the Rs themselves (or, equivalently, the Os) is important. Van Eeden derived her results assuming that the Os constituted

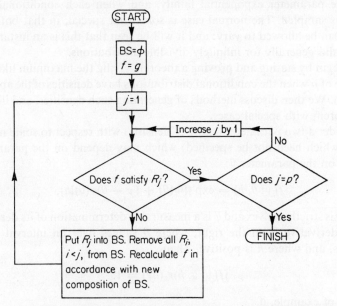

Figure 2.9   Flow-chart for an algorithm to calculate isotonic regression by van Eeden's method

what she called the set of "essential restrictions" of "$\lesssim$": by this she meant that there was no smaller set of restrictions which defined "$\lesssim$". The above results have been derived without making this assumption, but obviously if some of the Rs are implied by other members of the set, the algorithm will waste time in needless comparisons; for this reason it is desirable that the Rs should all correspond to essential restrictions of "$\lesssim$".

2.4 MAXIMUM LIKELIHOOD AND BAYESIAN ESTIMATION OF ORDERED PARAMETERS IN EXPONENTIAL FAMILIES

Brief mention is made in Section 1.1 of problems of estimating isotonic regression in which, for each $x \in X$, the conditional distribution of $\tilde{y}$ is normal, and each conditional distribution is sampled. Another problem mentioned (see also Section 2.1) is of the same character, except that the conditional distributions are Bernoullian: $\Pr(\tilde{y} = 1|\tilde{x} = x) = \mu(x)$, $\Pr(\tilde{y} = 0|\tilde{x} = x) = 1 - \mu(x)$. The present section is devoted primarily to the problem of estimating the regression function $\mu(x) = E(\tilde{y}|\tilde{x} = x)$ when it is known to be isotonic with respect to a given quasi-order, when the conditional distributions belong to a one parameter exponential family, and when each conditional distribution is sampled. The normal case is somewhat special, in that both parameters can be allowed to vary: and it will be seen that this is an instance of a result valid generally for infinitely divisible distributions.

We begin by stating and proving a theorem giving the maximum likelihood estimate of μ when the conditional distributions have densities of the appropriate form. We then discuss methods of generating such densities, and illustrate the theorem with special cases.

Consider a two parameter family of densities with respect to some measure $\nu_h(dy)$ (which need not be specified) which may depend on the parameter h but not on the parameter θ:

$$f(y; \theta, h) = \exp\{[\Phi(\theta) + (y - \theta)\varphi(\theta)]h\}, \qquad (2.7)$$

where Φ is strictly convex and φ is a measurable determination of its derivative, say its derivative from the right, where θ ranges over an interval of real numbers, and where h is positive. Then

$$\int f(y; \theta, h)\nu_h(dy) = 1.$$

By way of example, if

$$\Phi(\theta) = \theta^2/2, \qquad \varphi(\theta) = \theta, \qquad h = 1/\sigma^2,$$

(2.7) gives the density with respect to

$$\nu_h(dy) = \left(\frac{h}{2\pi}\right)^{\frac{1}{2}} \exp(-hy^2/2) \, dy$$

of the normal distribution with mean θ and variance σ^2. If

$$\Phi(\theta) = \theta \log \theta + (1 - \theta) \log(1 - \theta),$$

$$\varphi(\theta) = \log[\theta/(1 - \theta)], \qquad h = 1,$$

(2.7) gives the density with respect to counting measure on $\{0, 1\}$ of the Bernoulli distribution with mean θ.

If φ has a derivative φ' (which is then positive since Φ is strictly convex) and if the formal differentiation is valid, it is found that

$$\int (y - \theta)f(y; \theta, h)\nu_h(dy) = 0, \qquad (2.8)$$

so that θ is the mean of the distribution with density (2.7). A further differentiation of (2.8) yields

$$\int (y - \theta)^2 f(y; \theta, h) \nu_h(dy) = 1/h\varphi'(\theta); \tag{2.9}$$

so that providing the formal calculations are justified, $h\varphi'(\theta)$ is the reciprocal of the variance, or the *precision* of the distribution with density (2.7).

It is noted in passing that the negative logarithm of the density (2.7) with respect to the measure $\exp[h\Phi(y)]\nu_h(dy)$ is h multiplied by the form (see (1.29))

$$\Delta(y, \theta) = \Phi(y) - \Phi(\theta) - (y - \theta)\varphi(\theta).$$

Also, the information $I(\theta_1, \theta_2) = E_{\theta_1}[-\log f(\tilde{y}; \theta_2, h) + \log f(\tilde{y}; \theta_1, h)]$ is equal to

$$h\Delta(\theta_1, \theta_2) = h[\Phi(\theta_1) - \Phi(\theta_2) - (\theta_1 - \theta_2)\varphi(\theta_2)].$$

Theorem 2.12

Let X be a finite set, quasi-ordered by "$\lesssim$". For each $x \in X$, let the conditional distribution of $\tilde{y}$ given $\tilde{x} = x$ have density given by (2.7) with $\theta = \mu(x) = E(\tilde{y}|\tilde{x} = x)$ and with $h = a\lambda(x)$ where $\lambda(x)$ is a known positive number for each x, but the positive number a may be unknown. Let independent random samples be taken from these conditional distributions, with sizes $m(x) > 0$, $x \in X$. Then the maximum likelihood estimate of $\mu(\cdot)$ given, that $\mu(\cdot)$ is isotonic, is furnished uniquely by the isotonic regression of the sample means, with weights $w(x) = \lambda(x)m(x)$, $x \in X$.

Proof. Let the sample values observed be $y_j(x)$, $j = 1, 2, \ldots, m(x)$, $x \in X$, and let $\bar{y}(x)$ denote the sample mean,

$$\bar{y}(x) = \sum_{j=1}^{m(x)} y_j(x)/m(x), \quad x \in X.$$

Then the log-likelihood function is

$$\sum_{x \in X} \sum_{j=1}^{m(x)} \{\Phi[\mu(x)] + [y_j(x) - \mu(x)]\varphi[\mu(x)]\} a\lambda(x)$$

$$= a \sum_{x \in X} \{\Phi[\mu(x)] + [\bar{y}(x) - \mu(x)]\varphi[\mu(x)]\} \lambda(x) m(x).$$

By Theorem 1.10 this is uniquely maximized in the class of isotonic $\mu(\cdot)$ by setting $\mu(\cdot)$ equal to the isotonic regression of $\bar{y}$ with weights $\lambda(x)m(x)$, $x \in X$. This completes the proof of the theorem. ∎

Bayesian estimation of isotonic regression

The calculations occurring in the proof of Theorem 2.12 can also be given a Bayesian interpretation. Suppose for each $x \in X$ the experimenter will make $m(x)$ observations on a distribution with density of form (2.7) with $\theta = \mu(x)$, $h = \lambda(x)$, where $\mu(\cdot)$ is unknown and $\lambda(\cdot)$ is known:

$$f(y; \mu(x), \lambda(x)) = \exp\{(\Phi[\mu(x)] + [y - \mu(x)]\varphi[\mu(x)])\lambda(x)\}. \quad (2.10)$$

Suppose it is known further that $\mu(\cdot)$ is isotonic with respect to a quasi-order on X. In adopting a Bayesian approach, the experimenter will select a prior distribution for the parameters $\mu(x)$, $x \in X$.

It is convenient to select the prior density in the family conjugate to (2.7). (The use of the term "conjugate" in "conjugate family" is not at all directly related to its use in "conjugate convex function".) Such a prior density, as a function of an argument θ with respect to some measure $d\rho(\theta)$, is of the form

$$\pi(\theta) \propto \exp\{[\Phi(\theta) + (\theta_0 - \theta)\varphi(\theta)]h\}. \quad (2.11)$$

Since $\Phi(\theta) + (\theta_0 - \theta)\varphi(\theta)$ assumes its maximum at $\theta = \theta_0$, the mode of this density is θ_0. If the parameter h is large, the density is sharply peaked about θ_0; $h = 0$ gives a flat (uniform) density. h will be called the *precision parameter*.

Suppose that for $x \in X$ the experimenter considers $\mu_0(x)$ a likely value of $\mu(x)$. He selects a precision parameter $h(x)$, and for fixed x would then consider a prior density with respect to $d\rho$ proportional to

$$\exp\{(\Phi(\theta) + [\mu_0(x) - \theta]\varphi(\theta))h(x)\}.$$

The regression function being known to be isotonic, the experimenter will select a modal function $\mu_0(\cdot)$ which is isotonic. If there were no order restrictions, he might now use a joint prior distribution according to which the $\mu(x)$ for distinct x are independent. The joint density with respect to the product measure would then be

$$\pi[\mu(\cdot)] \propto \prod_{x \in X} \exp\{(\Phi[\mu(x)] + [\mu_0(x) - \mu(x)]\varphi[\mu(x)])h(x)\}.$$

In the presence of the order restrictions, it will be convenient to use a density proportional to this one, but vanishing in the proscribed region:

$$\pi[\mu(\cdot)] \propto \exp\{\sum_{x \in X}(\Phi[\mu(x)] + [\mu_0(x) - \mu(x)]\varphi[\mu(x)])h(x)\} \quad (2.12)$$

for isotonic $\mu(\cdot)$,

$$\pi[\mu(\cdot)] = 0$$

if $\mu(\cdot)$ is not isotonic on X. The mode of this density is the isotonic function $\mu_0(\cdot)$, and the precision function $h(\cdot)$ determines its peakedness about the mode; $h \equiv 0$ corresponds to a "flat" or uniform prior.

Now suppose that for each $x \in X$ the experimenter makes $m(x)$ observations on the distribution with density (2.10) and that the observed sample means are $\bar{y}(x)$, $x \in X$. Then the posterior density of $\mu(\cdot)$ is proportional to

$$\prod_{x \in X} \exp\{(\Phi[\mu(x)] + [\bar{y}(x) - \mu(x)]\varphi[\mu(x)])\lambda(x)m(x)\}\pi(\mu(\cdot)),$$

or

$\pi(\mu(\cdot)|\bar{y}(\cdot))$

$$\propto \exp\{\sum_{x \in X}(\Phi[\mu(x)] + [\hat{\mu}(x) - \mu(x)]\varphi[\mu(x)])(m(x)\lambda(x) + h(x))\}$$

for isotonic $\mu(\cdot)$,

$$\pi(\mu(\cdot)|\bar{y}(\cdot)) = 0$$

otherwise, where

$$\hat{\mu}(x) = \frac{m(x)\lambda(x)\bar{y}(x) + h(x)\mu_0(x)}{m(x)\lambda(x) + h(x)}, \qquad x \in X.$$

The posterior mode maximizes the sum appearing in the exponent in the class of isotonic $\mu(\cdot)$. But this is precisely the extremizing problem of Section 1.4, and the solution is the isotonic regression $\hat{\mu}^*(\cdot)$ of the function $\hat{\mu}(\cdot)$, with weights $m\lambda + h$. The following result has thus been arrived at.

Theorem 2.13

Let X be a finite set, quasi-ordered by "$\lesssim$". Let there be associated with each x in X a distribution having density (2.7) with $\theta = \mu(x)$ unknown, and with $h = \lambda(x)$, known. Let independent random samples be taken from these distributions with sizes $m(x) > 0$, $x \in X$. Let $\mu(\cdot)$ have prior distribution over the class of isotonic $\mu(\cdot)$ with density (2.12), with isotonic mode $\mu_0(\cdot)$ (specified) and precision function $h(\cdot)$ (specified). Then the posterior distribution of $\mu(\cdot)$ has density of the same form with mode $\hat{\mu}^*$, the isotonic regression with weights $m\lambda + h$, of the weighted average

$$\hat{\mu}(x) = [m(x)\lambda(x)\bar{y}(x) + h(x)\mu_0(x)]/[m(x)\lambda(x) + h(x)], \qquad x \in X,$$

of $\bar{y}(\cdot)$ and $\mu_0(\cdot)$. In particular, if $h(x) \equiv 0$ (flat prior) the posterior mode coincides with the maximum likelihood estimate.

Exponential families

The distributions (2.7) are certainly special, yet they include a large class of distributions. In particular, as is seen below, with $h = 1$, (2.7) gives the one parameter family determined by a distribution F which is arbitrary non-degenerate, except that its moment generating function is assumed to converge in a neighbourhood of the origin.

Let $\Theta(\tau)$ denote the cumulant generating function of $\bar{F}$:

$$e^{\Theta(\tau)} = \int e^{y\tau} F(dy). \tag{2.13}$$

Then

$$g(y; \tau) = \exp\{y\tau - \Theta(\tau)\} \tag{2.14}$$

is a one parameter family of probability densities with respect to the measure dF. For τ interior to the interval of convergence, $\Theta(\tau)$ is analytic. Let $\theta(\tau) = d\Theta(\tau)/d\tau$. Differentiation of (2.13) shows that $\theta(\tau)$ is the mean of the distribution with density $g(y; \tau)$:

$$\theta(\tau) = \int y g(y; \tau) F(dy). \tag{2.15}$$

A further differentiation of (2.15) shows that

$$d^2\Theta(\tau)/d\tau^2 = \int y(y - \theta(\tau)) g(y; \tau) F(dy)$$
$$= \int [y - \theta(\tau)]^2 g(y; \tau) F(dy) > 0,$$

so that $\Theta(\cdot)$ is strictly convex. Its derivative $\theta(\cdot)$ is then strictly increasing.

Let $\Phi(\cdot)$ denote the convex conjugate of $\Theta(\cdot)$:

$$\Phi(y) = \max_\tau \{y\tau - \Theta(\tau)\}.$$

This maximum is attained at the solution of $\theta(\tau) = y$, i.e., for $\tau = \varphi(y)$, where φ is the inverse function of $\theta(\cdot)$. We observe that

$$\Phi(y) = y\varphi(y) - \Theta[\varphi(y)]. \tag{2.17}$$

On differentiating (2.17) it is also found that

$$\varphi(y) = d\Phi(y)/dy \tag{2.18}$$

Making a change of parameter in (2.14), replace τ by $\varphi(\theta)$. In view of (2.17),

$$f(y; \theta) = g(y; \tau(\theta)) = \exp\{\Phi(\theta) + (y - \theta)\varphi(\theta)\}, \tag{2.19}$$

of the form (2.7) with $h = 1$.

If h is a positive integer and if $\tilde{y}$ is the average of h independent random variables with density (2.19), then its density is (2.7) with respect to a measure independent of θ. To see this note that the moment generating function of the distribution (2.19) is

$$\int \exp\left[yt + \Phi(\theta) + (y - \theta)\varphi(\theta)\right]F(dy) = \exp\left[\Theta(t + \tau) - \Theta(\tau)\right]$$

where $\tau = \varphi(\theta)$, using (2.13) and (2.17). Then the moment generating function of $\tilde{\tilde{y}}$ is

$$\exp\{h[\Theta(\tau + t/h) - \Theta(\tau)]\}.$$

Setting $\tau = 0$ the moment generating function of the average of h independent random variables with distribution F is found to be $\exp\{h\Theta(t/h)\}$. Let $\bar{F}_h$ denote the distribution function of this average, so that

$$\exp\{h\Theta(t/h)\} = \int \exp(ut)\bar{F}_h(du).$$

Then the moment generating function of $\tilde{\tilde{y}}$ is

$$\exp\{h[\Theta(\tau + t/h) - \Theta(\tau)]\}$$
$$= \int \exp\{h[u(t/h + \tau) - \Theta(\tau)]\}\bar{F}_h(du)$$
$$= \int \exp(ut)\exp\{h[\Phi(\theta) + (u - \theta)\varphi(\theta)]\}\bar{F}_h(du).$$

Thus the density of $\tilde{\tilde{y}}$ with respect to $\bar{F}_h$ is given by (2.7).

That there are densities of the form (2.7) with arbitrary positive h associated with an infinitely divisible distribution F may be seen as follows. Let $M(t) = \exp[\Theta(t)]$ be the moment generating function, convergent in a neighbourhood of the origin, of an infinitely divisible distribution F. Then for every positive integer n there exists a moment generating function M_n such that $M(t) = [M_n(t)]^n$. Given $h > 0$, let $\{m(n)\}$ be a sequence of positive integers such that

$$\lim_{n \to \infty} m(n)/n = h.$$

For each n, $[M(t)]^{m(n)/n} = [M_n(t)]^{m(n)}$ is the moment generating function of the sum of $m(n)$ independent random variables with moment generating function M_n. It can be shown to follow that the limit as $n \to \infty$, $[M(t)]^h$, is then the moment generating function of a distribution. (cf. Chung (1968), Theorem 7.6.5.) Thus if F is infinitely divisible and has moment generating function $\exp[\Theta(t)]$ for t in a neighbourhood of the origin, for each $h > 0$ there is a distribution G_h such that

$$e^{h\Theta(\tau)} = \int e^{y\tau}G_h(dy)$$

and thus a distribution F_h such that

$$e^{h\Theta(\tau)} = \int e^{h\tau s}F_h(ds). \tag{2.20}$$

Following a development similar to that of the case $h = 1$ it is found that $f(y; \theta, h)$ given by (2.7) is a density with respect to the measure F_h.

Example 2.4. Estimation of ordered means and coefficients of variation of normal distributions.

Results obtained in discussion of this example will be used in Section 3.2. Let F_h have density $(h/2\pi)^{\frac{1}{2}} \exp(-hy^2/2)$ with respect to Lebesgue measure. Then $\Theta(\tau)$ defined by (2.20) is $\Theta(\tau) = \tau^2/2$. Also $\theta(\tau) = \tau$, $\varphi(\theta) = \theta$, $\Phi(\theta) = \theta^2/2$, and

$$f(y; \theta, h)F_h(dy) = (h/2\pi)^{\frac{1}{2}} \exp\{-h(y-\theta)^2/2\}\, dy,$$

so that a random variable with this distribution is normally distributed with mean θ and variance $1/h$ (precision h).

Suppose $\bar{y}(x)$ is the mean of a sample of $m(x)$ from a normal distribution with mean $\theta = \mu(x)$ and precision $h = a\lambda(x)$ (variance $1/[a\lambda(x)]$), where $\lambda(x)$ is known, $x \in X$. Suppose $\mu(\cdot)$ is known to be isotonic with respect to a quasi-order "$\lesssim$" on X. By Theorem 2.12, *the maximum likelihood estimate of $\mu(\cdot)$ is the isotonic regression $\hat{\mu}^*(\cdot)$ of $\bar{y}(\cdot)$ with weights $m(x)\lambda(x)$, whether a is known or unknown.*

Suppose that $a = 1$ and that the prior distribution of $\mu(\cdot)$ has the truncated multivariate normal distribution with density

$$\pi[\mu(\cdot)] \propto \exp\{-\tfrac{1}{2} \sum_x [\mu(x) - \mu_0(x)]^2 h(x)\}$$

for isotonic $\mu(\cdot)$,

$$\pi[\mu(\cdot)] = 0 \text{ otherwise,}$$

with modal function $\mu_0(\cdot)$ and precision function $h(\cdot)$. Then *the posterior mode is the isotonic regression $\hat{\mu}^*(\cdot)$ of*

$$\hat{\mu}(\cdot) = [\lambda(\cdot)m(\cdot)\bar{y}(\cdot) + h(\cdot)\mu_0(\cdot)]/[\lambda(\cdot)m(\cdot) + h(\cdot)],$$

with weights $\lambda(\cdot)m(\cdot) + h(\cdot)$, and the posterior precision function is $\lambda(\cdot)m(\cdot) + h(\cdot)$, the posterior distribution being the truncated multivariate normal with density

$$\pi[\mu(\cdot)|\bar{y}(\cdot)] \propto \exp\{-\tfrac{1}{2} \sum_x [\mu(x) - \hat{\mu}(x)]^2 [\lambda(x)m(x) + h(x)]\}$$

for isotonic $\mu(\cdot)$,

$$\pi[\mu(\cdot)|\bar{y}(\cdot)] = 0 \text{ for other } \mu(\cdot).$$

Let us determine also the maximum likelihood estimate of a in the case in which a is unknown. Let the observations be $y_{xj}, j = 1, 2, \ldots, m(x)$, $x \in X$, and let

$$n = \sum_{x \in X} m(x)$$

be the total sample size. Then the likelihood L satisfies

$$-2 \log L = \text{const.} - n \log a + a\{S^2 + \sum_x [s(x) - \mu(x)]^2 m(x) \lambda(x)\}$$

where

$$S^2 = \sum_x \lambda(x) \sum_{j=1}^{m(x)} [y_{xj} - s(x)]^2$$

and

$$s(x) = \sum_{j=1}^{m(x)} y_{xj}/m(x), \qquad x \in X.$$

It is clear from this formula and the definition of isotonic regression that the maximum likelihood estimate of $\mu(\cdot)$ is the isotonic regression $\hat{\mu}^*(\cdot)$ of $s(\cdot)$, even without an appeal to Theorem 2.12. The maximum likelihood estimate of a is

$$n\{S^2 + \sum_x [s(x) - \hat{\mu}^*(x)]^2 \lambda(x) m(x)\}^{-1},$$

and the maximum likelihood estimate of $1/a$ is

$$\{S^2 + \sum_x [s(x) - \hat{\mu}^*(x)]^2 \lambda(x) m(x)\}/n.$$

Suppose $c(x) > 0$ is given for $x \in X$, and suppose that the function $\mu(\cdot)/c(\cdot)$, rather than $\mu(\cdot)$, is known to be isotonic. Set

$$g_1(x) = \bar{y}(x)/c(x), \qquad x \in X,$$

and let g_1^* be the isotonic regression of g_1 with weights $[c(x)]^2 m(x) \lambda(x)$, $x \in X$. By the lemma before Theorem 1.2, g_1^* is the maximum likelihood estimate of $\mu(\cdot)/c(\cdot)$ given that $\mu(\cdot)/c(\cdot)$ is isotonic. (The proof of this lemma is valid for quasi-orders as well as for simple orders.) In particular ($c(x) = \sigma(x) = 1/(\lambda(x))^{\frac{1}{2}}$, $a = 1$), if the standard deviations $\sigma(x)$ are known for $x \in X$, then the maximum likelihood estimate of the coefficient of variation $\mu(\cdot)/\sigma(\cdot)$, known to be isotonic, is the isotonic regression of $\bar{y}(\cdot)/\sigma(\cdot)$ with weights $m(x)$, $x \in X$.

Example 2.5. The gamma distribution.

For $h > 0$, set

$$F_h(dy) = h(hy)^{h-1} e^{-hy} \, dy/\Gamma(h), \qquad y > 0.$$

Then $\Theta(\tau)$ defined by (2.20) is

$$\Theta(\tau) = -\log(1 - \tau), \qquad \tau < 1.$$

Also,

$$\theta(\tau) = 1/(1 - \tau), \qquad \varphi(\theta) = 1 - 1/\theta, \qquad \Phi(\theta) = \theta - \log \theta - 1$$

for $\theta > 0$, and

$$f(y; \theta, h) F_h(dy) = (hy/\theta)^{h-1} e^{-hy/\theta} (h\, dy/\theta)/\Gamma(h), \qquad y > 0.$$

This is the density of the gamma distribution with mean θ and precision h/θ^2 (variance θ^2/h).

Suppose $\bar{y}(x)$ is the mean of a sample of $m(x)$ from the gamma distribution with mean $\theta = \mu(x)$ and precision $a\lambda(x)/[\mu(x)]^2$, where $\lambda(x)$ is known but a may be unknown. By Theorem 2.12, *the maximum likelihood estimate of* $\mu(\cdot)$, *known to be isotonic with respect to a quasi-order on* X, *is the isotonic regression* $\hat{\mu}^*(\cdot)$ *of* $\bar{y}(\cdot)$ *with weights* $m(x)\lambda(x)$, *whether a is known or unknown*.

Example 2.6. Estimation of ordered variances of normal distributions with known means.

Subtracting known means yields distributions with zero means, so that it may be supposed without loss of generality that for $j = 1, 2, \ldots, m(x)$, $y_j(x)$ are independent observations on a normal distribution with mean 0 and variance $\sigma^2(x)$, $x \in X$. The statistics

$$\tilde{t}_j(x) = \tilde{y}_j^2(x), \qquad j = 1, 2, \ldots, m(x),$$

are sufficient for $\sigma^2(x)$, $x \in X$, and $\tilde{t}_j(x)$ has the Gamma distribution with mean $\sigma^2(x)$, variance $2\sigma^4(x)$, precision $1/2\sigma^4(x)$. In applying the result of Example 2.5 set $\mu(x) = \sigma^2(x)$. Then the precision is

$$1/2\sigma^4(x) = a\lambda(x)/[\mu(x)]^2 = a\lambda(x)/\sigma^4(x),$$

so that $a\lambda(x) = 1/2$. *Then the maximum likelihood estimate of* $\mu = \sigma^2$ *known to be isotonic with respect to a quasi-order on* X *is the isotonic regression* $\tilde{t}^*$ *of* $\tilde{t}$ *with weights* $m(x)$, where

$$\tilde{t}(x) = \frac{1}{m(x)} \sum_{j=1}^{m(x)} y_j^2(x), \qquad x \in X.$$

Herbach (1959) encounters an instance of this in connection with maximum likelihood estimation of variances in Model II ANOVA. Let

$$\tilde{x}_{ij} = \mu + \tilde{\eta}_i + \tilde{\varepsilon}_{ij}, \qquad i = 1, 2, \ldots, I, \qquad i = 1, 2, \ldots, J,$$

where μ is constant, and $\{\tilde{\eta}_i\}$ and $\{\tilde{\varepsilon}_{ij}\}$ are independent and normally distributed with

$$\mathrm{E}\,\tilde{\eta}_i = \mathrm{E}\,\tilde{\varepsilon}_{ij} = 0, \qquad \mathrm{Var}\,\tilde{\eta}_i = \sigma_1^2, \qquad \mathrm{Var}\,\tilde{\varepsilon}_{ij} = \sigma_2^2,$$
$$i = 1, 2, \ldots, I, \qquad j = 1, 2, \ldots, J.$$

An appropriate orthogonal transformation transforms $\{\tilde{x}_{ij}\}$ into $\{\tilde{z}_{ij}\}$ with joint density proportional to

$$\lambda_1^{-I/2} \lambda_2^{-I(J-1)/2} \exp\{-\tfrac{1}{2}(s_1/\lambda_1 + s_2/\lambda_2)\},$$

where

$$s_1 = (z_{11} - \sqrt{N}\mu)^2 + \sum_{i=2}^{I} z_{i1}^2, \qquad N = IJ,$$

$$s_2 = \sum_{j=2}^{J} \sum_{i=1}^{I} z_{ij}^2, \qquad \lambda_1 = J\sigma_1^2 + \sigma_2^2, \qquad \lambda_2 = \sigma_2^2.$$

Maximum likelihood estimates of μ, λ_1, λ_2, subject to the restrictions

$$\lambda_1 \geq \lambda_2 \geq 0,$$

are required.

Obviously the maximum likelihood estimate of μ is

$$\hat{\mu} = z_{11}/\sqrt{N}.$$

Substituting this for μ in the likelihood yields

$$\lambda_1^{-I/2} \lambda_2^{-I(J-1)/2} \exp\{-\tfrac{1}{2}(s_1'/\lambda_1 + s_2/\lambda_2)\}$$

where

$$s_1' = \sum_{i=1}^{I} \zeta_i^2, \qquad \zeta_1 = 0, \qquad \zeta_i = z_{i1}, \qquad i = 2, 3, \ldots, I.$$

This is precisely the likelihood that would be obtained if independent samples of sizes I and $I(J - 1)$ were taken from two normal distributions with means 0 and variances λ_1 and λ_2 respectively, and observed values $\zeta_1, \ldots, \zeta_I$ from the first and values z_{ij}, $i = 1, 2, \ldots, I$, $j = 2, 3, \ldots, J$ from the second. Thus an instance of the present example is obtained, with $X = \{1, 2\}$, $m(1) = I$, $m(2) = I(J - 1)$, $\lambda_1 = \sigma^2(1)$, $\lambda_2 = \sigma^2(2)$. The prescribed quasi-order on X is a simple order: $1 \gtrsim 2$. *The maximum likelihood estimate of $\lambda_x = \sigma^2(x)$, $x = 1, 2$, is the isotonic regression of $\bar{\imath}(\cdot)$ with weights $m(\cdot)$ where*

$$\bar{\imath}(1) = \frac{1}{I} \sum_{i=1}^{I} \zeta_i^2 = s'/I,$$

$$\bar{\imath}(2) = \frac{1}{I(J-1)} \sum_{j=2}^{J} \sum_{i=1}^{I} z_{ij}^2 = \frac{s_2}{I(J-1)}.$$

Example 2.7 Maximum likelihood estimation of ordered binomial parameters.

Let F in (2.13) be the distribution function of the Bernoulli distribution with probability 1/2 attached to each of the values 0 and 1. Then

$$e^{\Theta(\tau)} = (1/2)(e^{0\tau} + e^{1\tau}) = (1 + e^{\tau})/2, \quad -\infty < \tau < \infty,$$

and

$$g(y; \tau) = 2e^{y\tau}/(1 + e^{\tau}), \quad y = 0, 1,$$

is a probability density with respect to dF. This is the Bernoulli distribution with probabilities $1/(1 + e^{\tau})$ and $e^{\tau}/(1 + e^{\tau})$ attached to the values 0 and 1 respectively. Its mean is

$$\theta(\tau) = \sum_{y=0}^{1} y g(y; \tau)(1/2) = e^{\tau}/(1 + e^{\tau}).$$

The inverse function φ of $\theta(\cdot)$ is

$$\varphi(y) = \log[y/(1 - y)], \quad 0 < y < 1.$$

The conjugate function Φ of Θ is

$$\Phi(y) = y \log y + (1 - y) \log(1 - y) + \log 2,$$

and the density (2.7) is

$$f(y; \theta) = g(y; \varphi(\theta)) = \exp\{\Phi(\theta) + (y - \theta)\varphi(\theta)\},$$

or

$$f(y; \theta) = 2\theta^y(1 - \theta)^{1-y}, \quad y = 0, 1,$$

the Bernoulli distribution with mean θ. This density is of form (2.7) with $h = 1$.

Suppose $\bar{y}(x)$ is the mean of a sample of $m(x)$ from a Bernoulli distribution with mean $\theta = \mu(x)$, $x \in X$; $m(x)\bar{y}(x)$ has the binomial distribution with parameters $(m(x), \mu(x))$. Suppose the parameters μ are known to satisfy order restrictions which require μ to be isotonic with respect to a prescribed partial order on X. In the absence of order restrictions $\bar{y}(x)$ would be used as a basic estimate of $\mu(x)$, $x \in X$. With the order restrictions, the isotonic regression of $\bar{y}$ with some prescribed weights may be used. In applying Theorem 2.12 by setting $h = 1 = \lambda(x) = 1$, $x \in X$, it is found that the *maximum likelihood estimate of μ is the isotonic regression of $\bar{y}$ with weights m, the sample sizes.*

In a Bayesian approach a prior distribution having density (2.11) is used:

$$\pi(\theta) \propto [\theta^{\theta_0}(1 - \theta)^{1-\theta_0}]^h$$

with respect to the measure $d\rho(\theta) = d\theta/\theta(1 - \theta)$ on $(0, 1)$. The prior density of $\mu(\cdot)$ with respect to the product measure is then

$$\pi[\mu(\cdot)] \propto \prod_{x \in X} \{[\mu(x)]^{\mu_0(x)}[1 - \mu(x)]^{1-\mu_0(x)}\}^{h(x)}$$

for isotonic $\mu(\cdot)$,

$$\pi[\mu(\cdot)] = 0 \text{ otherwise,}$$

where $\mu_0(\cdot)$ is the prior mode function and $h(\cdot)$ the prior precision function. *The posterior mode is the isotonic regression $\hat{\mu}^*(\cdot)$ of*

$$\hat{\mu}(\cdot) = [m(\cdot)\bar{y}(\cdot) + h(\cdot)\mu_0(\cdot)]/[m(\cdot) + h(\cdot)],$$

with weights $m(\cdot) + h(\cdot)$. The posterior density with respect to the product measure

$$\prod_{x \in X} d\mu(x)/\{\mu(x)[1 - \mu(x)]\}$$

is

$$\pi[\mu(\cdot)|\bar{y}(\cdot)] \propto \prod_{x \in X} \{[\mu(x)]^{\hat{\mu}(x)}[1 - \mu(x)]^{1-\hat{\mu}(x)}\}^{m(x) + h(x)}$$

for isotonic $\mu(\cdot)$,

$$\pi[\mu(\cdot)|\bar{y}(\cdot)] = 0 \text{ for other } \mu(\cdot).$$

This is a truncated Beta distribution, nonzero only at isotonic $\mu(\cdot)$.

Example 2.8   Bioassay

This is the example mentioned at the beginning of Section 2.1. For each of various levels x of a stimulus (e.g., dose of an insecticide) the probability of a response (e.g., death of the insect) is $\mu(x)$, known to be nondecreasing in x. If X is the finite set of stimulus levels, it is simply ordered by the natural order. Suppose for $x \in X$ there are $m(x)$ independent trials at stimulus level x, $n(x)$ responses occur, and $\bar{y}(x) = n(x)/m(x)$ is the average number of responses per trial. This is an instance of estimation of ordered binomial parameters, Example 2.7. *The maximum likelihood estimate of μ is furnished by the isotonic regression of $\bar{y}$ with weights m (or any weights proportional to sample sizes).*

Example 2.9   Estimation of safe distance; a traffic problem

Suppose cars travelling along a one-way street A arrive at an intersection with another one-way street B (see Figure 2.10). The car in A will cross the intersection if the distance d of the nearest car in B is at least $\tilde{s}$, the minimum safe distance as judged by the driver in A. An observer of successive cars in A notes values $d_1, d_2, \ldots, d_k$ of $\tilde{d}$, but the values of $\tilde{s}$ associated with the drivers in A are unknown. All that is observable is whether $s \le d_i$ or $s_i > d_i$. Let the ordered observed distances be $x_1 < x_2 < \ldots < x_k$ and let x_i be observed m_i times: $y_{ij} = 1$ if the jth driver observing a car in B at distance x_i went

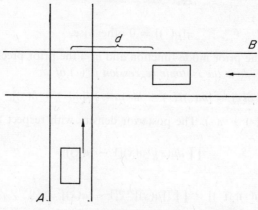

Figure 2.10

through (his safe distance was at least as large as x_i), otherwise $y_{ij} = 0$. It is desired to estimate $p_i = F(x_i) = \Pr(\tilde{s} \le x_i)$. A natural estimate of p_i is

$$\bar{y}_i = \sum_{j=1}^{n_i} y_{ij}/n_i,$$

but this may violate the condition $p_1 \le p_2 \le \ldots \le p_k$. The situation here is the same as in the case of bioassay (Example 2.8) and *the isotonic regression of* $\{\bar{y}_i\}$ *with weights* $\{m_i\}$ *gives also the maximum likelihood estimate.*

Example 2.10 *Maximum likelihood estimation of a discrete distribution with increasing failure rate*

Suppose objects on test fail only at discrete times $x_1 < x_2 < \ldots$. Let $p(x)$ denote the probability that an object chosen at random will fail at time x, and $\rho(x)$ the conditional probability that it will fail at time x given that it has not failed before. The conditional probability $\rho(x)$ is the (discrete) failure rate; suppose it is assumed to be nondecreasing in x.

Let n objects be put on test, and let $n(x_i) \ge 0$ denote the number that fail at time x_i, $i = 1, 2, \ldots, k$,

$$\sum_{i=1}^{k} n(x_i) = n.$$

The likelihood is
$$\prod_{i=1}^{k} [p(x_i)]^{n(x_i)}$$
and the maximizing probabilities will have $p(x_i) = 0$ for $i > k$, i.e.,
$$\sum_{i=1}^{k} p(x_i) = 1.$$
Also
$$\rho(x_i) = p(x_i) / \sum_{j \geq i} p(x_j)$$
and
$$p(x_i) = \rho(x_i) \prod_{j < i} [1 - \rho(x_j)], \qquad i = 1, 2, \ldots, k.$$
The condition
$$\sum_{i=1}^{k} p(x_i) = 1$$
is equivalent to $\rho(x_k) = 1$, and then the likelihood may be written
$$\prod_{i=1}^{k-1} [\rho(x_i)]^{n(x_i)} [1 - \rho(x_i)]^{m(x_i) - n(x_i)},$$
where
$$m(x_i) = \sum_{j=i}^{k} n(x_j).$$

Thus it is as if, first, n objects are put on test and $n(x_1)$ found to fail at the first possible time, x_1; the remaining $n - n(x_1) = m(x_2)$ represent independent trials of an event with probability $\rho(x_2)$; $n(x_2)$ fail at time x_2, etc. The failure rates satisfy $0 < \rho(x_1) \leq \rho(x_2) \leq \ldots \leq \rho(x_{k-1}) \leq 1$. In the absence of the order restrictions the natural estimates would be $\bar{y}(x_i) = n(x_i)/m(x_i)$. Formally the maximum likelihood estimation problem is precisely that discussed in Example 2.7, estimation of ordered binomial parameters. *The solution is the isotonic regression of $\bar{y}(x_i)$ with weights $m(x_i)$, $i = 1, 2, \ldots, k - 1$.*

2.5 MAXIMUM LIKELIHOOD ESTIMATION OF STOCHASTICALLY ORDERED DISTRIBUTIONS

Let a random variable $\tilde{y}$ be known to be stochastically smaller than a random variable $\tilde{z}$; i.e.,
$$F(y) = \Pr\{\tilde{y} \leq y\} \geq G(y) = \Pr\{\tilde{z} \leq y\}$$
for all real y. Suppose all $m + n$ observations of random samples of size m from F and n from G are arranged in increasing order. Some may coincide; when y and z observations occur at the same point, list the y observations

first. Reading from left to right, let a_1 y observations precede all z observations. It will be convenient to assume $a_1 > 0$; the contrary case can easily be reduced to this one. Let b_1 z observations precede the next group of y observations; $b_1 > 0$. Let a_2 uninterrupted y observations follow; $a_2 > 0$ (unless $a_1 = m$). Following these are b_2 z observations; again $b_2 > 0$ unless $b_1 = n$, etc. Suppose in this way the observations are separated into r groups. Again it is assumed that $b_r > 0$; the contrary case can be reduced to this one. Let I_1 be an interval containing only the first a_1 y observations and the first b_1 z observations; I_2 a contiguous interval containing only the second a_2 y observations and the second b_2 z observations, etc. The maximum likelihood estimate of the distribution function F, assumed discrete, will assign equal (unknown) probabilities p_i to all y observations in I_i, $i = 1, 2, \ldots, r$. The maximum likelihood estimate of G will assign equal probabilities q_i to the z observations in I_i, $i = 1, 2, \ldots, r$. The stochastic ordering requirement will be satisfied if and only if

$$\sum_{j=1}^{i} a_j p_j \geq \sum_{j=1}^{i} b_j q_j, \quad i = 1, 2, \ldots, r.$$

Thus, formulating the problem of maximum likelihood estimation in terms of multinomial parameters:

$$a_i p_i = \Pr\{\tilde{y} \in I_i\}, \quad b_i q_i = \Pr\{\tilde{z} \in I_i\}, \quad i = 1, 2, \ldots, r.$$

The problem is to maximize

$$\prod_{i=1}^{r} p_i^{a_i} q_i^{b_i} \tag{2.21}$$

subject to

$$\sum_{j=1}^{i} a_j p_j \geq \sum_{j=1}^{i} b_j q_j, \quad i = 1, 2, \ldots, r-1, \tag{2.22}$$

and

$$\sum_{j=1}^{r} a_j p_j = \sum_{j=1}^{r} b_j q_j = 1. \tag{2.23}$$

Inequalities (2.22) and Equation (2.23) are reminiscent of restrictions which occurred in Section 1.5 in the discussion of dual cones; it will be seen that Proposition 1.1 yields the solution of the present problem. Set $X_1 = \{1, 2, \ldots, r\}$, and define functions p, q, a and b on X_1 by

$$p(x) = p_x, \quad q(x) = q_x, \quad a(x) = a_x, \quad b(x) = b_x, \quad x \in X_1;$$

a and b are assumed to be positive functions. Therefore

$$\sum_{x \in X_1} a(x) = m, \quad \sum_{x \in X_1} b(x) = n.$$

Also, using the argument of the Remark in Section 1.5, we see that (2.22) and (2.23) together are equivalent to

$$\left. \begin{array}{l} \sum_{x \in X_1} [a(x)p(x) - b(x)q(x)]z(x) \leq 0 \\ \text{for every function } z \text{ on } X_1 \text{ iso-} \\ \text{tonic with respect to the natural} \\ \text{order} \end{array} \right\} \quad (2.24)$$

and

$$\sum_{x \in X_1} a(x)p(x) = \sum_{x \in X_1} b(x)q(x) = 1. \quad (2.25)$$

The problem may now be written as follows: to maximize

$$\sum_{x \in X_1} [a(x) \log p(x) + b(x) \log q(x)] \quad (2.26)$$

subject to (2.24) and (2.25).

Set

$$N = m + n = \sum_{x \in X_1} [a(x) + b(x)].$$

Now consider X_1 with the natural order

$$1 \prec 2 \prec \ldots \prec r,$$

and the function

$$g_q(x) = [a(x) + b(x)]/Nb(x), \quad x \in X_1.$$

Let $q^*(x)$, $x \in X_1$, denote the isotonic regression of g_q on X_1 with respect to the natural order, with weights $Nb(x)$, $x \in X_1$. Consider also X_1 with the reversed order:

$$1 \succ 2 \succ \ldots \succ r,$$

and the function

$$g_p(x) = [a(x) + b(x)]/Na(x), \quad x \in X_1.$$

Let $p^*(x)$, $x \in X_1$, denote the isotonic regression of g_p on X_1 with respect to the reversed order, with weights $Na(x)$, $x \in X_1$. Proposition 1.1 will be used to prove the following theorem.

Theorem 2.14

The unique maximum of (2.26) *subject to* (2.24) *and* (2.25) *is furnished by* p^* *and* q^*.

It follows that maximum likelihood estimates of F and G are obtained by assigning probability $p^*(i)$ to each y observation in I_i and probability $q^*(i)$ to each z observation in I_i, $i = 1, 2, \ldots, r$.

The following lemma is useful in proving Theorem 2.14.

Lemma

For $x \in X_1$,

$$\frac{1}{p^*(x)} + \frac{1}{q^*(x)} = N.$$

Proof. With p^* as defined above, it must be shown that $1/(N - 1/p^*)$ is the isotonic regression of $g_q = (a + b)/Nb$ with respect to the natural order on X_1 with weights b. By Robertson's Theorem, Theorem 1.9, it suffices to show that $n - 1/p^*$ is the isotonic regression of $Nb/(a + b)$ with respect to the reverse order n on X_1, with weights $(a + b)/N$; or equivalently, with weights $(a + b)$. By Theorem 1.4, it is enough to show that

$$\sum_{x \in X_1} \left[\frac{Nb(x)}{a(x) + b(x)} - \left(N - \frac{1}{p^*(x)} \right) \right] \left(N - \frac{1}{p^*(x)} \right) [a(x) + b(x)] = 0 \quad (2.27)$$

and

$$\left. \sum_{x \in X_1} \left[\frac{Nb(x)}{a(x) + b(x)} - \left(N - \frac{1}{p^*(x)} \right) \right] z(x)[a(x) + b(x)] \leq 0 \right\} \quad (2.28)$$

for every function z on X_1 isotonic with respect to the reverse order.

Equation (2.27) can be rewritten as

$$\sum_{x \in X_1} \left[\frac{Na(x)}{a(x) + b(x)} - \frac{1}{p^*(x)} \right] \left[\frac{1}{p^*(x)} \right] [a(x) + b(x)]$$

$$= N \sum_{x \in X_1} \left[\frac{Na(x)}{a(x) + b(x)} - \frac{1}{p^*(x)} \right] [a(x) + b(x)].$$

But by Robertson's Theorem, $1/p^*$ is the isotonic regression of $Na/(a + b)$ with respect to the natural order, with weights $(a + b)$. Therefore both members are 0 by Theorem 1.4, completing the verification of (2.27). Also (2.28) is equivalent to

$$\sum_{x \in X_1} \left[\frac{Na(x)}{a(x) + b(x)} - \frac{1}{p^*(x)} \right] [-z(x)][a(x) + b(x)] \leq 0.$$

ESTIMATION OF ORDERED DISTRIBUTIONS

This is immediate by Theorem 1.4, since $-z$ is isotonic with respect to the natural order on X_1. Thus (2.28) is verified, and the proof of the lemma complete. ∎

Proof of Theorem 2.14. In applying Proposition 1.1, set

$$k = 2r, \quad X = K = \{1, 2, \ldots, k\},$$
$$w(x) = a(x) \quad \text{if} \quad 1 \leq x \leq r,$$
$$w(x) = b(x - r) \quad \text{if} \quad r < x \leq 2r = k,$$

and

$$g_1(x) = 0, \quad x \in X.$$

For given functions p and q on X_1, define

$$f(x) = p(x) \quad \text{for} \quad 1 \leq x \leq r,$$
$$f(x) = q(x - r) \quad \text{for} \quad r < x \leq k.$$

In effect, p is the restriction of f to X_1, and a translation of the domain carries the restriction of f to $X - X_1$ into q. Set

$$\Phi(u) = -\log u, \quad \varphi(u) = -1/u \quad \text{for} \quad u > 0.$$

Then the negative of (2.26) may be written:

$$\sum_{x \in X} \Phi[f(x)]w(x). \tag{2.29}$$

Let $\mathscr{K}$ be the class of functions f on X satisfying (2.24) and (2.25). Then the problem is to minimize (2.29) in $\mathscr{K}$.

Let $g_2(x)$ be the function on X determined by g_p and g_q on X_1:

$$g_2(x) = [a(x) + b(x)]/Na(x) \quad \text{if} \quad 1 \leq x \leq r,$$
$$g_2(x) = [a(x - r) + b(x - r)]/Nb(x - r) \quad \text{if} \quad r < x \leq k.$$

Set

$$g_2^*(x) = p^*(x) \quad \text{for} \quad 1 \leq x \leq r,$$
$$g_2^*(x) = q^*(x - r) \quad \text{for} \quad r < x \leq k.$$

Endow the translate $X - X_1$ of X_1 with the natural order, and X_1 itself with the reverse order, so that X bears the partial order

$$1 \gtrsim 2 \gtrsim \ldots \gtrsim r, \quad r + 1 \lesssim r + 2 \lesssim \ldots \lesssim k.$$

It is easy to see that g_2^* is the isotonic regression of g_2 with respect to this partial order on X, with weights w. It will be shown that $u = g_2^*$ satisfies

(1.61) and 1.62). It will then be concluded from Proposition 1.1 that g_2^* minimizes (2.29) in $\mathscr{K}$ and thus that p^* and q^* furnish the unique maximum to (2.26) subject to (2.24) and (2.25).

Since $g_1 = 0$, the equation (1.61):

$$\sum_x [u(x) - g_2(x)]\{\varphi[u(x)] - g_1(x)\}w(x) = 0$$

is immediate from Theorem 1.7. It remains to verify (1.62):

$$\left.\begin{array}{r}\sum_x [f(x) - g_2(x)]\{\varphi[u(x)] - g_1(x)\}w(x) \geq 0 \\ \text{for all } f \in \mathscr{K}.\end{array}\right\}$$

Here this becomes

$$\sum_{x \in X_1} \left[q(x) - \frac{a(x) + b(x)}{Na(x)}\right]\left[-\frac{1}{p^*(x)}\right]a(x)$$
$$+ \sum_{x \in X_1} \left[q(x) - \frac{a(x) + b(x)}{Nb(x)}\right]\left[-\frac{1}{q^*(x)}\right]b(x) > 0$$

for all p, q satisfying (2.24) and (2.25). Using the above lemma the left member reduces to

$$-\sum_{x \in X_1} \{[a(x)p(x) - b(x)q(x)]\left[\frac{1}{p^*(x)}\right] + Nb(x)q(x) - a(x) - b(x)\}.$$

By (2.25)
$$\sum_{x \in X_1} b(x)q(x) = 1,$$

so that
$$\sum_{x \in X_1} [Nb(x)q(x) - a(x) - b(x)] = 0.$$

Thus it remains to show that

$$\sum_{x \in X_1} [a(x)p(x) - b(x)q(x)]\left[\frac{1}{p^*(x)}\right] \leq 0.$$

But since p^* is isotonic with respect to the reverse order on X_1, $1/p^*$ is isotonic with respect to the natural order on X_1, and the inequality is immediate from (2.24). ∎

2.6 COMPLEMENTS

The squared difference $[\mu(x) - g(x)]^2$ in Theorem 2.1 can be interpreted as a "loss" sustained if $g(x)$ is taken as estimate of the unknown true $\mu(x)$. In these terms Theorem 2.1 states that if μ is isotonic, the (weighted) average loss is not greater when the estimate g is replaced by its isotonic regression g^*. This conclusion is valid for many other loss functions as well. For example, suppose $\Delta(\cdot, \cdot)$ is defined by

$$\Delta(u, v) = \Phi(u) - \Phi(v) - (u - v)\Phi'(v),$$

where Φ is a given convex function (cf. Section 1.4). It follows from Theorem 1.10 that the conclusion of Theorem 2.1 holds when squared error loss $[\mu(x) - g(x)]^2$ is replaced by $\Delta[g(x), \mu(x)]$ (which coincides with squared error loss when $\Phi(x) = x^2/2$).

Maximum likelihood estimation of ordered multinomial parameters, the subject of Example 2.1, is discussed by Chacko (1966).

The cumulative isotonic regression function is studied in Brunk (1970). The lemma in Section 2.2 is due to Marshall (1970).

In Section 2.2 the estimation of a regression function μ, known to be isotonic, by the isotonized sample regression function has been discussed briefly. Even if μ is not isotonic, one may consider estimating the (unknown) isotonic regression function μ^*, defined as the isotonic regression of μ. Of course if μ is isotonic, μ and μ^* coincide. It should be noted that μ^* is not defined until a weight function w on X is specified. If weights w on X are specified, one may take the isotonic regression of $\bar{y}$ with weights w as estimator of μ^*. A consistency theorem is immediate from the strong law of large numbers.

Theorem 2.15

The sample isotonic regression function $\bar{y}^*$ with weights $w(x)$, $x \in X$, converges with probability 1 to the isotonic regression μ^* at each $x \in X$ as

$$\min_{x \in X} m(x) \to \infty.$$

In particular, if the regression μ is isotonic, the sample isotonic regression function will converge almost surely to $\mu = \mu^*$, whatever positive weights $w(x)$ are used, as long as they are fixed, independently of sample size.

Proof of theorem. By definition, $\bar{y}^*$ is the isotonic regression of $\bar{y}$ and μ^* is the isotonic regression of μ, both with weights $w(x)$, $x \in X$. It is shown (in a more general context) in Chapter 7 that

$$\sum_x [\mu^*(x) - \bar{y}^*(x)]^2 w(x) \leq \sum_x [\mu(x) - \bar{y}(x)]^2 w(x),$$

But by the strong law of large numbers, $\bar{y}(x)$ converges with probability 1 to $\mu(x)$ for each $x \in X$ as

$$\min_{x \in X} m(x) \to \infty.$$

It follows that $\bar{y}^*(x)$ converges with probability 1 to $\mu^*(x)$ for each $x \in X$. ∎

In Section 2.2 it is observed that if X is held fixed while sample size $m(x) \to \infty$ for each $x \in X$, the sample regression function $\bar{y}$ and the isotonic regression $\bar{y}^*$ are both strongly consistent estimators of an isotonic μ. In the case of a simple order, consistency theorems for the isotonic regression $\bar{y}^*$ have been proved in which $m(x)$ may even be only 1 for each x, provided there are enough observation points x. The first theorem of this kind known to the authors was given by Ayer and coworkers (1955). A generalization appears in Brunk (1955), and both are subsumed by Theorem 6.2 in Brunk (1958), of which a corrected version appears as Theorem 4.1 in Brunk (1970). This theorem is stated here. For convenience, suppose it is desired to estimate the regression function μ on $[0, 1]$. Let $\{x_r\}$ be a sequence of "observation points" in $[0, 1]$, not necessarily distinct. Let $D(x_r)$ be a distribution with mean $\mu(x_r)$ associated with the observation point x_r, $r = 1, 2, \ldots$. Let $\{y_r(x_r)\}$ be a sequence of independent random variables such that $y_r(x_r)$ has distribution $D(x_r)$, $r = 1, 2, \ldots$; in particular, $Ey_r(x_r) = \mu(x_r)$. Let $a(\cdot)$ be a given bounded positive function on $[0, 1]$, bounded away from 0; $a(x_r)$ is to be interpreted as a weighting factor to be applied to $y_r(x_r)$. For a fixed positive integer r, let $0 \leq x_{r,1} \leq \ldots < x_{r,k}$ be the $k = k(r)$ distinct numbers among $x_1, \ldots, x_r$, arranged in increasing order. Let $m_i = m_{r,i}$ denote the number of numbers among $x_1, \ldots, x_r$ which are equal to $x_{r,i}$, and let $w_i = w_{r,i}$ denote the sum of the weights at $x_{r,i}$:

$$m_i = \sum_{j: x_j = x_{r,i}} 1, \quad w_i = m_i a(x_{r,i}).$$

Set

$$\bar{y}_r(x_{r,i}) = \sum_{j: x_j = x_{r,i}} y_j(x_{r,i})/m_i.$$

Let $\bar{y}^*(x_{r,i})$ be the isotonized sample regression with weights w_i, $i = 1, 2, \ldots, k$. The function $\bar{y}^*(x)$ may be defined for $x_{r,1} \leq x \leq x_{r,k}$ by linear interpolation between adjacent observation points, or to be constant on, say, each half-open interval $[x_{r,i}, x_{r,i+1})$. For each r, let $N_r(J)$ denote the number of numbers among $x_1, \ldots, x_r$ which lie in an interval J:

$$N_r(J) = \text{card } \{i: 1 \leq i \leq r, x_i \in J\}.$$

Theorem: Let $a(\cdot)$ be a bounded positive function on $[0, 1]$, bounded away from 0. Let $\mu(\cdot)$ be continuous and nondecreasing on $[0, 1]$. Let $\{x_r\}$ be a sequence of observation points, not necessarily distinct, such that for each interval $J \subset (0, 1)$,

$$\limsup_r r/N_r(J) < \infty.$$

Let the variances of the observed independent random variables $y_r(x_r)$ be bounded. If $0 < \alpha < \beta < 1$, then

$$\Pr\{\lim_r \sup_{\alpha \leq x \leq \beta} |\bar{y}^*(x) - \mu(x)| = 0\} = 1.$$

The Up-and-Down Blocks algorithm (Section 2.3) was developed by J. B. Kruskal (1964b), and the Minimum Violator algorithm by W. A. Thompson, Jr. (1962). The Minimum Lower Sets algorithm is given in Brunk, Ewing, Utz (1957) and in Brunk (1955). For the case of simple order Miles (1959) discusses the Pool-Adjacent-Violators algorithm (Ayer and coworkers (1955)) and the Minimum Lower Sets and Maximum Upper Sets algorithms.

Van Eeden (1956, 1957a, 1957b, 1958) dealt with a range of estimation problems for which isotonic regression provides a solution. Her methods can be used in cases where estimates are required which not only satisfy certain order restrictions, but also lie in specified closed subsets of the real line. They are, however, also applicable when no restrictions of the latter type are imposed. One of van Eeden's theorems forms the basis of the algorithm described in Section 2.3. Another, which should be considered before undertaking computations in the case of a partial order which is not simple is this: if there are two disjoint subsets of X such that no element of one is order-related to any element of the other, then the isotonic regression may be computed for the two subsets separately, i.e. as if they were distinct spaces.

Other algorithms have recently come to the authors' attention. Hoadley (1971) gives an algorithm for simple order. Gebhardt (1970) gives algorithms for simple order and for a partial order arising from the requirement that a regression function should be monotonic nondecreasing in each of two independent variables (cf. Example 2.2); the algorithm can, however, be generalized to deal with any partial order.

Ayer and coworkers (1955) obtained the sample isotonic regression as maximum likelihood estimate of simply ordered probabilities in the stimulus-response (bioassay) situation discussed in Section 2.1 and in Example 2.8. Viewed as the problem of estimating the distribution of the minimum stimulus that will lead to the response, this coincides with Moran's (1966) traffic problem described in Example 2.9. The bioassay problem was also described in van Eeden (1960).

The problem discussed in Example 2.9 is posed in Moran (1966). In their paper on estimating distributions with monotonic failure rate, Marshall and Proschan (1965, page 76) note that when the distribution is discrete the problem is formally equivalent to that of maximum likelihood estimation of ordered binomial parameters (Example 2.7).

W. T. Reid and, independently, Brunk, Ewing and Utz (1957) (see also Brunk (1955, 1958)) recognized that the maximum likelihood estimates of ordered means of distributions belonging to an arbitrary one parameter exponential family are given by the isotonic regression of the sample means with weights proportional to sample sizes. This is the case $a\lambda(x) \equiv 1$ of Theorem 2.12. Consistency theorems appear in Ayer and coworkers (1955), Brunk (1955), Brunk (1958) and in Brunk (1970, Chapter 4).

The class of normal distributions is not, of course, a one parameter exponential family, yet van Eeden (1957a) and Bartholomew (1959a) showed that isotonic regression furnishes maximum likelihood estimates of ordered means in sampling from normal distributions with distinct variances also, provided the ratios of the variances are known. This is the case $\Phi(\theta) = \frac{1}{2}\theta^2$ of Theorem 2.12.

In Theorem 2.12 the problem of maximizing

$$\sum_x \{\Phi[f(x)] + [\tilde{y}(x) - f(x)]\varphi[f(x)]\}\lambda(x)m(x)$$

is interpreted in terms of estimating an isotonic regression function given independent observations on distributions with densities (2.7). But this extremum problem can also be interpreted as a problem of maximum likelihood estimation of means of dependent random variables. Suppose that the random variables $\tilde{y}(x)$, $x \in X$, are linearly related: there exists a known function ψ and known positive numbers $m(x)$, $x \in X$, such that

$$\sum_x [\tilde{y}(x) - \mu(x)]\psi[\mu(x)]m(x) \equiv 0, \qquad (2.30)$$

where $\mu(x) = E\tilde{y}(x)$ (unknown). Suppose further that the likelihood is proportional to

$$\exp \sum_x \{\Phi[\mu(x)] + [y(x) - \mu(x)]\varphi[\mu(x)]\}m(x). \qquad (2.31)$$

Then again, by the corollary to Theorem 1.10, the maximum likelihood estimate of μ, known to be isotonic in X, is just the isotonic regression of $\{y(x), x \in X\}$ with weights $\{m(x), x \in X\}$. Example 2.1, estimation of ordered multinomial parameters, gives an instance of this, in which $\Phi(u) = u \log u$, $\varphi(u) = \Phi'(u) = \log u + 1$, $\mu > 0$ and $\psi(u) = 1$. Set $X = K = \{1, 2, \ldots, k\}$, and identify $\tilde{y}(x)$ with $\tilde{r}_x$, $x \in X$, and $\mu(x) = E\tilde{y}(x) = E\tilde{r}_x = np_x$, $x \in X$. The random variables $\tilde{y}(x) = \tilde{r}_x$ are linearly related by (2.30), in which $\psi(\cdot) \equiv 1$, $m(x) = 1$, $x \in X$, which becomes

$$\sum_x [\tilde{r}_x - np_x] = 0,$$

or

$$\sum_x \tilde{r}_x = n.$$

The joint discrete density of $\tilde{r}_1, \ldots, \tilde{r}_2$ on the hyperplane

$$\sum_x r_x = n$$

is

$$n! \prod_x p_x^{r_x} / \prod_x r_x!.$$

The likelihood is proportional to

$$\exp\{\sum_x r_x \log p_x\}$$

and, since

$$\sum_x (r_x - np_x) = 0,$$

to

$$\exp\{\sum_x [np_x \log np_x + (r_x - np_x)(1 + \log np_x)]\} \qquad (2.32)$$

which is (2.31).

It is interesting to note also that, since

$$\sum_x r_x = n,$$

(2.32) is also proportional to

$$\exp\{\sum_x (r_x \log \mu_x - \mu_x)\},$$

which is the likelihood function for a sample from independent Poisson distributions.

A more direct but long derivation of the maximum likelihood estimates of two stochastically ordered distributions is given by Brunk and coworkers (1966). Another derivation, based on the dual theory of isotonic regression, (see Section 1.5) is given in Barlow and Brunk (1972).

CHAPTER 3

Testing the Equality of Ordered Means: Likelihood Ratio Tests in the Normal Case

3.1 INTRODUCTION

A substantial part of statistical theory is concerned with making inferences about the means of normal populations. When several such means have to be compared, it is often useful to begin the analysis by making a test of the null hypothesis that they are equal. Most of the standard tests which are used in this situation are not designed to detect any specific kind of differences between the population means. However, it has already been seen that there are many practical situations in which the means are subject to order restrictions; and, when this is so, it would be expected that more powerful tests could be devised. In fact, a simple example where this is the case has already been met when, in Chapter 1, the one-tailed test for a pair of means was mentioned. One of the most important applications of isotonic regression theory is to testing hypotheses about means. It is the principal object of this and the following chapter to present the methods which have been developed for this purpose.

In order to set the problem in the context of the previous chapters it will be described in the following terms. A random variable $\tilde{y}$ has a normal distribution with mean $\mu(x)$ and variance $\sigma^2(x)$. Independent observations have been made on $\tilde{y}$ at k different levels of x denoted by $x_1, x_2, \ldots, x_k$. We wish to test the null hypothesis

$$H_0: \mu(x_1) = \mu(x_2) = \ldots = \mu(x_k),$$

using prior knowledge which enables us to impose order restrictions on the $\mu(x_i)$s. It will be found necessary to assume, in most cases, either that $\sigma^2(x_i)$ is known for each x_i in the experiment, or that it is known up to a multiplicative constant. If the form of the dependence of $\sigma^2(x)$ on x is known, it may be possible to make a transformation from the variable $\tilde{y}$ to a variable with approximately constant variance. The reader will recognize that the set-up which has been described is the same as that on which the standard statistical tests for means are based.

Two main approaches have been made to the problem of testing H_0 against ordered alternatives. One is to convert it into a standard regression problem by the introduction of artificial variables, or "scores", in place of the xs. This method is described in the following chapter. The second approach is by way of the likelihood ratio principle, which yields a test criterion provided that the likelihood of the sample under the alternative hypothesis can be maximized. In the present chapter the theory of this approach will be developed and its applications discussed.

The use of the likelihood ratio principle is usually justified by its intuitive appeal and by the desirable asymptotic properties which tests derived by this method possess. In this case it will be seen that the asymptotic justification is lacking: nevertheless it will be found that the tests obtained are intuitively acceptable, are closely related to the standard χ^2 and F tests for the unordered case, and have power characteristics which compare favourably with their competitors.

In the unordered case, normal theory tests are often used as approximations in circumstances when the random variable is not exactly normal. The situation concerning ordered tests is similar: in particular, the tests may be used for testing the equality of a set of proportions.

3.2 THE $\bar{\chi}^2$ AND $\bar{E}^2$ TESTS

The $\bar{\chi}^2$ test

Suppose that, for each i in $\{1, 2, \ldots, k\}$, n_i values of $\tilde{y}$ are observed when x has the value x_i. The resulting data may then be set out as in Table 3.1. It will be noted that, in order to simplify the notation in the remainder of this chapter, the "tilde" previously used to distinguish random variables has been omitted, and μ_i has been written for $\mu(x_i)$, and σ_i^2 for $\sigma^2(x_i)$.

Table 3.1

Population	1	2	...	k
Observations	y_{11}	y_{21}	...	y_{k1}
	y_{12}	y_{22}	...	y_{k2}
	.	.		.
	.	.		.
	.	.		.
	y_{1n_1}	y_{2n_2}	...	y_{kn_k}
Sample means	$\bar{y}_1$	$\bar{y}_2$	...	$\bar{y}_k$
Population means	μ_1	μ_2	...	μ_k
Population variances	σ_1^2	σ_2^2	...	σ_k^2

In many situations the object, given the above data, may be to test the null hypothesis

$$H_0: \mu_1 = \mu_2 = \ldots = \mu_k$$

against the alternative hypothesis that these means are not all equal—denoted by $\bar{H}_0$. For such situations tests are well known. Tests will be considered against the members of a class $\mathcal{H}$ of alternative hypotheses, each of which restricts the population means, not only to satisfy $\bar{H}_0$, but also to be such that μ_i is an isotonic function of i with respect to some partial ordering on the set $K = \{1, 2, \ldots, k\}$. Thus a member of $\mathcal{H}$ restricts the population means so that they satisfy a number of inequalities of the form $\mu_i \leq \mu_j$, and are not all equal. A particular member of $\mathcal{H}$ to which considerable attention will be devoted is

$$H_1: \mu_1 \leq \mu_2 \leq \ldots \leq \mu_k,$$

not all μs equal, referred to as the case of simple order. However, in this section full generality will be retained.

In what follows, the notation H will denote both an alternative hypothesis of the class $\mathcal{H}$ and the region to which the population means are restricted by the corresponding defining partial order—that is, the union of the regions corresponding to null and alternative hypotheses. Similarly, the notation H_0 will be used to denote the region within which the population means are all equal.

The likelihood ratio test for testing H_0 against an alternative $H \in \mathcal{H}$ rejects H_0 for small values of the statistic

$$\lambda = \frac{\max_{\mu \in H_0} L(\mathbf{y}_1, \mathbf{y}_2, \ldots, \mathbf{y}_k | \boldsymbol{\mu}, \boldsymbol{\sigma})}{\max_{\mu \in H} L(\mathbf{y}_1, \mathbf{y}_2, \ldots, \mathbf{y}_k | \boldsymbol{\mu}, \boldsymbol{\sigma})}, \tag{3.1}$$

where $\mathbf{y}'_i = [y_{i1}, y_{i2}, \ldots, y_{in_i}]$, $\boldsymbol{\mu}' = [\mu_1, \mu_2, \ldots, \mu_k]$, $\boldsymbol{\sigma}' = [\sigma_1, \sigma_2, \ldots, \sigma_k]$, and $L(\cdot)$ is the likelihood function. For the moment $\boldsymbol{\sigma}$ is supposed known.

Under H_0, it is well known that the maximum likelihood estimators of the population means are given by

$$\hat{\mu}_1 = \hat{\mu}_2 = \ldots = \hat{\mu}_k = \bar{y} = \hat{\mu}, \tag{3.2}$$

say, where

$$\bar{y} = \sum_{i=1}^{k} w_i \bar{y}_i / \sum_{i=1}^{k} w_i \text{ and } w_i = n_i \sigma_i^{-2}.$$

And, under H, it has been shown in Example 2.4 that the maximum likelihood estimates are $\hat{\mu}_1^*, \hat{\mu}_2^*, \ldots, \hat{\mu}_k^*$, where $\hat{\mu}_i^*$ is the isotonic regression of $\bar{y}_i$, with weights w_i, with respect to the partial ordering on K which defines H.

THE $\bar{\chi}^2$ AND E^2 TESTS

The likelihood function is as follows:

$$L(y_1, y_2, \ldots, y_k | \mu, \sigma)$$

$$= (2\pi)^{-(N/2)} \prod_{i=1}^{k} \sigma_i^{-n_i} \exp\left\{-\tfrac{1}{2} \sum_{i=1}^{k} \frac{1}{\sigma_i^2} \sum_{j=1}^{n_i} (y_{ij} - \mu_i)^2\right\}, \quad (3.3)$$

where N is the total sample size. The likelihood ratio test rejects H_0, therefore, for large values of the statistic

$$-2 \log \lambda = \sum_{i=1}^{k} w_i n_i^{-1} \sum_{j=1}^{n_i} (y_{ij} - \hat{\mu})^2 - \sum_{i=1}^{k} w_i n_i^{-1} \sum_{j=1}^{n_i} (y_{ij} - \hat{\mu}_i^*)^2$$

$$= \sum_{i=1}^{k} w_i (\hat{\mu}_i^* - \hat{\mu})^2 + 2 \sum_{i=1}^{k} w_i (\bar{y}_i - \hat{\mu}_i^*)(\hat{\mu}_i^* - \hat{\mu})$$

$$= \sum_{i=1}^{k} w_i (\hat{\mu}_i^* - \hat{\mu})^2, \quad (3.4)$$

the second term vanishing, as can be seen from application of Theorem 1.7 to the isotonic regression $\hat{\mu}_i^*$.

The test statistic (3.4) is similar in form to the χ^2 statistic calculated to test H_0 against $\bar{H}_0$, and will be denoted by $\bar{\chi}_k^2$. The subscript k is introduced only to denote the number of means being compared; it does not define the null hypothesis distribution of the test statistic, which will be seen to depend on the alternative H and the weights $w_1, w_2, \ldots, w_k$.

In the usual model for the one-way analysis of variance all populations have the same variance σ^2, and the likelihood ratio test rejects H_0 for large values of

$$\bar{\chi}_k^2 = \sum_{i=1}^{k} n_i (\hat{\mu}_i^* - \hat{\mu})^2 / \sigma^2. \quad (3.5)$$

In this case, an alternative form of the statistic is often useful for purposes of calculation. It has been seen in Chapter 2 that the process by which the $\hat{\mu}_i^*$s are obtained involves the amalgamation of some of the original samples into level sets, within each of which $\hat{\mu}_i^*$ is constant and equal to the mean value of y in the amalgamated sample. If the number of level sets is l, the amalgamated samples being of sizes $N_1, N_2, \ldots, N_l$, and having means $Y_1, Y_2, \ldots, Y_l$, then

$$\bar{\chi}_k^2 = \sum_{i=1}^{l} N_i (Y_i - \hat{\mu})^2 / \sigma^2 = \sum_{i=1}^{l} N_i (Y_i - \bar{Y})^2 / \sigma^2, \quad (3.6)$$

where

$$\bar{Y} = \sum_{i=1}^{l} N_i Y_i / \sum_{i=1}^{l} N_i.$$

Thus $\sigma^2 \bar{\chi}_k^2$ is identical with the between groups sum of squares after amalgamation has taken place. The calculations are illustrated by the following example.

Example 3.1

Assume $\sigma^2 = 10$; given the following data we wish to test H_0 against H_1.

i	1	2	3	4
n_i	5	12	10	14
$\bar{y}_i$	15.0	13.4	17.2	16.4

To find the estimates $\hat{\mu}_i^*$ under H_1, the Pool-Adjacent-Violators algorithm or any of the other algorithms described in Chapter 2 is applied, and samples 1 and 2, and samples 3 and 4, are pooled, giving estimates

$$\hat{\mu}_1^* = \hat{\mu}_2^* = \frac{5 \times 15.0 + 12 \times 13.4}{17} = 13.871,$$

and

$$\hat{\mu}_3^* = \hat{\mu}_4^* = \frac{10 \times 17.2 + 14 \times 16.4}{24} = 16.733.$$

So the value of $\bar{\chi}_4^2$ is found by substituting into (3.6), with $l = 2$, $N_1 = 17$, $Y_1 = 13.781$, $N_2 = 24$, $Y_2 = 16.733$. We obtain

$$\bar{\chi}_4^2 = \frac{1}{10}\left[17(13.871)^2 + 24(16.733)^2 - 41\left\{\frac{17 \times 13.871 + 24 \times 16.733}{41}\right\}^2\right]$$

$$= \frac{1}{10}(9990.791 - 9909.238)$$

$$= 8.155.$$

The $\bar{E}^2$ test

In most applications the variances, σ_i^2, are unknown, and have to be estimated from the data. If this is the case the $\bar{\chi}^2$ test can be extended provided that the variances are of the form

$$\sigma_i^2 = a_i \sigma^2 \quad (i = 1, 2, \ldots, k)$$

where the a_i are known constants and σ^2 is unknown.

In this case the maximum likelihood estimators of the population means under H_0 are again as given in (3.2), the weights now being $w_i = n_i a_i^{-1}$, $i = 1, 2, \ldots, k$; and the estimator of σ^2 is

$$\hat{\sigma}_0^2 = \sum_{i=1}^{k} a_i^{-1} \sum_{j=1}^{n_i} (y_{ij} - \hat{\mu})^2 / N. \tag{3.7}$$

The problem of finding maximum likelihood estimators under H has been solved in Example 2.4. The population means are estimated by $\hat{\mu}_1^*, \hat{\mu}_2^*, \ldots, \hat{\mu}_k^*$, the values of the isotonic regression of $\bar{y}_i$ with the above weights; and the corresponding estimator of σ^2 is

$$\hat{\sigma}^2 = \left[\sum_{i=1}^{k} a_i^{-1} \sum_{j=1}^{n_i} (y_{ij} - \bar{y}_i)^2 + \sum_{i=1}^{k} n_i a_i^{-1} (\bar{y}_i - \hat{\mu}_i^*)^2 \right] / N$$

$$= \sum_{i=1}^{k} a_i^{-1} \sum_{j=1}^{n_i} (y_{ij} - \hat{\mu}_i^*)^2 / N. \tag{3.8}$$

The likelihood function in this case is

$$(2\pi\sigma^2)^{-(N/2)} \prod_{i=1}^{k} a_i^{-(n_i/2)} \exp\left\{ -\frac{1}{2\sigma^2} \sum_{i=1}^{k} a_i^{-1} \sum_{j=1}^{n_i} (y_{ij} - \mu_i)^2 \right\}, \tag{3.9}$$

and it follows that the likelihood ratio test rejects H_0 for small values of the statistic

$$\lambda = \left(\frac{\hat{\sigma}^2}{\hat{\sigma}_0^2} \right)^{N/2}. \tag{3.10}$$

In the case most frequently encountered in practice, when $a_i = 1$ for all i, this statistic is given by

$$\lambda^{2/N} = \sum_{i=1}^{k} \sum_{j=1}^{n_i} (y_{ij} - \hat{\mu}_i^*)^2 / \sum_{i=1}^{k} \sum_{j=1}^{n_i} (y_{ij} - \hat{\mu})^2.$$

It is more convenient to take the complement of this quantity, and H_0 is accordingly rejected for large values of

$$\bar{E}_k^2 = \sum_{i=1}^{k} n_i (\hat{\mu}_i^* - \hat{\mu})^2 / \sum_{i=1}^{k} \sum_{j=1}^{n_i} (y_{ij} - \hat{\mu})^2$$

$$= \sigma^2 \bar{\chi}_k^2 / S_T, \tag{3.11}$$

where S_T denotes the total sum of squares for the data.

In analysis of variance terminology $\bar{E}_k^2$ is the ratio of the between groups sum of squares, calculated after amalgamation, and the total sum of squares. Thus the notation chosen is a natural extension of that used by Welch (1937) for the case when there are no order restrictions on the means. Another form in which the statistic can be written is

$$\bar{E}_k^2 = \frac{l-1}{N-k} F \bigg/ \left(1 + \frac{l-1}{N-k} F\right),$$

where F is the usual mean square ratio for the one-way analysis of variance, calculated after amalgamation. It is important, however, to recognize that a critical region of the form

$$\bar{E}_k^2 \geq \text{constant}$$

is not equivalent to one of the form

$$F \geq \text{constant,}$$

as in the ordinary analysis of variance. This is because l, in the equation connecting $\bar{E}_k^2$ and F, is a random variable in the present problem. This point will be met again in Section 3.5.

Example 3.2

This example is based on data given in Bartholomew (1959a), with $k = 5$ and $n_i = 4$ for all i. We wish to test H_0 against the alternative H_1.

i	1	2	3	4	5
$\bar{y}_i$	27.50	30.75	31.50	35.00	24.00
$\sum_{j=1}^{4} y_{ij}^2$	3380	3601	3712	3984	3201

$$\sum_{i=1}^{5}\sum_{j=1}^{4} y_{ij} = 595, \quad \text{and} \quad \sum_{i=1}^{5}\sum_{j=1}^{4} y_{ij}^2 = 17\,878.$$

Thus, under H_0, the common mean is estimated by

$$\hat{\mu} = 595/20 = 29.75.$$

Application of any of the estimation algorithms described in the previous chapter gives the following estimates of the population means under H_1:

$$\hat{\mu}_1^* = 27.50; \quad \hat{\mu}_2^* = \hat{\mu}_3^* = \hat{\mu}_4^* = \hat{\mu}_5^* = 30.31.$$

The total sum of squares is

$$17\,878 - 595^2/20 = 176.75;$$

and the between groups sum of squares when groups 2–5 have been amalgamated is

$$110^2/4 + 485^2/16 - 595^2/20 = 25.31.$$

Therefore

$$\bar{E}_5^2 = 25.31/176.75 = 0.143.$$

Methods for testing the significance of both the $\bar{\chi}_k^2$ and $\bar{E}_k^2$ statistics will be developed in Section 3.3.

The $\bar{E}^2$ test for main effects in orthogonal designs

Shorack (1967) has extended the above theory to the case of the two-way cross-classification with one observation per cell, and to the Latin square design, and points out that such an extension can be made to any orthogonal design. A general method of testing against ordered alternatives will be given for the linear model in Chapter 4, but because of their simplicity orthogonal designs will be dealt with briefly at this point, using the balanced two-way layout as an illustration.

Consider first the case when the error variance is known. The standard parametric model is

$$y_{tij} = \mu + \alpha_t + \beta_i + \gamma_{ti} + e_{tij} \qquad (3.12)$$

$$(t = 1, 2, \ldots, k; i = 1, 2, \ldots, m; j = 1, 2, \ldots, n)$$

where

$$\sum_{t=1}^{k} \alpha_t = \sum_{i=1}^{m} \beta_i = \sum_{t=1}^{k} \gamma_{ti} = \sum_{i=1}^{m} \gamma_{ti} = 0,$$

and the es are mutually independent normal random variables with zero means and common variance σ^2. Since all cell frequencies are equal

$$\bar{y}_{t..} = \mu + \alpha_t + \bar{e}_{t..} \qquad (t = 1, 2, \ldots, k),$$

in the usual notation for means.

If it is desired to test the null hypothesis

$$H_0: \alpha_1 = \alpha_2 = \ldots = \alpha_k = 0$$

against an alternative H restricting the αs to satisfy inequalities of the form $\alpha_i \leq \alpha_j$, the problem is essentially the same as that discussed earlier in this section for the case of the one-way classification when σ^2 is known. To establish the equivalence, the following substitutions are made in the earlier results: for each i, $i = 1, 2, \ldots, k$, μ_i is replaced by $\mu + \alpha_i$; $\bar{y}_i$ by $\bar{y}_{i..}$; and n_i by mn. The test statistic now has the form

$$\bar{\chi}_k^2 = mn \sum_{t=1}^{k} (\hat{\mu}_t^* - \bar{y}_{...})^2 / \sigma^2,$$

where $\hat{\mu}_t^*$ is obtained as the isotonic regression of $\bar{y}_{t..}$ with equal weights. A similar test can obviously be constructed for the βs. In any other orthogonal design we proceed in the same way using the sample means for the main effect in question.

If σ^2 is unknown the appropriate extension of the foregoing test can be obtained by the likelihood ratio method, as in Shorack (1967). It has already been shown that, for the one-way classification, the likelihood ratio is given by

$$\lambda^{2/N} = \hat{\sigma}^2 / \hat{\sigma}_0^2,$$

where $\hat{\sigma}_0^2$ and $\hat{\sigma}^2$ are the maximum likelihood estimators of σ^2 under the null and alternative hypotheses respectively. This result applies, in fact, to more general linear models, and in particular to any orthogonal design: thus, in the present example, these estimators of σ^2 have to be determined when the model (3.12) applies. If $\mu_{ti} = \mu + \alpha_t + \beta_i + \gamma_{ti}$, then

$$N\hat{\sigma}_0^2 = \min_{H_0} S_s,$$

and

$$N\hat{\sigma}^2 = \min_{H} S_s,$$

where

$$S_s = \sum_{t=1}^{k} \sum_{i=1}^{m} \sum_{j=1}^{n} (y_{tij} - \mu_{ti})^2.$$

For an orthogonal design the two minimizations are easily effected because S_s can be decomposed into parts each involving only one set of parameters. In the present example, it is well known that

$$S_s = S_\mu + S_\alpha + S_\beta + S_\gamma + S_R, \qquad (3.13)$$

where

$$S_\mu = kmn(\hat{\mu} - \mu)^2,$$

$$S_\alpha = mn \sum_{t=1}^{k} (\hat{\alpha}_t - \alpha_t)^2,$$

$$S_\beta = kn \sum_{i=1}^{m} (\hat{\beta}_i - \beta_i)^2,$$

$$S_\gamma = n \sum_{t=1}^{k} \sum_{i=1}^{m} (\hat{\gamma}_{ti} - \gamma_{ti})^2,$$

$$S_R = \sum_{t=1}^{k} \sum_{i=1}^{m} \sum_{j=1}^{n} (y_{tij} - \hat{\mu}_{ti})^2$$

(the residual sum of squares) and the use of a circumflex denotes the ordinary unrestrained least squares estimator of the corresponding parameter. In this case, the least squares estimators are as follows:

$$\hat{\mu} = \bar{y}_{...},$$
$$\hat{\alpha}_t = \bar{y}_{t..} - \bar{y}_{...},$$
$$\hat{\beta}_i = \bar{y}_{.i.} - \bar{y}_{...},$$
$$\hat{\gamma}_{ti} = \bar{y}_{ti.} - \bar{y}_{t..} - \bar{y}_{.i.} + \bar{y}_{...},$$

and

$$\hat{\mu}_{ti} = \hat{\mu} + \hat{\alpha}_t + \hat{\beta}_i + \hat{\gamma}_{ti} = \bar{y}_{ti.}.$$

On minimizing (3.13) subject to the restraints H it is found that the terms involving the βs and the γs vanish, so that

$$N\hat{\sigma}^2 = \min_H S_\alpha + S_R.$$

Similarly, after putting $\alpha_t = 0$ for all t,

$$N\hat{\sigma}_0^2 = mn \sum_{t=1}^{k} \hat{\alpha}_t^2 + S_R.$$

Hence the likelihood ratio test rejects H_0 for large values of the statistic

$$\bar{E}_k^2 = 1 - \lambda^{2/N} = \frac{mn \sum_{t=1}^{k} (\bar{y}_{t..} - \bar{y}_{...})^2 - mn \sum_{t=1}^{k} (\bar{y}_{t..} - \hat{\mu}_t^*)^2}{mn \sum_{t=1}^{k} (\bar{y}_{t..} - \bar{y}_{...})^2 + S_R}.$$

Making use of (3.4), with $w_i = mn$, $n_i = 1$, $i = 1, 2, \ldots, k$, it is found that

$$E_k^2 = \frac{mn \sum_{t=1}^{k} (\hat{\mu}_t^* - \bar{y}_{...})^2}{mn \sum_{t=1}^{k} (\bar{y}_{t..} - \bar{y}_{...})^2 + S_R}. \tag{3.14}$$

The numerator of this expression is $\sigma^2 \bar{\chi}_k^2$, and the denominator is the sum of the relevant main effect sum of squares and the residual sum of squares. This result applies in any situation concerned with testing the equality of ordered main effects in an orthogonal design. The test statistic (3.11) already developed for the case of the one-way classification can be seen to arise from the application of this general result.

Two-sided versions of $\bar{\chi}_k^2$ and $\bar{E}_k^2$

In some applications the direction of the ordering under H may be unknown. Suppose, for example, that it is known only that $\mu_1, \mu_2, \ldots, \mu_k$ form a monotonic sequence, without knowledge as to whether it increases or decreases as the subscript runs from 1 to k. The likelihood ratio method may still be applied. The likelihood is first maximized under the increasing alternative, and then under the decreasing alternative. The larger of the two maxima is the one which is substituted in λ. Since λ is a monotonic function of $\bar{\chi}_k^2$ (or $\bar{E}_k^2$) this procedure is equivalent to calculating $\bar{\chi}_k^2$ ($\bar{E}_k^2$) under each assumption about the direction of the ordering, and using the larger of the two values as the value of the test statistic. When $k = 2$, this reduces to the standard two-tail test. The same method can clearly be used if the alternative hypothesis is a union of several hypotheses of the class $\mathcal{H}$.

3.3 THE NULL HYPOTHESIS DISTRIBUTIONS OF $\bar{\chi}^2$ AND $\bar{E}^2$

Basic result for any alternative of the class $\mathcal{H}$

The basic distributional results concerning $\bar{\chi}_k^2$ and $\bar{E}_k^2$ are contained in the following theorems.

Theorem 3.1

If H_0 is true then

$$\Pr\{\bar{\chi}_k^2 \geq C\} = \sum_{l=2}^{k} P(l, k) \Pr\{\chi_{l-1}^2 \geq C\}, \qquad C > 0$$

$$\Pr\{\bar{\chi}_k^2 = 0\} = P(1, k),$$

where $P(l, k)$ is the probability that the isotonic regression function $\hat{\mu}_i^*$ takes exactly l distinct values, and the notation χ_ν^2 is used to denote a random variable having the χ^2 distribution with ν degrees of freedom.

Theorem 3.2

If H_0 is true then

$$\Pr\{\bar{E}_k^2 \geq C\} = \sum_{l=2}^{k} P(l,k) \Pr\{B_{\frac{1}{2}(l-1),\frac{1}{2}(N-l)} \geq C\}, \qquad C > 0$$

$$\Pr\{\bar{E}_k^2 = 0\} = P(1,k),$$

where $P(l, k)$ is as in the previous theorem, and the notation $B_{a,b}$ is used to denote a random variable having the Beta distribution with parameters a and b.

The density functions of $\bar{\chi}_k^2$ and $\bar{E}_k^2$ are thus weighted sums of well-known densities; their dependence on the alternative hypothesis envisaged is through the weights $P(l, k)$. Distributions of the types given in Theorems 3.1 and 3.2 will be referred to subsequently as $\bar{\chi}^2$ and $\bar{E}^2$ distributions respectively. Theorem 3.1 shows that the asymptotic distribution of $-2 \log \lambda = \bar{\chi}_k^2$ is not that of χ_{k-1}^2: an illustration of the fact mentioned earlier that the usual asymptotic results concerning likelihood ratio tests do not hold in the situation under consideration. The reason for this is that an open region cannot be found lying in the region of the restricted parameter space and containing the points corresponding to the null hypothesis.

Theorem 3.1 was first proved for the case of simple order in Bartholomew (1959a). An extension to cover any alternative of the class $\mathscr{H}$ was given in Bartholomew (1961a); but this latter proof has been found to be deficient in that it is not valid for some members of $\mathscr{H}$. More recently, Shorack (1967) has published an alternative proof. Like Bartholomew's this proof is based on induction on the number of restrictions imposed by the alternative hypothesis. Both proofs depend for their validity on the assumption that the addition of further restrictions to the alternative hypothesis either leaves the estimates $\hat{\mu}_1^*, \ldots, \hat{\mu}_k^*$ unchanged, or results in a new set of estimates formed by making further amalgamations of the groups. That this assumption is valid for the alternative of simple order can be seen from consideration of the Pool-Adjacent-Violators algorithm. It is not, however, valid for a general partial order. A new proof is therefore given here, valid for any alternative of the class $\mathscr{H}$, and covering Theorems 3.1 and 3.2. The new proof is somewhat longer than its predecessors, but the authors have not been able to shorten it without loss of generality.

It will be recalled that $\bar{\chi}_k^2$, and the numerator of $\bar{E}_k^2$, are calculated as the between groups sum of squares for the one-way layout formed by the amalgamation process. If the groups to be amalgamated were to be selected at random the conditional distribution of $\bar{\chi}_k^2$, given l, would be that of χ_{l-1}^2; similarly $\bar{E}_k^2$ would have the distribution of $B_{\frac{1}{2}(l-1),\frac{1}{2}(N-l)}$. The amalgamations, however, are not random, but are a consequence of inequalities involving the group means $\bar{y}_i$. Thus in order to establish the theorems it must be shown that the existence of the restraints does not affect the conditional distribution. The proof is in three parts: the first specifies the two kinds of restraint which the amalgamation procedure imposes; and in the second and third parts it is shown that they can be disregarded for the purposes of finding the distributions of $\bar{\chi}_k^2$ and $\bar{E}_k^2$.

To simplify the proof, attention is confined to the case when $\sigma_i^2 = \sigma^2$ for all i. The extension to the case when the σs are not all equal is straightforward. It will be found convenient to use the following notation, similar to that introduced in Chapter 1. Denoting the set $\{1, 2, \ldots, k\}$ by K, for any non-empty subset A of K, define

$$N_A = \sum_{i \in A} n_i,$$

and

$$\text{Av } A = \sum_{i \in A} n_i \bar{y}_i / N_A.$$

The following three lemmas will be required in the proof.

Lemma A

If A and B are non-empty subsets of K such that $A \supset B$, $A \neq B$, then the covariance of Av A and Av A − Av B is zero.

Proof.

$$\begin{aligned} &\text{Cov } (\text{Av } A, \text{Av } A - \text{Av } B) \\ &= \text{Var } (\text{Av } A) - \text{Cov } (\text{Av } A, \text{Av } B) \\ &= \frac{\sigma^2}{N_A} - \text{Cov}\left\{\frac{N_{A-B} \text{ Av } (A - B) + N_B \text{ Av } B}{N_A}, \text{Av } B\right\} \\ &= \frac{\sigma^2}{N_A} - 0 - \frac{\sigma^2}{N_A}. \end{aligned}$$ ∎

Lemma B

If $u_1, u_2, \ldots, u_r$ are independent normal random variables with zero means and unit variances, and R is a set of restrictions on the us, of the form

$$\sum_{i=1}^r a_i u_i \geq 0,$$

such that the probability that R is satisfied is non-zero, then the conditional distribution of

$$\sum_{i=1}^{r} u_i^2,$$

given R, is that of a χ^2 random variable with r degrees of freedom.

Proof. Make a polar transformation to random variables $\rho, \theta_1, \theta_2, \ldots, \theta_{r-1}$, given by

$$u_1 = \rho^{\frac{1}{2}} \cos \theta_1 \cos \theta_2 \ldots \cos \theta_{r-1}$$
$$u_i = \rho^{\frac{1}{2}} \cos \theta_1 \cos \theta_2 \ldots \cos \theta_{r-i} \sin \theta_{r-i+1}, \quad i = 2, 3, \ldots, r-1$$
$$u_r = \rho^{\frac{1}{2}} \sin \theta_1.$$

Then

$$\sum_{i=1}^{r} u_i^2 = \rho,$$

and it is well-known—see, for example, Kendall and Stuart (1958, p. 247)—that ρ is distributed as χ_r^2, independently of the θs. Since, in terms of the new variables, the restrictions R involve only the θs, it follows that the conditional distribution of ρ is also that of χ_r^2. ∎

Lemma C

If $z_1, z_2, \ldots, z_r$ are independent normally distributed random variables with common mean and variances $b_1^{-1}, b_2^{-1}, \ldots, b_r^{-1}$, then the conditional distribution of

$$\sum_{i=1}^{r} b_i (z_i - \bar{z})^2,$$

given that $z_1 \leq z_2 \leq \ldots \leq z_r$, is that of χ^2 with $r - 1$ degrees of freedom, where

$$\bar{z} = \sum_{i=1}^{r} b_i z_i / \sum_{i=1}^{r} b_i.$$

Proof. Make an orthogonal transformation from the variables $b_1^{\frac{1}{2}} z_1, b_2^{\frac{1}{2}} z_2, \ldots, b_r^{\frac{1}{2}} z_r$ to variables $u_1, u_2, \ldots, u_r$, such that

$$u_r = (\sum_{i=1}^{r} b_i)^{\frac{1}{2}} \bar{z}.$$

Then it is easily shown that

$$\sum_{i=1}^{r} b_i (z_i - \bar{z})^2 = \sum_{i=1}^{r-1} u_i^2,$$

and that the us are independent normally distributed random variables, all with unit variances and all but u_r having zero means. The restraints

$$z_{i+1} - z_i \geq 0 \qquad (i = 1, 2, \ldots, r-1)$$

take the form

$$\sum_{j=1}^{r-1} d_{ij} u_j \geq 0 \qquad (i = 1, 2, \ldots, r-1),$$

and thus Lemma B can be applied to give the required result. ∎

Proof of Theorems 3.1 *and* 3.2. The proof is based on the Minimum Lower Sets algorithm, described in Section 2.3. It will be recalled that this algorithm determines the isotonic regression function $\hat{\mu}_i^*$ by partitioning K into l level sets $B_1, B_2, \ldots, B_l$ such that $\mathrm{Av}\, B_1 < \mathrm{Av}\, B_2 < \ldots < \mathrm{Av}\, B_l$. These are the sets of constancy of $\hat{\mu}_i^*$, which is given by $\hat{\mu}_i^* = \mathrm{Av}\, B_j$, for all $i \in B_j$ ($j = 1, 2, \ldots, l$). Also define

$$L_i = \bigcup_{j=1}^{i} B_j, \qquad i = 1, 2, \ldots, l;$$

and let L_0 denote the empty set ϕ. In the remainder of this section any set denoted by L, whether subscripted or not, will be understood to be a lower set.

The ith stage of the algorithm leads to the determination of B_i as the largest level set of minimum average among the level sets of the form $L - L_{i-1}$. In searching for this level set L is considered to range over the set of all possible non-empty lower sets: but it is easily seen that attention may be restricted to cases when L contains L_{i-1} as a proper subset. For suppose $L \not\supset L_{i-1}$: then $L' = L \cup L_{i-1}$ is a lower set containing L_{i-1}, and $L' - L_{i-1} = L - L_{i-1}$. Thus the level set $L - L_{i-1}$ has not been excluded from consideration if attention is restricted to lower sets containing L_{i-1}. The case $L = L_{i-1}$ can also be excluded, as then the set $L - L_{i-1}$ is empty.

Necessary and sufficient conditions can now be stated for $\hat{\mu}_i^*$ to be the isotonic function defined as above by some partition of K into level sets $B_1, B_2, \ldots, B_l$. These are:

(A.1) If $L \supset L_{i-1}$, $L - L_{i-1} \neq \phi$, then

$$\mathrm{Av}\, B_i \leq \mathrm{Av}\, (L - L_{i-1}) \qquad (i = 1, 2, \ldots, l).$$

(A.2) If $L \supset L_i$, $L - L_i \neq \phi$, then

$$\mathrm{Av}\, B_i < \mathrm{Av}\, (L - L_{i-1}) \qquad (i = 1, 2, \ldots, l-1).$$

The conditions A.1 arise from the fact that, for each i, B_i is a level set of minimum average amongst all level sets under consideration at the ith stage; conditions A.2 arise from the fact that B_i is the largest such set.

In order to find the conditional distributions, given $B_1, B_2, \ldots, B_l$, of the $\bar{\chi}^2$ and $\bar{E}^2$ statistics it is first shown that conditions A.1 and A.2 can be replaced by an equivalent set of conditions, as follows:

(B.1) If $L_i \supset L \supset L_{i-1}$, $L_i - L \neq \phi$, $L - L_{i-1} \neq \phi$, then

$$\operatorname{Av} B_i \leq \operatorname{Av}(L - L_{i-1}) \qquad (i = 1, 2, \ldots, l).$$

(B.2) $\qquad \operatorname{Av} B_i < \operatorname{Av} B_{i+1} \qquad (i = 1, 2, \ldots, l-1).$

It is straightforward to show that conditions A imply B. A.1 implies B.1 since the conditions of B.1 are simply a subset of those of A.1. To show that A.2 implies B.2, let $L = L_{i+1}$ in A.2 obtaining, for each value of i,

$$\operatorname{Av} B_i < \operatorname{Av}(B_i \cup B_{i+1}),$$

which implies

$$\operatorname{Av} B_i < \operatorname{Av} B_{i+1}.$$

To complete the proof of equivalence it must now be shown that conditions B imply A. That they imply A.1 may be shown as follows. Let i have any value in $\{1, 2, \ldots, l\}$, and let $L \supset L_{i-1}$, $L - L_{i-1} \neq \phi$. Now let j be the smallest integer such that $L_j \supset L$ (j exists because $L_l \supset L$ is always true). Then

$$L - L_{i-1} = \bigcup_{h=i}^{j} (L \cap B_h).$$

Now let h be an integer, $i \leq h \leq j$, which is such that $L \cap B_h$ is non-empty (there must be at least one such h, namely $h = j$). Then either $L \supset B_h$, or $L \cap B_h \neq B_h$. It will be shown that, in either case, $\operatorname{Av} B_h \leq \operatorname{Av}(L \cap B_h)$. In the first case, $B_h = L \cap B_h$, and the result is trivially satisfied. Otherwise define

$$L' = (L \cap B_h) \cup L_{h-1}.$$

Then $L_h \supset L' \supset L_{h-1}$, $L_h - L' \neq \phi$, and $L' - L_{h-1} \neq \phi$. Applying B.1 gives

$$\operatorname{Av} B_h \leq \operatorname{Av}(L' - L_{h-1}) = \operatorname{Av}(L \cap B_h),$$

so that, in either case, $\operatorname{Av} B_h \leq \operatorname{Av}(L \cap B_h)$. Then applying B.2 gives

$$\operatorname{Av} B_i \leq \operatorname{Av}(L \cap B_h),$$

the inequality being strict unless $h = i$. Thus, as $\operatorname{Av}(L - L_{i-1})$ is a weighted average of the quantities $\operatorname{Av}(L \cap B_h)$ over all h such that $i \leq h \leq j$ and $L \cap B_h \neq \phi$, it follows that

$$\operatorname{Av} B_i \leq \operatorname{Av}(L - L_{i-1}),$$

the inequality being strict unless $i = j$. Thus it has been proved that conditions B imply A.1.

To show that conditions B also imply A.2, let i have any value in $\{1, 2, \ldots, l - 1\}$, and let $L \supset L_i$, $L - L_i \neq \phi$. Then, as it has already been shown that conditions B imply A.1, it follows that

$$\operatorname{Av} B_{i+1} \leq \operatorname{Av}(L - L_i),$$

and then, applying B.2, that

$$\operatorname{Av} B_i < \operatorname{Av}(L - L_i).$$

Now, as $L - L_{i-1}$ may be expressed as the union of the disjoint sets $L - L_i$ and B_i, $\operatorname{Av}(L - L_{i-1})$ must be the weighted average of $\operatorname{Av}(L - L_i)$ and $\operatorname{Av} B_i$. Thus it follows that

$$\operatorname{Av} B_i < \operatorname{Av}(L - L_{i-1}),$$

proving that conditions B imply A.2.

Conditions A and B have now been proved to be equivalent, and hence conditions B can be used to determine the conditional distributions of the $\bar{\chi}_k^2$ and $\bar{E}_k^2$ statistics, given $B_1, B_2, \ldots, B_l$. To do this, consider the joint conditional distribution of the following components into which the total sum of squares

$$\sum_{i=1}^{k} \sum_{j=1}^{n_i} (y_{ij} - \bar{y})^2$$

may be partitioned:

$$Q_1 = \sum_{i=1}^{k} n_i (\hat{\mu}_i^* - \hat{\mu})^2 = \sum_{i=1}^{l} N_{B_i} (\operatorname{Av} B_i - \bar{y})^2$$

and

$$Q_2 = \sum_{i=1}^{k} \sum_{j=1}^{n_i} (y_{ij} - \hat{\mu}_i^*)^2 = \sum_{i=1}^{l} \sum_{h \in B_i} \sum_{j=1}^{n_h} (y_{hj} - \operatorname{Av} B_i)^2.$$

Let

$$\mathbf{y}' = [y_{11}, y_{12}, \ldots, y_{1n_1}, y_{21}, \ldots, y_{kn_k}]$$

be the vector of the N observations. An orthogonal transformation to $\mathbf{z} = \mathbf{Hy}$ is made such that the elements h_{ij} in the first l rows of $\mathbf{H}$ are given by:

$$h_{ij} = \begin{cases} N_{B_i}^{-\frac{1}{2}}, & \text{if } p \in B_i \\ 0, & \text{otherwise} \end{cases}$$

where p is the first subscript of the jth element of $\mathbf{y}$. If, for example, $B_1 = \{1, 2, \ldots, r\}$, then the first row of $\mathbf{H}$ would be

$$\Big[\underbrace{\Big(\sum_{i=1}^{r} n_i\Big)^{-\frac{1}{2}}, \ldots \Big(\sum_{i=1}^{r} n_i\Big)^{-\frac{1}{2}}}_{\sum_{i=1}^{r} n_i \text{ elements}}, \underbrace{0, \ldots\quad 0}_{\sum_{i=r+1}^{k} n_i \text{ elements}}\Big].$$

The remaining rows of $\mathbf{H}$ can be chosen in any way subject to the orthogonality requirement. As

$$z = N_{B_i}^{\frac{1}{2}} \operatorname{Av} B_i, \quad (i = 1, 2, \ldots, l)$$

it follows that

$$\sum_{i=1}^{l} z_i^2 = \sum_{i=1}^{l} N_{B_i} (\operatorname{Av} B_i)^2;$$

the orthogonality of the transformation then ensures that

$$Q_2 = \sum_{i=1}^{k} \sum_{j=1}^{n_i} y_{ij}^2 - \sum_{i=1}^{l} z_i^2 = \sum_{i=l+1}^{N} z_i^2.$$

We may now appeal to standard results concerning orthogonal transformations of normal random variables to deduce that $\mathbf{z}$ has a multivariate normal distribution with dispersion matrix $\sigma^2 \mathbf{I}_N$; on the null hypothesis,

$$E(\mathbf{z}') = \mu[B_{B_1}^{\frac{1}{2}}, \ldots N_{B_l}^{\frac{1}{2}}, 0, \ldots 0],$$

where μ is the common population mean.

The effect of imposing the restraints $\mathbf{B}$ on the distribution of Q_1 and Q_2 must now be examined. Consider first the conditions B.1. These are of the form $\operatorname{Av} B_i - \operatorname{Av}(L - L_{i-1}) \leq 0$, where $B_i \supset L - L_{i-1}$ and $B_i \neq L - L_{i-1}$; Lemma A can therefore be applied to show that, for any i, the conditions can be expressed in terms of random variables which are uncorrelated with, and therefore independent of, $\operatorname{Av} B_i = N_{B_i}^{-\frac{1}{2}} z_i$. As these random variables are functions only of the ys corresponding to the non-zero elements in the ith row of **H**, they are also independent of $z_1, z_2, \ldots, z_{i-1}, z_{i+1}, \ldots, z_l$. It then follows that the conditions B.1 can be expressed as a set of conditions of the form

$$\sum_{i=l+1}^{N} a_i z \geq 0.$$

Turning next to the conditions B.2 it is seen that these involve only the variables $z_1, z_2, \ldots, z_l$. Now, since Q_1 can be expressed in terms of $z_1, z_2, \ldots, z_l$ alone, its distribution is unaffected by the conditions B.1. Its distribution is thus that of

$$\sum_{i=1}^{l} N_{B_i}(\text{Av } B_i - \bar{y})^2,$$

subject to the restrictions $\text{Av } B_1 < \text{Av } B_2 < \ldots < \text{Av } B_l$. By Lemma C, this distribution is that of $\sigma^2 \chi^2_{l-1}$. The distribution of

$$Q_2 = \sum_{i=l+1}^{N} z_i^2,$$

given the conditions B.1, is, by Lemma B, that of $\sigma^2 \chi^2_{N-l}$. Further, since the zs are independent, Q_1 and Q_2 are independently distributed.

These results enable us to find, for any partition of K into level sets $B_1, B_2, \ldots, B_l$, the conditional distributions of $\bar{\chi}^2_k$ and $\bar{E}^2_k$, given that the maximum likelihood estimates of the population means are obtained by making the amalgamations defined by these level sets. These distributions are those of χ^2_{l-1} and $B_{\frac{1}{2}(l-1),\frac{1}{2}(N-l)}$ respectively. As these distributions do not depend on the sets $B_1, B_2, \ldots, B_l$ except through their number, it follows that they are also the conditional distributions of the two test statistics given that the estimates, under H, of the population means take l distinct values. These results lead directly to Theorems 3.1 and 3.2. ∎

The probabilities $P(l, k)$

The theorems just proved give the general forms of the null hypothesis distributions of the $\bar{\chi}^2_k$ and $\bar{E}^2_k$ statistics. In order to determine completely the distribution in any particular situation, it is necessary to find the probabilities $P(l, k)$. It is not possible in all cases to determine expressions for these probabilities in a simple closed form. In the following pages we shall describe a general method, and some other methods appropriate in particular cases. The general method can be regarded as being in two parts: the first is to find $P(k, k)$, the second to find a recurrence formula for $P(l, k)$, $(l < k)$, in terms of $P(i, j)$, $(i \leq j < k)$. These results, combined with the fact that

$$\sum_{l=1}^{k} P(l, k) = 1,$$

enable all the probabilities to be calculated.

For any given k, the probabilities will obviously depend on the alternative hypothesis considered. This dependence will not, however, be made explicit in the notation, as it will be clear from the context which hypothesis is being discussed. The probabilities will also depend on the weights w_i, and this fact will sometimes be made explicit in the notation, viz. $P(l, k; w_1, w_2, \ldots, w_k)$, or $P(l, k; \mathbf{w})$: where there is no risk of confusion, or where the weights are equal, they will be omitted.

In what follows it will be found convenient, in dealing with certain partially ordered alternatives, to adopt the following convention. When an inequality sign separates two groups of quantities, each enclosed in square brackets, the inequality will be understood to hold between each member of one group and each member of the other. Thus, for example, the alternative defined by $\mu_1 \leq \mu_2$; $\mu_1 \leq \mu_3$; $\mu_2 \leq \mu_4$; $\mu_2 \leq \mu_5$; $\mu_3 \leq \mu_4$; $\mu_3 \leq \mu_5$; $\mu_4 \leq \mu_6$; $\mu_5 \leq \mu_6$ may be written in the simpler form $\mu_1 \leq [\mu_2, \mu_3] \leq [\mu_4, \mu_5] \leq \mu_6$. Such alternatives may arise when there is an *a priori* ordering of the populations, but there are ties in the ordering. The simplest case, which will be encountered later, is the alternative that will be referred to as that of "simple loop order":

$$\mu_1 \leq [\mu_2, \mu_3] \leq \mu_4.$$

The method given for finding $P(k, k)$ is applicable in all situations, although it will be seen that a simple explicit solution exists only in certain cases. It is observed that $P(k, k)$ is the probability that no amalgamation is necessary, and that this will happen if the means $\bar{y}_i$ satisfy the inequalities imposed by the alternative hypothesis. As an example, consider the case when $k = 4$ and the alternative hypothesis is that of simple loop order as defined above. Then

$$P(4, 4) = \Pr\{\bar{y}_1 < [\bar{y}_2, \bar{y}_3] < \bar{y}_4\},$$

where the $\bar{y}_i$ are mutually independent and $\bar{y}_i \sim N(\mu, \sigma_i^2/n_i)$, $i = 1, 2, 3, 4$. Now the event $\bar{y}_1 < [\bar{y}_2, \bar{y}_3] < \bar{y}_4$ is the union of the two disjoint events

(a) $\bar{y}_1 < \bar{y}_2 < \bar{y}_3 < \bar{y}_4$ and   (b) $\bar{y}_1 < \bar{y}_3 \leq \bar{y}_2 < \bar{y}_4$.

It follows that $P(4, 4) = \Pr\{(a)\} + \Pr\{(b)\}$. Thus, by expressing the above partial ordering on the $\bar{y}_i$s as a union of simple orderings, $P(4, 4)$ may be expressed as a sum of probabilities of the form $\Pr\{\bar{y}_{\alpha_1} < \bar{y}_{\alpha_2} < \bar{y}_{\alpha_3} < \bar{y}_{\alpha_4}\}$, where $(\alpha_1, \alpha_2, \alpha_3, \alpha_4)$ is some permutation of the integers $(1, 2, 3, 4)$. Similar methods may be employed for any alternative hypothesis of the class $\mathscr{H}$, reducing the problem of finding $P(k, k)$ to that of finding probabilities of the form $\Pr\{\bar{y}_1 < \bar{y}_2 < \ldots < \bar{y}_k\}$. Such probabilities may be expressed in a more familiar form by making the transformation

$$z_i = \bar{y}_{i+1} - \bar{y}_i \qquad (i = 1, 2, \ldots, k-1).$$

The z_is so obtained have a multivariate normal distribution with correlation matrix with the following elements:

$$\left. \begin{aligned} \rho_{i,i+1} &= -\left[\frac{w_i w_{i+2}}{(w_i + w_{i+1})(w_{i+1} + w_{i+2})}\right]^{\frac{1}{2}}, & (i = 1, 2, \ldots, k-2) \\ \rho_{ii} &= 1, & (i = 1, 2, \ldots, k-1) \\ \rho_{ij} &= 0, & |i - j| > 1. \end{aligned} \right\} \quad (3.15)$$

The foregoing considerations enable $P(k, k)$ to be expressed as a sum of probabilities, each having the form

$$\Pr\{z_1 > 0, z_2 > 0, \ldots, z_{k-1} > 0\} = P_{k-1}, \qquad (3.16)$$

say. It is unnecessary to specify the variances of the zs since the required probability does not depend on their scale.

There is a vast literature on the determination of the probability (3.16). For papers prior to 1963 the reader may consult Gupta (1963). Simple explicit expressions are available for $k = 2, 3$ and 4 (see, for example, Childs (1967)) as follows. (For P_3, as in subsequent results for P_4, the expression is presented in a form modified by making use of the knowledge that $\rho_{ij} = 0$ for $|i - j| > 1$.)

$$\left. \begin{aligned} P_1 &= \frac{1}{2} \\ P_2 &= \frac{1}{4} + \frac{1}{2\pi} \sin^{-1} \rho_{12} \\ P_3 &= \frac{1}{8} + \frac{1}{4\pi} (\sin^{-1} \rho_{12} + \sin^{-1} \rho_{23}). \end{aligned} \right\} \quad (3.17)$$

For $k > 4$ there are no closed expressions for P_{k-1}, except in special cases, the most important of which is the case when the weights are equal. In this case P_{k-1} is the probability that the ranks of the $\bar{y}$s are in increasing order, and is therefore $1/k!$, as all permutations of the ranks are equally likely. It should be observed that this result also holds for any set of symmetrically dependent $\bar{y}$s which are such that ties occur with zero probability: in particular it holds if the $\bar{y}$s are continuous, independently and identically distributed, random variables.

In the case of general weights several methods are available for finding numerical values of P_{k-1} for $k > 4$. Kendall (1941) and Moran (1948) have given series expansions in terms of the ρs, but these do not converge fast enough for practical use unless the ρs are small—in the present application, provided the weights are not too unequal, the ρs will be in the neighbourhood of $-\frac{1}{2}$. Plackett (1954) gave a reduction formula for the multivariate normal integral, and also an approximation for $k = 5$ quoted in (3.19) below. McFadden (1960), Abrahamson (1964) and Childs (1967) also give methods for the case $k = 5$. These include the formula

$$P_4 = \frac{1}{16} + \frac{1}{8\pi}(\sin^{-1}\rho_{12} + \sin^{-1}\rho_{23} + \sin^{-1}\rho_{34})$$
$$+ \left(\frac{2}{\pi}\right)^2 \int_0^{-\rho_{34}} \int_0^{-\rho_{12}} \{(1-u^2)(1-v^2) - \rho_{23}^2\}^{-\frac{1}{2}} du\, dv. \quad (3.18)$$

A table is given by Abrahamson (1964) from which values of P_4 can easily be obtained. Discussion of the general problem of finding P_{k-1} for $k > 4$ is outside the scope of this book. In addition to (3.18), Plackett's approximation for P_4 will be quoted and some simple upper bounds for P_{k-1} derived, valid for all k. These will be sufficient for the present purpose.

By assuming that the ρs were small, Plackett (1954) obtained the approximation

$$P_4 \approx \frac{1}{16} + \frac{1}{8\pi}(\sin^{-1}\rho_{12} + \sin^{-1}\rho_{23} + \sin^{-1}\rho_{34})$$
$$+ \frac{1}{4\pi^2}\sin^{-1}\rho_{12}\sin^{-1}\rho_{34}. \quad (3.19)$$

Note that the first two terms of (3.19) are the same as the first two appearing in the exact expression of (3.18). An idea of the accuracy of (3.19) can be obtained from applying it in the case of equal weights: here we have an exact value of 1/120 whereas the approximation gives 1/144. It was shown in Bartholomew (1959b) that a discrepancy of this order has a negligible effect on significance levels.

Simple upper bounds for P_{k-1} can be obtained as follows.

P_{k-1}
$= \Pr\{\bar{y}_1 < \bar{y}_2 < \ldots < \bar{y}_k\}$
$= \Pr\{(\bar{y}_1 < \bar{y}_2 < \ldots < \bar{y}_j) \text{ and } (\bar{y}_j < \bar{y}_{j+1} < \ldots < \bar{y}_k)\}$
$= \Pr\{\bar{y}_1 < \bar{y}_2 < \ldots < \bar{y}_j\} \Pr\{\bar{y}_j < \bar{y}_{j+1} < \ldots \bar{y}_k | \bar{y}_1 < \bar{y}_2 \ldots < \bar{y}_j\}$
$\quad < \Pr\{\bar{y}_1 < \bar{y}_2 < \ldots < \bar{y}_j\} \Pr\{\bar{y}_j < \bar{y}_{j+1} < \ldots < \bar{y}_k\}, \quad (3.20)$
$(j = 2, 3, \ldots, k-1).$

The last step follows from the fact that the conditioning event $\{\bar{y}_1 < \bar{y}_2 < \ldots < \bar{y}_j\}$ makes $\bar{y}_j$ the largest of j normal random variables, and so decreases the probability of the event $\{\bar{y}_j < \bar{y}_{j+1} < \ldots < \bar{y}_k\}$. This method gives a set of bounds—one for each value of j. In particular, if $j = 2$, $P_{k-1} < \frac{1}{2}P_{k-2}$, where P_{k-2} must be understood as referring to the last $(k - 2)$ $\bar{y}$s. The values of the bound for $k = 5$ and equal weights are as follows.

j	2	3	4	Exact value
Bound	$\frac{1}{48}$	$\frac{1}{36}$	$\frac{1}{48}$	$\frac{1}{120}$

The bound is a poor one in this case but can be approached under suitable limiting conditions. If it is supposed that the smallest weight of the set $\{\bar{y}_1, \bar{y}_2, \ldots, \bar{y}_j\}$ is d times as large as the largest weight of the set $\{\bar{y}_{j+1}, \bar{y}_{j+2}, \ldots, \bar{y}_k\}$, and $d \to \infty$, then $\Pr\{\bar{y}_j < \bar{y}_{j+1} < \ldots < \bar{y}_k | \bar{y}_1 < \bar{y}_2 < \ldots < \bar{y}_j\} \to \Pr\{\bar{y}_j < \bar{y}_{j+1} < \ldots < \bar{y}_k\}$. A cruder bound still, which has the merit of being distribution-free, may be found by applying (3.20) repeatedly with $j = 2$ (or $k - 1$). This gives $P_{k-1} < 1/2^{k-1}$.

The second stage in finding the $P(l, k)$s is to establish a recurrence relation between those for a given value of k and those for smaller values. In order to do this we recall the proof of Theorems 3.1 and 3.2, where we derived conditions B.1 and B.2 which are necessary and sufficient for the level sets $B_1, B_2, \ldots, B_l$ to define the amalgamation giving the maximum likelihood estimates of $\mu_1, \mu_2, \ldots, \mu_k$. To find $P(l, k)$, we first find the probability that these conditions are satisfied for a given partition of K into l level sets and then sum over the set $\mathscr{L}_{lk}$ of all such partitions. The following notation is required. For any non-empty subset A of K let

$$W_A = \sum_{i \in A} w_i,$$

and C_A be the number of elements in A. As in Chapter 1, let

$$\text{Av } A = \sum_{i \in A} w_i \bar{y}_i / W_A;$$

and let $\mathbf{w}(A)$ be the vector whose elements are the weights w_i, $i \in A$. As the proof of Theorems 3.1 and 3.2 was presented for the case when all population variances are the same the results quoted from these must be modified by the substitution of w_i for n_i where necessary.

To find the probability that $B_1, B_2, \ldots, B_l$ satisfy conditions B.1 and B.2 we observe first that, as before, Lemma A shows that the events defined by these conditions are independent. They are therefore considered separately, beginning with B.2. This requires that the means Av B_1, Av B_2, ..., Av B_l form an increasing set, and therefore

$$\Pr\{\text{B.2}\} = P(l, l; W_{B_1}, W_{B_2}, \ldots, W_{B_l}). \tag{3.21}$$

Turning now to B.1, it is seen that these conditions may be split into l sets, one for each value of i, $1 \leq i \leq l$. If the conditions corresponding to a particular value of i are now considered, it is seen that these involve only the quantities $\bar{y}_\alpha$, $\alpha \in B_i$. We therefore examine the reduced problem of estimating the means of the populations in B_i subject to the restrictions imposed by H within that group. It is then seen that the conditions in B.1 corresponding to the value of i being dealt with are merely those which are necessary and sufficient for the estimates in the reduced problem to be all equal. The probability of this happening is $P(1, C_{B_i}; \mathbf{w}(B_i))$, and hence

$$\Pr\{\text{B.1}\} = \prod_{i=1}^{l} P(1, C_{B_i}; \mathbf{w}(B_i)). \tag{3.22}$$

Combining (3.21) and (3.22), and summing over $\mathscr{L}_{lk}$, the result

$$P(l, k; \mathbf{w}) = \sum_{\{B_1, B_2, \ldots, B_l\} \in \mathscr{L}_{lk}} P(l, l; W_{B_1}, W_{B_2}, \ldots, W_{B_l}) \prod_{i=1}^{l} P(1, C_{B_i}; \mathbf{w}(B_i)) \tag{3.23}$$

is obtained.

It must be emphasized that each probability appearing in the sum has to be determined under those restraints of H which apply to the group(s) in question. It may be noted that setting $l = k$ results in the method already described of finding $P(k, k)$ for any alternative, given $P(k, k)$ for the alternative of simple order. Thus, with knowledge of P_i, $i = 1, 2, \ldots, k-1$, (3.23) enables all the probabilities $P(l, k)$ to be found.

It has already been pointed out that $P_{k-1} = 1/k!$ when the weights are equal and that this result is, provided ties occur with zero probability, distribution-free. Thus, for any alternative and equal weights $P(k, k)$ is, again provided ties have zero probability, distribution-free; and this will later be seen to be the case for all the $P(l, k)$s when the alternative is that of simple order. It has already been seen in Chapter 2 that isotonic regression is used to find restricted maximum likelihood estimators for parameters of other distributions besides the normal. Some results of this section therefore have application to a much wider class of problems. The distribution-free character of the $P(l, k)$ also has implications for the robustness of the $\bar{\chi}^2$ and $\bar{E}^2$ tests to which attention will be drawn in Chapter 4.

The probabilities $P(l, k)$ in some special cases

The case of simple order

Perhaps the most important special case is that of simple order where the alternative, denoted by H_1, specifies

$$\mu_1 \leq \mu_2 \leq \ldots \leq \mu_k.$$

We shall begin by taking general weights, for which case, as we have already seen, a simple exact solution can be found only for $k \leq 4$. For the case when the weights are equal, a solution valid for all k will be derived.

In order to obtain results for general weights, (3.23) will be applied, taking account of the fact that the level sets in this case are blocks of consecutive integers; thus the enumeration of all members of $\mathscr{L}_{lk}$ is, in this case, quite a simple operation. For example, when $k = 3$, $l = 2$, there are only two cases to consider. These are

$$B_1 = \{1\},\ B_2 = \{2, 3\} \quad \text{and} \quad B_1 = \{1, 2\},\ B_2 = \{3\}.$$

Application of (3.23) gives

$$\begin{aligned}P(2, 3; \mathbf{w}) &= P(2, 2; w_1, w_2 + w_3)P(1, 2; w_2, w_3) \\ &\quad + P(2, 2; w_1 + w_2, w_3)P(1, 2; w_1, w_2) \\ &= \tfrac{1}{2} \cdot \tfrac{1}{2} + \tfrac{1}{2} \cdot \tfrac{1}{2} = \tfrac{1}{2}.\end{aligned} \quad (3.24)$$

$P(3, 3; \mathbf{w})$ is given by P_2 of (3.18), with ρ_{12} as given by (3.16). Finally, by subtraction, it is found that

$$P(1, 3; \mathbf{w}) = \frac{1}{4} - \frac{1}{2\pi} \sin^{-1} \rho_{12}. \quad (3.25)$$

When $k = 4$, $P(4, 4; \mathbf{w})$ is given by P_3 of (3.18). The calculation of $P(3, 4)$ proceeds as follows.

$$\begin{aligned}P(3, 4; \mathbf{w}) &= P(3, 3; w_1, w_2, w_3 + w_4)P(1, 2; w_3, w_4) \\ &\quad + P(3, 3; w_1, w_2 + w_3, w_4)P(1, 2; w_2, w_3) \\ &\quad + P(3, 3; w_1 + w_2, w_3, w_4)P(1, 2; w_1, w_2).\end{aligned}$$

The second factor in each of these terms is $\tfrac{1}{2}$. The first factors follow by substituting the appropriate weights in $P(3, 3)$. Thus, for example,

$$P(3, 3; w_1, w_2, w_3 + w_4) = \frac{1}{4} + \frac{1}{2\pi} \sin^{-1} \rho,$$

where

$$\rho = -\left[\frac{w_1(w_3 + w_4)}{(w_1 + w_2)(w_2 + w_3 + w_4)}\right]^{\frac{1}{2}}$$

$$= \frac{\rho_{12}}{[1 - \rho_{23}^2]^{\frac{1}{2}}}.$$

Since $\rho_{13} = 0$, it follows that ρ is the partial correlation coefficient between z_1 and z_2 given z_3, which will be denoted by $\rho_{12.3}$. A similar argument for the other two terms gives

$$P(3, 4; \mathbf{w}) = \frac{3}{8} + \frac{1}{4\pi}(\sin^{-1}\rho_{12.3} + \sin^{-1}\rho_{13.2} + \sin^{-1}\rho_{23.1}). \quad (3.26)$$

Now

$$P(2, 4; \mathbf{w}) = P(2, 2; w_1, w_2 + w_3 + w_4)P(1, 3; w_2, w_3, w_4)$$
$$+ P(2, 2; w_1 + w_2 + w_3, w_4)P(1, 3; w_1, w_2, w_3)$$
$$+ P(2, 2; w_1 + w_2, w_3 + w_4)P(1, 2; w_1, w_2)P(1, 2; w_3, w_4)$$
$$= \frac{1}{2}\left(\frac{1}{4} - \frac{1}{2\pi}\sin^{-1}\rho_{23} + \frac{1}{4} - \frac{1}{2\pi}\sin^{-1}\rho_{12}\right) + \frac{1}{8}$$
$$= \frac{1}{2} - P(4, 4; \mathbf{w}). \quad (3.27)$$

Finally, on subtraction it is found that

$$P(1, 4; \mathbf{w}) = \tfrac{1}{2} - P(4, 4; \mathbf{w}). \quad (3.28)$$

When $k > 4$ progress is hindered by the lack of simple closed expressions for P_{k-1}, although some results for $k = 5, 6, 7$ (specifically the values of $P(l, k)$ satisfying $k - 3 \leq l \leq 4$) depend only on the results for $k \leq 4$. Thus, for $k = 5$, the process of finding the $P(l, k)$s is relatively simple. $P(5, 5; \mathbf{w})$ may be found exactly from Abrahamson's (1964) table, or approximately from Plackett's formula (3.19). Then application of (3.23) in the same manner as above gives

$$P(4, 5; \mathbf{w}) = \frac{1}{4} + \frac{1}{8\pi}(\sin^{-1}\rho_{12} + \sin^{-1}\rho_{34}) + \frac{1}{8\pi}(\sin^{-1}\rho_{12.3}$$
$$+ \sin^{-1}\rho_{23.1} + \sin^{-1}\rho_{34.2} + \sin^{-1}\rho_{13.2}$$
$$+ \sin^{-1}\rho_{23.4} + \sin^{-1}\rho_{24.3})$$

$$P(3, 5; \mathbf{w}) = \frac{1}{2} + \frac{1}{8\pi} (\sin^{-1} \rho_{12.34} + \sin^{-1} \rho_{13.24} + \sin^{-1} \rho_{14.23}$$
$$+ \sin^{-1} \rho_{23.14} + \sin^{-1} \rho_{24.13} + \sin^{-1} \rho_{34.12})$$
$$- \frac{1}{8\pi} (\sin^{-1} \rho_{12} + \sin^{-1} \rho_{23} + \sin^{-1} \rho_{34})$$
$$- \frac{1}{4\pi^2} (\sin^{-1} \rho_{12} \sin^{-1} \rho_{34.12} + \sin^{-1} \rho_{23} \sin^{-1} \rho_{14.23}$$
$$+ \sin^{-1} \rho_{34} \sin^{-1} \rho_{12.34})$$
$$P(2, 5; \mathbf{w}) = \tfrac{1}{2} - P(4, 5; \mathbf{w}). \tag{3.29}$$

Finally, $P(1, 5; \mathbf{w})$ is found by subtraction.

In principle there is no difficulty in obtaining numerical values of the $P(l, k)$s for $k > 5$, but the procedure becomes increasingly complicated as k increases. More research is required to see whether useful approximations can be found.

The above-mentioned difficulties do not apply if the weights are equal. In this case it is possible to find a complete solution for all k, which is given by the following theorem. Its two corollaries lead to methods of obtaining the $P(l, k)$s which are, in practice, more convenient than the direct use of the theorem. The statement of the theorem, and its proof, requires the following definitions. By an *l*-composition of k is meant a representation of k as the sum of l ordered positive integers $c_1, c_2, \ldots, c_l$; this will be referred to as the *l*-composition $(c_1, c_2, \ldots, c_l)$. Thus $(3, 3, 4, 2, 1)$ and $(4, 3, 1, 3, 2)$ are distinct 5-compositions of 13. An *l*-partition of k is a partition of k into l unordered positive integral parts. Such a partition will be denoted by $b_1^{\pi_1} b_2^{\pi_2} \ldots b_\lambda^{\pi_\lambda}$, where

$$\sum_{i=1}^{\lambda} \pi_i = l$$

—the notation indicates that, in the partition, the integer b_i occurs π_i times. To ensure uniqueness of this representation, the convention that $b_1 > b_2 > \ldots > b_\lambda$ may be adopted. Thus each of the above 5-compositions of 13 is a representation of the 5-partition $4 \; 3^2 \; 2 \; 1$.

Theorem 3.3

The probabilities $P(l, k)$ satisfy

$$P(l, k) = \sum_{\mathscr{P}_{lk}} \prod_{i=1}^{\lambda} \frac{\{P(1, b_i)\}^\pi}{\pi_i!},$$

where $\mathscr{P}_{lk}$ denotes the set of all l-partitions $b_1^{\pi_1} b_2^{\pi_2} \ldots b_\lambda^{\pi_\lambda}$ of k.

Proof. We make use of (3.23), and observe that, in the present case each member of $\mathscr{L}_{lk}$ is equivalent to an l-composition of k. Thus, denoting by $\mathscr{C}_{lk}$ the set of all l-compositions $(c_1, c_2, \ldots, c_l)$ of k, (3.23) can be written in the form

$$P(l, k) = \sum_{\mathscr{C}_{lk}} \prod_{i=1}^{l} P(1, c_i) P(l, l; \mathbf{c}), \tag{3.30}$$

where $\mathbf{c} = (c_1, c_2, \ldots, c_l)'$. The summation is now split into two parts, enumerating all l-compositions by first summing over all l-partitions, and then over the $l!/(\pi_1! \pi_2! \ldots \pi_\lambda!)$ permutations of each partition. Denoting the latter summation by Σ', (3.30) becomes

$$P(l, k) = \sum_{\mathscr{P}_{lk}} \prod_{i=1}^{\lambda} P(1, b_i)^{\pi_i} \Sigma' P(l, l; \mathbf{c}). \tag{3.31}$$

Now $P(l, l; c_1, c_2, \ldots, c_l) = \Pr\{x_1 < x_2 < \ldots < x_l\}$, where the x_is are mutually independent random variables distributed as $N(0, \sigma^2/c_i)$; and this result still holds on making any permutation of the subscripts $1, 2, \ldots, l$ on both sides of the equation. Thus, denoting summation over all $l!$ permutations by Σ'', it follows that

$$\Sigma'' P(l, l; c_1, c_2, \ldots, c_l) = 1, \tag{3.32}$$

as exactly one permutation of $x_1, x_2, \ldots, x_l$ arranges their values in ascending order. The summation Σ'' is not, however, the same as Σ', because, in general, there will be repetitions among the cs. In general, any particular l-composition occurring in Σ' will occur $\pi_1! \pi_2! \ldots \pi_\lambda!$ times in Σ''. Thus $\Sigma' P(l, l; \mathbf{c})$ in (3.31) may be replaced by $1/(\pi_1! \pi_2! \ldots \pi_\lambda!)$, and the theorem is proved. ∎

Corollary A

The probabilities $P(l, k)$ are given by

$$P(l, k) = |S_k^l|/k! \qquad (l = 1, 2, \ldots, k), \tag{3.33}$$

where $|S_k^l|$ is the coefficient of z^l in the expansion of $z(z + 1) \ldots (z + k - 1)$. (The notation arises from the fact that the above function generates the moduli of the Stirling numbers of the first kind.)

Proof. The result is proved by induction on k. Assuming that (3.33) holds for $k = 1, 2, \ldots, m - 1$, it will be shown, by substitution in the result of Theorem 3.3, that it holds for $k = m$. It will be sufficient to consider the results for $P(l, m)$, $l = 2, 3, \ldots, m$, as, if they are as given by the corollary, so also must be $P(1, m)$ as it can easily be verified that

$$\sum_{l=1}^{k} |S_k^l| = k!$$

In applying the theorem in these cases, we observe that the largest possible value of b_i appearing on the right hand side is $(m - l + 1)$, and the induction hypothesis can therefore be applied to make the substitution $P(1, b_i) = b_i^{-1}$, obtaining

$$P(l, m) = \sum_{\mathscr{P}_{lm}} \prod_{i=1}^{\lambda} \frac{1}{b_i^{\pi_i} \pi_i!}$$

$$= \frac{1}{l!} \sum_{\mathscr{C}_{lm}} \frac{1}{c_1 c_2 \ldots c_l}. \tag{3.34}$$

The final stage in the proof is to show that the last expression is equal to

$$\frac{1}{m} \sum_{\mathscr{U}_{l-1,m}} \frac{1}{c_1 c_2 \ldots c_{l-1}}, \tag{3.35}$$

where $\mathscr{U}_{l-1,m}$ is the set of all $(c_1, c_2, \ldots, c_{l-1})$ satisfying $1 \leq c_1 < c_2 \ldots < c_{l-1} < m$. As this is one form of the coefficient of z^l in $z(z + 1) \ldots (z + m - 1)/m!$, the proof will then be complete.

This last step is established by induction on l.

$$m \sum_{\mathscr{C}_{lk}} \frac{1}{c_1 c_2 \ldots c_l} = \sum_{\mathscr{C}_{lk}} \frac{c_1 + c_2 + \ldots + c_l}{c_1 c_2 \ldots c_l}$$

$$= l \sum_{c_1 + c_2 + \ldots + c_{l-1} < m} \frac{1}{c_1 c_2 \ldots c_{l-1}}.$$

$$= l \sum_{h=l-1}^{m-1} \sum_{\mathscr{C}_{l-1,h}} \frac{1}{c_1 c_2 \ldots c_{l-1}}. \tag{3.36}$$

Assuming (3.34) is equal to (3.35) for $l - 1$ and lesser values the last expression in (3.36) may be written.

$$l \sum_{h=l-1}^{k-1} \frac{(l-1)!}{h} \sum_{\mathscr{U}_{l-2,h}} \frac{1}{c_1 c_2 \ldots c_{l-2}} = l! \sum_{\mathscr{U}_{l-1,k}} \frac{1}{c_1 c_2 \ldots c_{l-1}}$$

Since the result is obviously true for $l = 2$ the foregoing argument shows that it is true for all l. ∎

Corollary B

The probabilities $P(l, k)$ satisfy

$$P(1, k) = \frac{1}{k}$$

$$P(k, k) = \frac{1}{k!}$$

$$P(l, k) = \frac{1}{k} P(l-1, k-1)$$
$$+ \frac{k-1}{k} P(l, k-1), \quad (l = 2, 3, \ldots, k-1).$$

Proof. These results follow directly from the generating function $z(z+1) \ldots (z+k-1)/k!$ of Corollary A, and are related to well-known properties of the Stirling numbers. ∎

Numerical values of the $P(l, k)$s, for $k \leq 12$, are given in the Appendix Table A.5.

The simple tree alternative

An important class of alternative hypotheses in the set $\mathscr{H}$ is that of alternatives specified by a partial ordering of the type

$$\mu_1 \leq [\mu_2, \mu_3, \ldots, \mu_k],$$

which will be referred to as "simple tree order". The application of (3.23) to find the probabilities $P(l, k)$ for such alternatives is straightforward as the set $\mathscr{L}_{lk}$ is easily defined. It consists of all partitions $(B_1, B_2, \ldots, B_l)$ where B_1 is composed of 1 and $k-l$ members of the set $\{2, 3, \ldots, k\}$, and $B_2, \ldots, B_l$ are, in some order, the remaining members of that set. As before, all the probabilities $P(l, k; \mathbf{w})$ can be found by using (3.23) if the probabilities $P(l, l)$ can be found for general weights and $l \leq k$. This may be done using the general method already described for $P(k, k)$, summing probabilities of the form P_{k-1} over all simple orderings consistent with the simple tree. This requires the summation of $(k-1)!$ probabilities, and the following direct method is preferable.

If $z_i = \bar{y}_{i+1} - \bar{y}_1$ $(i = 1, 2, \ldots, k-1)$, then

$$P(k, k; \mathbf{w}) = \Pr\{z_1 > 0, z_2 > 0, \ldots, z_{k-1} > 0\}.$$

This is the probability P_{k-1} as before, but now the correlation coefficients on which its value depends are as follows:

$$\rho_{ij} = \left[\frac{w_{i+1} w_{j+1}}{(w_1 + w_{i+1})(w_1 + w_{j+1})}\right]^{\frac{1}{2}}, \quad i \neq j \quad (3.37)$$
$$= 1, \quad i = j.$$

The expressions (3.17) may now be used to find $P(k, k; \mathbf{w})$, $k \leq 4$, with the exception that the expression for P_3 must be replaced by

$$\frac{1}{8} + \frac{1}{4\pi}(\sin^{-1}\rho_{12} + \sin^{-1}\rho_{13} + \sin^{-1}\rho_{23}).$$

Then, applying (3.23), expressions can be obtained for all the $P(l, k; \mathbf{w})$ for these values of k.

For this alternative, unlike that of simple order, a relatively straightforward approach is available for finding $P(k, k; \mathbf{w})$ for $k > 4$. This arises as the correlation coefficients ρ_{ij} ($i \neq j$) are of the form $\lambda_i \lambda_j$. For a multivariate normal distribution with mean zero and such a correlation structure, Dunnett and Sobel (1955), amongst others, have shown that the probability in the all-positive orthant can be written

$$P_{k-1} = \int_{-\infty}^{\infty} \left\{ \prod_{i=1}^{k-1} \Phi\left(\frac{\lambda_i x}{\sqrt{(1-\lambda_i^2)}}\right) \right\} \phi(x) \, dx, \qquad (3.38)$$

where $\Phi(\cdot)$ and $\phi(\cdot)$ are the distribution and density functions respectively of a standard normal random variable. This result does not lead to simple expressions for the probabilities $P(l, k; \mathbf{w})$, but they may be obtained by numerical integration.

If the weights are equal further simplification is possible. All the terms in (3.23) now have the same form and, if the common weight is w,

$$P(l, k) = \binom{k-1}{k-l} P(1, k-l+1) P(l, l; (k-l+1)w, w, \ldots, w). \qquad (3.39)$$

The last probability on the right hand side is the probability in the all-positive orthant of an $(l - 1)$-dimensional normal distribution with mean zero and all off-diagonal elements of the correlation matrix equal to $1/(k - l + 2)$. This probability has been tabulated by Ruben (1954) for positive values of k, l such that $k \leq 50$, $k - 10 \leq l \leq k$. In Ruben's notation

$$P(l, k) = \binom{k-1}{l-1} P(1, k-l+1) \bar{u}_{l-1}(k-l+2) \qquad (3.40)$$

$$(l = 2, 3, \ldots, k-1).$$

Numerical values of the $P(l, k)$s, for $k \leq 12$, are given in Table A.6.

The simple loop alternative

The two alternatives discussed above are, perhaps, the most important members of $\mathscr{H}$ in practice. The other members of $\mathscr{H}$, even for moderate k, are too numerous to consider individually: only one further illustration of the application of the general method will be given. We shall consider the simple loop alternative, for which $k = 4$:

$$\mu_1 \leq [\mu_2, \mu_3] \leq \mu_4.$$

First, as remarked when discussing this alternative at the beginning of this section, we see that

$$P(4, 4; \mathbf{w}) = \Pr\{\bar{y}_1 < \bar{y}_2 < \bar{y}_3 < \bar{y}_4\} + \Pr\{\bar{y}_1 < \bar{y}_3 < \bar{y}_2 < \bar{y}_4\},$$

and its value can therefore be found, as each of the probabilities on the right hand side is of the form P_3. To find $P(3, 4; \mathbf{w})$ the four partitions belonging to $\mathscr{L}_{34}$ are listed. These are:

$\{1, 2\}, \{3\}, \{4\}; \quad \{1, 3\}, \{2\}, \{4\}; \quad \{1\}, \{2\}, \{3, 4\}; \quad \text{and} \quad \{1\}, \{3\}, \{2, 4\}.$

The first of these contributes

$$P(1, 2; w_1, w_2)P(3, 3; w_1 + w_2, w_3, w_4) = \tfrac{1}{2}P(3, 3; w_1 + w_2, w_3, w_4)$$

to $P(3, 4; \mathbf{w})$, and the other three are treated similarly. $P(2, 4; \mathbf{w})$ is the sum of components belonging to the four partitions of $\mathscr{L}_{24}$, which are:

$\{1, 2\}, \{3, 4\}; \quad \{1, 3\}, \{2, 4\}; \quad \{1, 2, 3\}, \{4\}; \quad \text{and} \quad \{1\}, \{2, 3, 4\}.$

The contribution to the total probability arising from each of the first two is easily found to be $\tfrac{1}{8}$. The method for the last two may be illustrated on the third. Here the probability required is

$$P(1, 3; w_1, w_2, w_3)P(2, 2; w_1 + w_2 + w_3).$$

The first factor must be found under the restrictions $\mu_1 \leq [\mu_2, \mu_3]$, and the methods already described for the simple tree alternative are applied; the second factor is $\tfrac{1}{2}$. Combining these results, the full solution obtained for this alternative is:

$$\left.\begin{aligned}
P(4, 4; \mathbf{w}) &= \frac{1}{4\pi}(\sin^{-1}\tau_{12,13} + \sin^{-1}\tau_{24,34}) \\
P(3, 4; \mathbf{w}) &= \frac{1}{2} + \frac{1}{4\pi}(\sin^{-1}\tau_{[12]3,34} + \sin^{-1}\tau_{[13]2,24} \\
&\quad + \sin^{-1}\tau_{13,3[24]} + \sin^{-1}\tau_{12,2[34]}) \\
P(2, 4; \mathbf{w}) &= \frac{1}{2} - P(4, 4; \mathbf{w}) \\
P(1, 4; \mathbf{w}) &= \frac{1}{2} - P(3, 4; \mathbf{w}),
\end{aligned}\right\} \quad (3.41)$$

where

$$\tau_{12,13} = \left[\frac{w_2 w_3}{(w_1 + w_2)(w_1 + w_3)}\right]^{\frac{1}{2}},$$

$$\tau_{[12]3,34} = -\left[\frac{(w_1 + w_2)w_4}{(w_1 + w_2 + w_3)(w_3 + w_4)}\right]^{\frac{1}{2}}, \text{etc.}$$

In the special case of equal weights these expressions simplify to

$$\left.\begin{aligned} P(4, 4) &= \frac{1}{12}; \\ P(3, 4) &= \frac{1}{2} - \frac{1}{\pi}\sin^{-1}\frac{1}{\sqrt{3}} = 0.3041 \\ P(2, 4) &= \frac{5}{12}; \end{aligned}\right\} \quad (3.42)$$

and $\quad P(1, 4) = 0.1959.$

Concluding remarks on the $P(l, k)$s

The task of finding the probabilities $P(l, k)$ may not always be as complex as (3.23) might suggest. In particular, it may be considerably simplified if the partial ordering on K which defines the alternative hypothesis can be split into several partial orderings on disjoint subsets of K. Then the isotonic regression calculation may be performed separately on each of these subsets. As an illustration of the general method, consider the alternative

$$\mu_1 \leq \mu_2 \leq \mu_3, \mu_4 \leq [\mu_5, \mu_6],$$

for $k = 6$. In this case, each of the subsets $\{1, 2, 3\}$ and $\{4, 5, 6\}$ may be dealt with separately, and the appropriate probabilities $P(l, 3)$ for them are already known: let these be $P'(l, 3)$, $P''(l, 3)$ respectively. Then the probabilities $P(l, 6)$ for this alternative are given by

$$P(l, 6) = \sum_{j=\max(1,l-3)}^{\min(3,l-1)} P'(j, 3)P''(l-j, 3), \quad l = 2, 3, \ldots, 6.$$

Note that the weights have been suppressed for simplicity.

Another case which may be dealt with by this method is that of the alternative

$$\mu_1 \leq \mu_2, \mu_3 \leq \mu_4, \ldots, \mu_{k-1} \leq \mu_k \quad (k \text{ even}),$$

for which the following result may be obtained;

$$P(l, k) = \binom{\frac{1}{2}k}{k-l} 2^{-\frac{1}{2}k}, \quad l \geq \tfrac{1}{2}k$$

$$= 0, \quad l < \tfrac{1}{2}k.$$

This result, which depends essentially on that for $k = 2$, is distribution-free and holds regardless of the weights.

Various tables of critical values of the test statistics have been constructed, and these are presented, along with the tables of $P(l, k)$s already mentioned, at the end of the book.

Distributions of the two-sided test statistics

At the end of Section 3.2 two-sided versions of the $\bar{\chi}^2$ and $\bar{E}^2$ tests against the simple order alternative were introduced. Derivation of the exact distributions of these statistics is complicated, but a very close approximation may be derived from results for the one-sided tests. The method will be illustrated by considering the two-sided $\bar{\chi}^2$ test. This is based on the statistic

$$\bar{\chi}^2 = \max\{\bar{\chi}^2(1), \bar{\chi}^2(2)\},$$

where $\bar{\chi}^2(1)$ and $\bar{\chi}^2(2)$ are the statistics calculated under the assumptions of increasing order and decreasing order respectively. Let A_i ($i = 1, 2$) denote the event that $\bar{\chi}^2(i) \geq C$ when the null hypothesis is true, for some positive constant C. Then the probability that the two-sided test statistic exceeds C is

$$\Pr\{A_1 \cup A_2\} = \Pr\{A_1\} + \Pr\{A_2\} - \Pr\{A_1 \cap A_2\}. \qquad (3.43)$$

It is easily seen that $\Pr\{A_1\} = \Pr\{A_2\}$, and, further, $\Pr\{A_1 \cap A_2\} < \Pr\{A_1\}$. Hence it follows from (3.43) that

$$\Pr\{A_1\} < \Pr\{A_1 \cup A_2\} \leq 2\Pr\{A_1\}. \qquad (3.44)$$

Thus, if C is the $100\alpha\%$ point of $\bar{\chi}^2$ for simple order, it follows that the size of the two-sided test using the same critical value lies between α and 2α.

We now argue that, if C is large (that is, if α is small), the upper bound is a very close approximation to the true size of the test. We may write

$$\Pr\{A_1 \cap A_2\} = \Pr\{A_1\}\Pr\{A_2|A_1\}.$$

Now the occurrence of A_1 indicates a significant departure from the null hypothesis in one direction, and it may be expected that the probability of a significant departure in the other direction is reduced by this conditioning event; thus

$$\Pr\{A_2|A_1\} < \Pr\{A_1\},$$

and

$$\Pr\{A_1 \cap A_2\} < [\Pr\{A_1\}]^2.$$

It now follows from (3.43) that $\Pr\{A_1 \cup A_2\}$ differs from its upper bound by at most $[\Pr\{A_1\}]^2$, which will be negligible for the significance levels usually chosen. Hence the tables provided for the one-sided test may be used if the tabulated significance levels are doubled.

When $k = 2$ the procedure recommended above is exact and gives the usual two-tail test. For $k = 3$, exact 5% and 1% points have been given for the two-sided test by Kudô and Fujisawa (1964). Table 3.2 gives a comparison of the exact percentage points with those taken from the 2.5% and 0.5% rows of Table A.1. The agreement is remarkably good in every case.

Table 3.2 Comparison of the Exact and Approximate 5% and 1% Points for the Two-sided Version of $\bar{\chi}_3^2$ for Simple Order

$-\rho$	5%		1%	
	Exact	Approx.	Exact	Approx.
0.10	5.48	5.46	8.53	8.54
0.30	5.29	5.29	8.35	8.36
0.50	5.11	5.10	8.12	8.15
0.70	4.84	4.85	7.84	7.87
0.90	4.45	4.47	7.40	7.41

Note. The nature of the approximation leads us to expect that "Approx." $\leq$ "Exact". The fact that this is not always so suggests that one or both sets of figures are not exact to 2 decimals. This does not affect our conclusion that the agreement is very good.

An approximation to the null hypothesis distribution of $\bar{\chi}^2$

The alternative of simple order and equal weights will be dealt with here. A complete solution for this case has already been given, but calculation of the $P(l, k)$s, and thence of the distribution function of $\bar{\chi}_k^2$, may prove tedious if k is large. An approximate method would therefore be useful.

The approximation is based on the moments (or cumulants) of $\bar\chi_k^2$. These may be obtained from the characteristic function, which is given by

$$\varphi(t) = \sum_{l=1}^{k} P(l, k)(1 - 2it)^{-\frac{1}{2}(l-1)}$$

$$= \frac{(z + 1)(z + 2) \ldots (z + k - 1)}{k!}, \quad (3.45)$$

where $z = (1 - 2it)^{-\frac{1}{2}}$. This result follows at once from Corollary A. The cumulant generating function is thus

$$\psi(t) = \log \varphi(t) = \sum_{j=1}^{k-1} \log(j + z) - \log k!. \quad (3.46)$$

Expansion yields the following expressions for the first four cumulants:

$$\left.\begin{aligned}
\kappa_1 &= \sum_{j=2}^{k} j^{-1}, \\
\kappa_2 &= \sum_{j=2}^{k} \{3j^{-1} - j^{-2}\}, \\
\kappa_3 &= \sum_{j=2}^{k} \{15j^{-1} - 9j^{-2} + 2j^{-3}\}, \\
\text{and}\quad \kappa_4 &= \sum_{j=2}^{k} \{105j^{-1} - 87j^{-2} + 36j^{-3} - 6j^{-4}\}.
\end{aligned}\right\} \quad (3.47)$$

For purposes of computation it is convenient to use the fact that the sums

$$\sum_{j=2}^{k} j^{-r}$$

can be found from tables of the polygamma functions, which may be represented in the form

$$\psi^{(r)}(x) = \frac{d^{r+1}}{dx^{r+1}} \log \Gamma(x) = (-1)^{r+1} r! \sum_{j=0}^{\infty} (x + j)^{-r-1}.$$

Suitable tables are in the British Association Mathematical Tables, Vol. 1, or Davis (1933, 1935). If k is very large

$$\varphi(t) \approx \frac{k^{z-1}}{\Gamma(z+1)},$$

from which it is found, in the limit, that

$$\kappa_r \approx 1 \cdot 3 \cdot 5 \ldots (2r - 1) \cdot \log k. \quad (3.48)$$

It follows that the distribution approaches normality as $k \to \infty$, but the limiting form is reached extremely slowly because the cumulants depend on k through log k. The normal approximation is therefore of no practical value.

A suitable approximating distribution may be found by examining the form of the distribution as indicated by the shape coefficients β_1 and β_2. Some numerical values are given in Table 3.3. The last row of the table is

Table 3.3   Values of β_1 and β_2 for the Null Hypothesis Distribution of $\bar{\chi}_k^2$ with Equal Weights

k	10	20	50	100	∞
β_1	4.14	3.10	2.30	1.94	0.00
β_2	8.88	2.78	6.28	5.75	3.00
$2\beta_2 - 3\beta_1 - 6$	0.65	0.52	0.33	0.31	0.00

included to show that the shape of the $\bar{\chi}_k^2$ distribution is close to that of the gamma or chi-square distribution. For this distribution $2\beta_2 - 3\beta_1 - 6 = 0$. If the values given in the table are plotted in the (β_1, β_2) plane it will be evident that they lie much nearer to the line $2\beta_2 - 3\beta_1 - 6 = 0$ than to the beta values of any other of the well-known distributions with range $(0, \infty)$ e.g. the lognormal or inverse gamma. It thus appears reasonable to try to approximate the $\bar{\chi}_k^2$ distribution by a gamma distribution, and to investigate the closeness of the approximation in the area of greatest practical interest i.e. the upper tail of the distribution.

The method adopted will be to approximate the $\bar{\chi}_k^2$ distribution by a gamma distribution with parameters r and λ chosen so that the first two cumulants of the two distributions are the same. As the mean and variance of the gamma distribution are r/λ and r/λ^2 respectively, the required parameter values are

$$\lambda = \frac{\kappa_1}{\kappa_2}, \qquad r = \frac{\kappa_1^2}{\kappa_2}, \tag{3.49}$$

where κ_1, κ_2 are as given in (3.47). To check the accuracy of the approximation, consider $k = 10$. From Table A.3 it is known that the upper 5% and 1% points of $\bar{\chi}_{10}^2$ are 6.560 and 10.216 respectively. For this distribution

$$\kappa_1 = 1.9290, \quad \text{and} \quad \kappa_2 = 5.2371.$$

The parameters of the fitted gamma distribution are then

$$\lambda = 0.3683, \qquad r = 0.7105.$$

To find the probability that a variable with this distribution exceeds the critical values of interest, the tables of Pearson (1934) will be used. In his notation the required probability is $1 - I(u, p)$ where, for the 5% point, $u = 6.560\lambda/\sqrt{r}$ and $p = r - 1$. Two-way linear interpolation gives $1 - I(2.866, -0.2895) = 0.0495$ which, to three decimals, is equal to the exact value. A similar calculation at the 1% point gives $1 - I(4.464, -0.2895) = 0.0117$ which, again, is remarkably close to the exact value. Other calculations for small k show a similar measure of agreement. Hence the use of a gamma distribution with parameters given by (3.49) and (3.47) is recommended for any values of k not covered by the tables. Equivalently the $\bar{\chi}_k^2$ distribution may be approximated by that of a multiple c of a χ_ν^2 random variable, where $c = \frac{1}{2}\kappa_2/\kappa_1$ and $\nu = 2\kappa_1^2/\kappa_2$.

3.4 THE POWER FUNCTION OF $\bar{\chi}^2$

Exact methods

The power functions of both $\bar{\chi}_k^2$ and $\bar{E}_k^2$ can be found by methods similar to those used to find their null hypothesis distributions: for each value of l, $2 \leq l \leq k$, the associated probability of the statistic exceeding the appropriate critical value is calculated, and summation gives the value of the power function. However, the complexity of the analysis required is such that the results on power are more limited in scope. In particular, no results will be given for $\bar{E}_k^2$ beyond noting that, in the limit as $N \to \infty$, it becomes equivalent to $\bar{\chi}_k^2$. Our primary object in computing power is to compare the $\bar{\chi}_k^2$ test with its competitors, namely the ordinary χ^2 test in this chapter, and various other tests against ordered alternatives in Chapter 4. For these purposes the restricted scope of our results is not a severe drawback.

The theory of this section is based on Bartholomew (1961a). The case $k = 3$ with unit weights will be discussed in some detail. This will be sufficient to demonstrate what is involved in exact calculations of power. Results for any "equal weights" situation may be derived from those for unit weights by multiplying the μs by a scale factor. For $k = 4$ numerical results only will be given, for three alternative hypotheses.

When $k = 3$, $\mathcal{H}$ consists of alternatives of three basic types, represented by the following:

$$H_1: \mu_1 \leq \mu_2 \leq \mu_3 \quad \text{(simple order)}$$
$$H_2: \mu_1 \leq [\mu_2, \mu_3] \quad \text{(simple tree)}$$
$$H_3: \mu_1 \leq \mu_2.$$

All other members of $\mathcal{H}$ can be related to these by rearrangement of suffices or changing the direction of the inequalities.

Beginning with H_1, let C denote the chosen critical value of $\bar{\chi}_3^2$ for that alternative. Then the power function may be written

$$\Pr\{\bar{\chi}_3^2 \geq C\} = \sum_{j=2}^{3} \Pr\{\bar{\chi}_3^2 \geq C, l = j\}. \tag{3.50}$$

Our method is to find each of the probabilities in the above sum by direct integration. Thus

$$\Pr\{\bar{\chi}_3^2 \geq C, l = 3\} = \Pr\{\bar{y}_2 - \bar{y}_1 > 0, \bar{y}_3 - \bar{y}_2 > 0, \sum_{i=1}^{3}(\bar{y}_i - \bar{y})^2 \geq C\}.$$

To evaluate this probability we make the orthogonal transformation

$$z_1 = (\bar{y}_2 - \bar{y}_1)/\sqrt{2}, \quad z_2 = (2\bar{y}_3 - \bar{y}_2 - \bar{y}_1)/\sqrt{6}.$$

The variables thus defined are normally and independently distributed with means

$$\lambda_1 = (\mu_2 - \mu_1)/\sqrt{2}, \quad \lambda_2 = (2\mu_3 - \mu_2 - \mu_1)/\sqrt{6}$$

respectively, and unit variances. The required probability thus becomes

$$\Pr\{z_1 > 0, \sqrt{3}z_2 - z_1 > 0, z_1^2 + z_2^2 \geq C\}.$$

A polar transformation to variables (r, θ) is now made, given by

$$z_1 = r\cos\theta, \quad z_2 = r\sin\theta.$$

In terms of the new variables we must find

$$\Pr\{\cos\theta > 0, \sqrt{3}\sin\theta - \cos\theta > 0, r^2 \geq C\}$$

$$= \frac{\exp(-\tfrac{1}{2}\Delta^2)}{2\pi} \int_{(\pi/6)+\beta}^{(\pi/2)+\beta} \int_{\sqrt{C}}^{\infty} r \exp(-\tfrac{1}{2}r^2 + r\Delta\sin\theta) \, dr \, d\theta,$$

where

$$\lambda_1 = \Delta\sin\beta, \quad \lambda_2 = \Delta\cos\beta.$$

Integrating with respect to r then gives

$$\Pr\{\bar{\chi}_3^2 \geq C, l = 3\} = \frac{\exp(-\tfrac{1}{2}\Delta^2)}{2\pi} \int_{(\pi/6)+\beta}^{(\pi/2)+\beta} \psi(\Delta\sin\theta, C) \, d\theta, \tag{3.51}$$

where

$$\psi(x, C) = \frac{x\Phi(x - \sqrt{C}) + \phi(x - \sqrt{C})}{\phi(x)},$$

$\Phi(x)$ and $\phi(x)$ being the standardized normal distribution and density functions respectively. As the function ψ in (3.51) depends on tabulated functions of the normal distribution, the integral is easily evaluated by numerical methods.

Next, $\Pr\{\bar{\chi}_3^2 \geq C, l = 2\}$ is evaluated. To do this two probabilities must be added—the first corresponding to cases where the first and second populations are pooled, the second corresponding to cases where the second and third populations are pooled. The first of these is

$$\Pr\{\bar{y}_2 - \bar{y}_1 \leq 0, 2\bar{y}_3 - \bar{y}_2 - \bar{y}_1 > 0, \frac{(2\bar{y}_3 - \bar{y}_2 - \bar{y}_1)^2}{6} \geq C\}$$

$$= \Pr\{z_1 \leq 0, z_2 \geq \sqrt{C}\}$$

$$= \Phi(-\lambda_1)\Phi(\lambda_2 - \sqrt{C}).$$

The second probability may be found by similar methods to be

$$\Phi(-\lambda_3)\Phi(\lambda_4 - \sqrt{C}),$$

where

$$\lambda_3 = (\mu_3 - \mu_2)/\sqrt{2}, \qquad \lambda_4 = (\mu_3 + \mu_2 - 2\mu_1)/\sqrt{6}.$$

Thus

$$\Pr\{\bar{\chi}_3^2 \geq C, l = 2\} = \Phi(-\lambda_1)\Phi(\lambda_2 - \sqrt{C}) + \Phi(-\lambda_3)\Phi(\lambda_4 - \sqrt{C}). \quad (3.52)$$

Combining (3.51) and (3.52), the final expression for the power function is

$$\Pr\{\bar{\chi}_3^2 \geq C\} = \frac{\exp(-\tfrac{1}{2}\Delta^2)}{2\pi} \int_{(\pi/6)+\beta}^{(\pi/2)+\beta} \psi(\Delta \sin\theta, C)\, d\theta$$

$$+ \Phi(-\lambda_1)\Phi(\lambda_2 - \sqrt{C})$$

$$+ \Phi(-\lambda_3)\Phi(\lambda_4 - \sqrt{C}). \quad (3.53)$$

At this point it may be observed that the power function of $\bar{\chi}_3^2$ depends on the two quantities Δ^2 and β; for the above result may be rewritten as

$$\Pr\{\bar{\chi}_3^2 \geq C\} = \frac{\exp(-\tfrac{1}{2}\Delta^2)}{2\pi} \int_{(\pi/6)+\beta}^{(\pi/2)+\beta} \psi(\Delta \sin\theta, C)\, d\theta$$

$$+ \Phi(-\Delta \sin\beta)\Phi(\Delta \cos\beta - \sqrt{C})$$

$$+ \Phi\left(-\Delta \sin\left(\frac{\pi}{3}-\beta\right)\right)\Phi\left(\Delta \cos\left(\frac{\pi}{3}-\beta\right) - \sqrt{C}\right).$$

$$(3.54)$$

In this respect the $\bar{\chi}_3^2$ test differs from the usual χ^2 test used against an unrestricted alternative, the power of which depends only on Δ^2 which may be regarded as a measure of the dispersion of the means, as it may be expressed in the form

$$\sum_{i=1}^{3}(\mu_i - \bar{\mu})^2,$$

where

$$\bar{\mu} = \tfrac{1}{3}\sum_{i=1}^{3}\mu_i.$$

The second parameter β depends on the relative values of the differences $\mu_2 - \mu_1$ and $\mu_3 - \mu_2$. If the μs satisfy the restrictions of H_1, β must lie between 0 and $\pi/3$. In order to make numerical comparisons between the power of $\bar{\chi}_3^2$ and that of χ^2, we investigate the maximum and minimum values taken by (3.54), regarded as a function of β. It is found that, for any value of Δ, the power of $\bar{\chi}_3^2$ takes its maximum value when $\beta = \pi/6$ (corresponding to $\mu_2 - \mu_1 = \mu_3 - \mu_2$) and has two equal minima, within the region of the alternative hypothesis, when $\beta = 0$ $(\mu_1 = \mu_2)$ or $\beta = \pi/3$ $(\mu_2 = \mu_3)$. These two extreme values of the power are given in Table 3.4. Results for other alternative hypotheses and greater values of k are similar. The power of the $\bar{\chi}_k^2$ test depends on

$$\Delta^2 = \sum_{i=1}^{k}(\mu_i - \bar{\mu})^2,$$

where

$$\bar{\mu} = \frac{1}{k}\sum_{i=1}^{k}\mu_i,$$

and on $k - 1$ "configuration parameters", and results will be given for what we know or conjecture to be upper and lower limits of the power.

The determination of the power for the alternatives H_2 and H_3 proceeds along exactly similar lines. The expression for the power against H_2 is

$$\Pr\{\bar{\chi}_3^2 \geq C\} = \frac{\exp(-\tfrac{1}{2}\Delta^2)}{2\pi}\int_{(-\pi/6)+\beta}^{(\pi/2)+\beta}\psi(\Delta\sin\theta, C)\,d\theta$$
$$+ \Phi(-\Delta\sin\beta)\Phi(\Delta\cos\beta - \sqrt{C})$$
$$+ \Phi\left(-\Delta\sin\left(\frac{\pi}{3}+\beta\right)\right)\Phi\left(\Delta\cos\left(\frac{\pi}{3}+\beta\right) - \sqrt{C}\right).$$
(3.55)

Here the alternative hypothesis corresponds to values of β between 0 $(\mu_1 = \mu_2)$ and $2\pi/3$ $(\mu_1 = \mu_3)$; within this region the power, for given Δ^2, takes its minimum value at the endpoints and its maximum at $\beta = \pi/3$ ($\mu_2 = \mu_3$). The power of the $\bar{\chi}_3^2$ test against H_3 is

$$\Pr\{\bar{\chi}_3^2 \geq C\} = \frac{\exp(-\tfrac{1}{2}\Delta^2)}{2\pi} \int_{(-\pi/2)+\beta}^{(\pi/2)+\beta} \psi(\Delta \sin\theta, C)\, d\theta$$

$$+ \Phi(-\Delta \sin\beta)\{\Phi(\Delta \cos\beta - \sqrt{C})$$

$$+ \Phi(-\Delta \cos\beta - \sqrt{C})\}. \qquad (3.56)$$

The region corresponding to this alternative hypothesis is $0 \leq \beta \leq \pi$; and within this region minimum power is achieved at the endpoints, at each of which μ_1 and μ_2 are equal, and maximum power at $\beta = \pi/2$, when μ_3 lies halfway between μ_1 and μ_2.

The above results enable a general conclusion to be drawn about the effect of the spacing of the μs, for a given value of Δ, on the power function. For each alternative hypothesis the power is least when, of the differences between μs which that alternative restricts to be non-negative, all but one (or all where this is possible) are zero; and the power is large when these differences are large, and other differences small. In other words, prior information is most useful when the means that it orders differ appreciably.

The power functions of the $\bar{\chi}_3^2$ tests are now compared with each other and with the power function of the χ^2 test. This is done by means of Table 3.4 which, for tests at the 5% significance level, gives values of each power function for equally spaced values of Δ. For each of the three $\bar{\chi}_3^2$ tests the maximum and minimum values of the power for each value of Δ are given, obtained by numerical evaluation of the appropriate integrals. For the χ^2 test the power was obtained by approximating the noncentral χ^2 distribution by a χ^2 (gamma) distribution as recommended by Patnaik (1949). In Table 3.4 the values of the population means are specified by the value of Δ and the integers in the column headed "Spacing". These integers are chosen to represent the configuration of the means, the differences between them being in the same ratio as, and having the same sign as, the differences between the means. Thus, for example, corresponding to the values $\{1, 2, 3\}$ in the "Spacing" column, the power tabulated for a given value of Δ is that for

$$\mu_i = \frac{i\Delta}{\sqrt{2}} + \theta, \qquad (i = 1, 2, 3),$$

for any value of θ. This convention will be used frequently in the tables of power given in this and the following chapter. A further point concerning this and subsequent tables of power values is that, though the situation where

Table 3.4 Power Functions of $\bar{\chi}_3^2$ and χ^2

Test	Alternative hypothesis	Spacing	\Delta				
			0	1	2	3	4
$\bar{\chi}_3^2$	Simple order, H_1 ($\mu_1 \leqslant \mu_2 \leqslant \mu_3$)	1, 2, 3	0.050	0.244	0.605	0.892	0.987
		1, 2, 2	0.050	0.221	0.569	0.872	0.983
	Simple tree, H_2 ($\mu_1 \leqslant [\mu_2, \mu_3]$)	1, 2, 2	0.050	0.210	0.543	0.857	0.979
		1, 1, 2	0.050	0.173	0.494	0.831	0.974
	H_3 ($\mu_1 \leqslant \mu_2$)	1, 1, 2	0.050	0.178	0.491	0.824	0.971
		1, 3, 2	0.050	0.142	0.443	0.796	0.965
χ^2	μ_1, μ_2, μ_3 not all equal	any	0.050	0.130	0.402	0.776	0.959

all weights are unity has been dealt with, the tables apply to any situation where the weights all have the same value w, setting

$$\Delta^2 = w \sum_{i=1}^{k} (\mu_i - \bar{\mu})^2.$$

The first observations that may be made on the values of Table 3.4 are that, for any value of Δ, the power of the $\bar{\chi}_3^2$ test exceeds that of the χ^2 test, no matter what the configuration of the means is and that, indeed, the dependence of the power on the mean configuration is relatively slight. Further, the gain in power of $\bar{\chi}_3^2$ over χ^2 is considerable, especially when the alternative is that of simple order. It may help to think in terms of the "restrictiveness" of an alternative hypothesis as some measure of the amount of information about the means provided by the inequalities imposed on them. In Table 3.4 the alternatives have been arranged in decreasing order of restrictiveness, as each implies its successor; and it is seen that, for any value of Δ, power increases with the restrictiveness of the alternative hypothesis. For $k = 3$, there are obvious advantages in the use of $\bar{\chi}_3^2$ rather than χ^2 when prior information is strong enough to lead to the choice of the most restrictive alternative H_1; but when prior information enables only the single restriction of H_3 to be imposed, it is debatable whether the gain in power of $\bar{\chi}_3^2$ over χ^2 is sufficient to justify the extra work involved.

When $k = 4$ there are nine basic types of alternative, as follows:

(i) $\mu_1 \leq \mu_2 \leq \mu_3 \leq \mu_4$ (simple order);

(ii) $\mu_1 \leq [\mu_2, \mu_3] \leq \mu_4$ (simple loop);

(iii) $\mu_1 \leq [\mu_2, \mu_3, \mu_4]$ (simple tree);
(iv) $\mu_1 \leq \mu_2 \leq \mu_3$;
(v) $\mu_1 \leq \mu_2$;
(vi) $\mu_1 \leq \mu_2, \mu_3 \leq \mu_4$;
(vii) $\mu_1 \leq [\mu_2, \mu_3]$;
(viii) $\mu_1 \leq \mu_2 \leq [\mu_3, \mu_4]$;
(ix) $[\mu_1, \mu_2] \leq [\mu_3, \mu_4]$.

Table 3.5 Power Functions of $\bar{\chi}_4^2$ and χ^2

Test	Alternative hypothesis	Spacing	Δ = 0	1	2	3	4
$\bar{\chi}_4^2$	Simple order ($\mu_1 \leq \mu_2 \leq \mu_3 \leq \mu_4$)	1, 2, 3, 4	0.050	0.238	0.594	0.885	0.985
		1, 2, 2, 2	0.050	0.202	0.531	0.849	0.977
	Simple loop ($\mu_1 \leq [\mu_2, \mu_3] \leq \mu_4$)	1, 2, 2, 3	0.050	0.229	0.574	0.871	0.982
		1, 2, 2, 2	0.050	0.198	0.518	0.838	0.974
	Simple tree ($\mu_1 \leq [\mu_2, \mu_3, \mu_4])$	1, 2, 2, 2	0.050	0.187	0.489	0.813	0.967
		1, 1, 1, 2	0.050	0.140	0.416	0.767	0.955
χ^2	$\mu_1, \mu_2, \mu_3, \mu_4$ not all equal	any	0.050	0.115	0.350	0.710	0.945

Table 3.5 gives values of the power functions for the first three alternatives above. These are calculated by methods similar to those used when $k = 3$. In each case the numerical evaluation of a bivariate integral and several univariate integrals is involved: for details of the calculations for simple order, see Bartholomew (1961a). In Table 3.5 the power function for each alternative has been calculated for two configurations of the means which, it is conjectured, yield maximum and minimum values of the power within the alternative hypothesis region. The conjecture is based on the results for $k = 3$ and is supported by such other power calculations as have been made. The power characteristics of the tests illustrated by this table are similar to those seen for $k = 3$. As before, each of the $\bar{\chi}_4^2$ tests is better than the χ^2 test. For the simple order and simple tree alternatives the ranges of power values are slightly broader than when $k = 3$. In this table, as in Table 3.4, the alternative hypotheses have been arranged in order of decreasing restrictiveness, and it is seen that, for any value of Δ, the power of the tests

increases as the alternative becomes more restrictive. The gain in power of the $\bar{\chi}_4^2$ test against the simple order alternative over the ordinary χ^2 test is more pronounced than was the case when $k = 3$.

Approximate methods

For $k > 4$ the exact methods given above become extremely cumbersome, involving the evaluation of $(k - 2)$-dimensional integrals. We therefore consider approximate methods, of which three give solutions in certain special cases of practical interest. These are capable of extension, but it is doubtful if the effort involved would be worthwhile.

The conditions for which the first method is developed are as follows. The value of k must be large, $k - 1$ of the population means must be equal, and the remaining mean must be involved in only one of the inequalities defining the alternative hypothesis. Thus the method may be used to give approximations to the power of the $\bar{\chi}_k^2$ test against the simple order alternative for the mean configurations which, it is conjectured, yield the smallest power value for a given value of Δ, viz. $\mu_1 = \mu_2 = \ldots = \mu_{k-1} < \mu_k$ and $\mu_1 < \mu_2 = \mu_3 = \ldots = \mu_k$. The former configuration will be dealt with; the result for the latter is identical. As in the previous subsection the result for the "unit weights" situation is derived but it applies to any "equal weights" situation if we define

$$\Delta^2 = w \sum_{i=1}^{k} (\mu_i - \bar{\mu})^2,$$

where w^{-1} is the common variance of the sample means.

The first step in the calculation of $\bar{\chi}^2$ is to obtain the maximum likelihood estimators of the μs by means of one of the methods described in Chapter 2. If the Pool-Adjacent-Violators method is used the order in which the populations are amalgamated into blocks is unimportant, and we may choose to perform the operation in two stages: first to find the maximum likelihood estimators $\hat{\mu}_1^{*\prime}, \hat{\mu}_2^{*\prime}, \ldots, \hat{\mu}_k^{*\prime}$ of the population means subject to the restrictions $\mu_1 \leq \mu_2 \ldots \leq \mu_{k-1}$, and then to complete the amalgamation process by introducing the further restriction $\mu_{k-1} \leq \mu_k$. Then $\hat{\mu}_k^{*\prime} = \bar{y}_k$, and the final stage of the amalgamation process depends on whether $\bar{y}_k \geq \hat{\mu}_{k-1}^{*\prime}$ or $\bar{y}_k < \hat{\mu}_{k-1}^{*\prime}$. In the former case

$$\bar{\chi}_k^2 = \sum_{i=1}^{k-1} (\hat{\mu}_i^{*\prime} - \bar{y})^2 + (\bar{y}_k - \bar{y})^2$$

$$= \sum_{i=1}^{k=1} (\hat{\mu}_i^{*\prime} - \bar{y}')^2 + \left(\frac{k-1}{k}\right)(\bar{y}_k - \bar{y}')^2, \qquad (3.57)$$

where
$$\bar{y}' = \frac{1}{k-1} \sum_{i=1}^{k-1} \bar{y}_i.$$

Otherwise sample k must be amalgamated with the samples in the preceding block. The first approximation consists in assuming that this amalgamation has a negligible effect on the mean of that block and hence on the value of $\bar{\chi}_k^2$: that is, we take

$$\bar{\chi}_k^2 \approx \sum_{i=1}^{k-1} (\hat{\mu}_i^{*\prime} - \bar{y}')^2 \quad \text{if} \quad \bar{y}_k < \hat{\mu}_{k-1}^{*\prime}. \tag{3.58}$$

A rough justification for this step lies in a result of Miles (1959) that the average size of a block is of order $k/\log k$. Hence the incorporation of population k into the preceding block will have an effect which is asymptotically negligible. Combining (3.57) and (3.58), we may write

$$\bar{\chi}_k^2 \approx \bar{\chi}_{k-1}^2 + \bar{\chi}_2^2, \tag{3.59}$$

where the first term on the right hand side is calculated from $\bar{y}_1, \bar{y}_2, \ldots, \bar{y}_{k-1}$, and the second term is calculated from $\bar{y}_k$ and $\bar{y}'$; this stage involves the further approximation that $\hat{\mu}_{k-1}^{*\prime} \approx \bar{y}'$.

It is obvious that the quantities $\bar{\chi}_{k-1}^2$ and $\bar{\chi}_2^2$ in (3.59) are independent. $\bar{\chi}_{k-1}^2$ has the same distribution as under the null hypothesis, as all the populations concerned have the same mean; and the distribution of $\bar{\chi}_2^2$ is that which it will be found convenient to call the non-central $\bar{\chi}_2^2$ distribution, as follows:

$$\left. \begin{array}{l} \Pr\{\bar{\chi}_2^2 = 0\} = \Phi(-\Delta) \\ \Pr\{\bar{\chi}_2^2 \geq C\} = \Phi(\Delta - \sqrt{C}), \quad C \geq 0, \end{array} \right\} \tag{3.60}$$

where
$$\Delta^2 = \sum_{i=1}^{k} (\mu_i - \bar{\mu})^2 = \left(\frac{k-1}{k}\right)(\mu_k - \mu)^2,$$

μ being the common value of $\mu_1, \mu_2, \ldots, \mu_{k-1}$. As the distribution of $\bar{\chi}_{k-1}^2 + \bar{\chi}_2^2$ is not easy to obtain explicitly, a gamma distribution will be fitted to its first two moments (or cumulants), which may be obtained by adding the corresponding cumulants of $\bar{\chi}_{k-1}^2$ and $\bar{\chi}_2^2$. The first two cumulants of $\bar{\chi}_{k-1}^2$ are given in (3.47), and those of $\bar{\chi}_2^2$ may be found from the characteristic function, which is

$$\varphi(t) = \Phi(-\Delta) + (1 - 2it)^{-\frac{1}{2}} \Phi[\Delta(1 - 2it)^{-\frac{1}{2}}] \exp[it\Delta^2(1 - 2it)^{-1}].$$

The cumulants are then

$$\begin{aligned}\kappa_1 &= (\Delta^2 + 1)\Phi(\Delta) + \Delta\phi(\Delta) \\ \kappa_2 &= (\Delta^4 + 6\Delta^2 + 3)\Phi(\Delta) + \Delta(\Delta^2 + 5)\phi(\Delta) - \kappa_1^2.\end{aligned} \quad (3.61)$$

A similar argument to the above holds if the alternative hypothesis is the simple tree, and the population means are in a configuration which seems likely to give the smallest power for any given value of Δ, viz. when μ_1 is equal to all but one of $\mu_2, \mu_3, \ldots, \mu_k$ and less than the remaining mean. The

Table 3.6 Comparison of Exact Power Functions of $\bar{\chi}^2$ Tests with Approximations, for Mean Configurations Yielding Minimum Power

k	Alternative hypothesis	Spacing	Method of calculation	Δ				
				0	1	2	3	4
3	Simple order $(\mu_1 \leqslant \mu_2 \leqslant \mu_3)$	1, 2, 2	Exact	0.050	0.221	0.569	0.872	0.983
			Approx.	0.059	0.211	0.561	0.892	0.992
4	Simple order $(\mu_1 \leqslant \mu_2 \leqslant \mu_3 \leqslant \mu_4)$	1, 2, 2, 2	Exact	0.050	0.202	0.531	0.849	0.977
			Approx.	0.063	0.196	0.520	0.864	0.988
	Simple tree $(\mu_1 \leqslant [\mu_2, \mu_3, \mu_4])$	1, 1, 1, 2	Exact	0.050	0.140	0.416	0.767	0.955
			Approx.	0.041	0.133	0.401	0.771	0.967

only difference in the result lies in the fact that the null hypothesis cumulants of $\bar{\chi}^2_{k-1}$ used must be those for the simple tree alternative rather than that of simple order. These may be obtained, for $k \leq 12$, using the probabilities of Table A.6. The adequacy, for small k, of both approximations may be judged from Table 3.6. The closeness of the approximation depends on the value of Δ, but not, in any obvious way, on k for the two values investigated. On a proportional basis the agreement is somewhat better when Δ is large, as might be expected on the grounds that the form of the noncentral $\bar{\chi}^2_2$ distribution approaches that of noncentral χ^2 as the noncentrality parameter becomes large. The maximum discrepancy between exact and approximate power values is two units in the second decimal. Although these results are not particularly good, we suggest that they are adequate to warrant the use of the approximation for larger values of k when no other method is available, especially as the approximation would be expected to give more accurate results as k increases.

The second situation in which an approximation to the power function can be obtained involves an alternative hypothesis which is *decomposable*. By this it is meant that the populations may be divided into groups such that none of the restrictions imposed by the alternative hypothesis involves two populations belonging to different groups. Such alternatives have already been examined briefly when dealing with the probabilities $P(l, k)$. For such an alternative, where there are m ($m < k$) such groups, an approximation to the power may be found when, within each group, all populations have the same mean. The method relies on the fact that the $\bar{\chi}_k^2$ statistic may be written as follows:

$$\bar{\chi}_k^2 = \bar{\chi}_{(1)}^2 + \bar{\chi}_{(2)}^2 + \ldots + \bar{\chi}_{(m)}^2 + S, \tag{3.62}$$

where $\bar{\chi}_{(i)}^2$ ($i = 1, 2, \ldots, m$) is the $\bar{\chi}^2$ statistic that would be used to test the equality of the means in group i, and S is the (appropriately weighted) sum of squares of the group mean estimators about their grand mean which, by (1.22), can be seen to be the usual between groups sum of squares for testing the equality of the m group means. The fact that a decomposable alternative is being considered implies that the isotonic regression calculations may be carried out on each group separately. For each i, therefore, as the population means in group i are all equal, the distribution of $\bar{\chi}_{(i)}^2$ is a central $\bar{\chi}^2$ distribution. Furthermore, all the $m - 1$ components on the right hand side of (3.62) are independently distributed.

Now let l_i ($i = 1, 2, \ldots, m$) be the number of distinct estimates from which $\bar{\chi}_{(i)}^2$ is calculated. Then the conditional distribution of $\bar{\chi}_{(i)}^2$, given l_i, is that of χ^2 with $l_i - 1$ degrees of freedom (where χ^2 with zero degrees of freedom is interpreted as the distribution of a random variable taking the value zero with probability one). It follows that the conditional distribution, given $l_1, l_2, \ldots, l_m$, of $\bar{\chi}_{(1)}^2 + \bar{\chi}_{(2)}^2 + \ldots + \bar{\chi}_{(m)}^2$, is that of χ^2 with $l_1 + l_2 + \ldots + l_m - m$ degrees of freedom; and therefore, as $l = l_1 + l_2 + \ldots + l_m$ with probability one, the conditional distribution of $\bar{\chi}_{(1)}^2 + \bar{\chi}_{(2)}^2 + \ldots + \bar{\chi}_{(m)}^2$, given l, is that of χ^2 with $l - m$ degrees of freedom. As S, the final component of $\bar{\chi}^2$, has a noncentral χ^2 distribution with $m - 1$ degrees of freedom and noncentrality parameter Δ^2, it follows that the conditional distribution of $\bar{\chi}^2$, given l, is noncentral χ^2 with $l - 1$ degrees of freedom and noncentrality parameter Δ^2. Thus the distribution of $\bar{\chi}^2$ may be written as

$$\Pr\{\bar{\chi}^2 \geq C\} = \sum_{l=m}^{k} P(l, k) \Pr\{\chi_{l-1; \Delta^2}^2 \geq C\}, \tag{3.63}$$

where $\chi_{\nu; \delta^2}^2$ denotes a random variable having the noncentral χ^2 distribution with ν degrees of freedom and noncentrality parameter δ^2.

This result is exact, and the only approximation involved in its use arises from the need to approximate the noncentral χ^2 distribution. The case that has been considered is one particularly unfavourable to the $\bar{\chi}^2$ test in that those differences between means which the alternative hypothesis restricts to take non-negative values are all zero.

In the third situation for which an approximation to the power of $\bar{\chi}^2$ has been obtained, the alternative is again decomposable, but the groups all consist of two populations only and the present concern is with the power when the group means are all the same. The value of k must therefore be even, and the alternative hypothesis may be written (with suitable re-ordering if necessary) as

$$\mu_1 \leq \mu_2, \mu_3 \leq \mu_4, \ldots, \mu_{k-1} \leq \mu_k.$$

The particular situation of interest here is when the means satisfy $\mu_1 = \mu_3 = \ldots = \mu_{k-1} < \mu_2 = \mu_4 = \ldots = \mu_k$, as it is conjectured that this configuration will give maximum power for given Δ^2. The method could easily be extended to deal with any configuration such that the group means are all equal.

The $\bar{\chi}_k^2$ statistic for this problem may be expressed in the form (3.62), where $m = \frac{1}{2}k$ and group i consists of populations $(2i - 1, 2i)$. But now the distribution of S is central χ^2 with $\frac{1}{2}k - 1$ degrees of freedom, while each of the components $\bar{\chi}_{(1)}^2, \ldots, \bar{\chi}_{(\frac{1}{2}k)}^2$ has the noncentral $\bar{\chi}_2^2$ distribution with non-centrality parameter $2\Delta^2/k$. The distribution of $\bar{\chi}_k^2$ may therefore be approximated by fitting a gamma distribution to the first two cumulants, which may be obtained from the expressions (3.61).

Tables 3.7 and 3.8 give the results of power calculations based on the approximations derived above.

Table 3.7 Approximate Values of the Power Functions of the $\bar{\chi}_8^2$ and χ^2 Tests

Test	Alternative hypothesis	Configuration of means	Δ				
			0	1	2	3	4
$\bar{\chi}_8^2$	Simple order $\mu_1 \leq \mu_2 \leq \mu_3$	$\mu_1 < \mu_2 = \mu_3 = \ldots = \mu_8$	0.057	0.191	0.456	0.800	0.973
	Simple tree $(\mu_1 \leq [\mu_2, \mu_3, \ldots, \mu_8])$	$\mu_1 = \mu_2 = \ldots = \mu_7 < \mu_8$	0.042	0.096	0.270	0.592	0.882
	$\mu_1 \leq \mu_2, \mu_3 \leq \mu_4,$ $\mu_5 \leq \mu_6, \mu_7 \leq \mu_8$	$\mu_1 = \mu_3 = \mu_5 = \mu_7 < \mu_2$ $= \mu_4 = \mu_6 = \mu_8$	0.050	0.150	0.385	0.712	0.933
		$\mu_1 = \mu_2, \mu_3 = \mu_4,$ $\mu_5 = \mu_6, \mu_7 = \mu_8$	0.050	0.095	0.273	0.597	0.885
	$\mu_1 \leq \mu_2$	$\mu_3 = \mu_4 = \ldots = \mu_8$ $= \frac{1}{2}(\mu_1 + \mu_2)$	0.050	0.107	0.283	0.595	0.876
		$\mu_1 = \mu_2$	0.050	0.091	0.254	0.564	0.861
χ^2	$\mu_1, \mu_2, \ldots, \mu_8$ not all equal	any	0.050	0.090	0.249	0.535	0.853

Table 3.8 Approximate Values of the Power Functions of the $\bar{\chi}^2_{12}$ and χ^2 Tests

Test	Alternative hypothesis	Configuration of means	Δ = 0	1	2	3	4
$\bar{\chi}^2_{12}$	Simple order $(\mu_1 \leq \mu_2 \leq \ldots \leq \mu_{12})$	$\mu_1 < \mu_2 = \mu_3 = \ldots = \mu_{12}$	0.067	0.159	0.414	0.766	0.963
	Simple tree $(\mu_1 \leq [\mu_2, \mu_3, \ldots, \mu_{12}])$	$\mu_1 = \mu_2 = \ldots = \mu_{11} < \mu_{12}$	0.043	0.084	0.218	0.492	0.802
	$\mu_1 \leq \mu_2, \mu_3 \leq \mu_4, \ldots,$ $\mu_{11} \leq \mu_{12}$	$\mu_1 = \mu_3 = \ldots = \mu_{11} < \mu_2$ $= \mu_4 = \ldots = \mu_{12}$	0.050	0.141	0.352	0.659	0.901
		$\mu_1 = \mu_2, \mu_3 = \mu_4, \ldots,$ $\mu_{11} = \mu_{12}$	0.051	0.084	0.217	0.506	0.816
	$\mu_1 \leq \mu_2$	$\mu_3 = \mu_4 = \ldots = \mu_{12}$ $= \frac{1}{2}(\mu_1 + \mu_2)$	0.050	0.093	0.230	0.500	0.800
		$\mu_1 = \mu_2$	0.050	0.081	0.208	0.473	0.780
χ^2	$\mu_1, \mu_2, \ldots, \mu_{12}$ not all equal	any	0.050	0.080	0.205	0.466	0.776

The power outside the alternative hypothesis region

The preceding calculations of power have all been for sets of mean values which satisfy the order restrictions imposed by the alternative hypothesis. While our primary concern is with the power of the $\bar{\chi}^2$ test in such situations, its power when the alternative hypothesis is not satisfied is also of interest. In particular we should like to know about its power at points just outside the alternative hypothesis region. For instance, in testing against an increasing trend, while it may be expected that the population means show a substantial tendency to be in increasing order, we may not be certain that all the restrictions of the simple order alternative are satisfied. In such a situation, before using the $\bar{\chi}^2$ test we should like to know that its power is not seriously affected by the presence of a small number of reversals in a predominantly increasing sequence of means.

Another question of interest is that of determining the set of all mean configurations for which the $\bar{\chi}^2$ test is consistent, in the sense that its power approaches unity as Δ increases. This question will be investigated before presenting further numerical results, and the following theorem will be proved. In stating and proving the theorem, the notation introduced in the first section of this chapter will be used, with the following additions. Let μ_i^* be the isotonic regression of μ_i, with weights w_i, with respect to the partial order which defines the alternative hypothesis H, and let $l\hat{u}$ be the number of distinct values taken by μ_i^*, $i = 1, 2, \ldots, k$. It will also be found convenient to use the notation $\bar{y}_i^*$ instead of $\hat{\mu}_i^*$ for the maximum likelihood estimators of the μ_i in the region H.

Theorem 3.4

If

$$\Delta^2 = \sum_{i=1}^{k} w_i(\mu_i - \bar{\mu})^2 \to \infty$$

in such a way that w_i/w_j and $(\mu_i - \bar{\mu})/(\mu_j - \bar{\mu})$ remain constant for all (i, j), the power of the $\bar{\chi}^2$ test of the null hypothesis H_0 against the alternative hypothesis H tends to unity if $l_\mu \geq 2$, and to zero if $(\mu_1, \mu_2, \ldots, \mu_k)$ lies in the interior of the region such that $l_\mu = 1$.

Proof. For the way in which $\Delta^2 \to \infty$ to satisfy the requirements of the theorem it must hold, for some constants $v_1, v_2, \ldots, v_k, \theta_1, \theta_2, \ldots, \theta_k$, that

$$w_i = av_i, \quad \mu_i - \bar{\mu} = b^{\frac{1}{2}}\theta_i, \quad i = 1, 2, \ldots, k,$$

and that $ab \to \infty$. As the power of the $\bar{\chi}^2$ test depends on the means only through the quantities $\mu_i - \bar{\mu}$, it is convenient to take $\bar{\mu} = 0$, so that $\mu_i = b^{\frac{1}{2}}\theta_i$, $i = 1, 2, \ldots, k$.

It may now be observed that the weights w_i in the isotonic regression calculations leading to μ_i^*, $\bar{y}_i^*$ may be replaced by the weights v_i; and defining θ_i^* to be the isotonic regression of θ_i with weights v_i, it is seen that $\mu_i^* = b^{\frac{1}{2}}\theta_i^*$, $i = 1, 2, \ldots, k$, and that l_μ is the number of distinct values taken by θ_i^*, $i = 1, 2, \ldots, k$.

The $\bar{\chi}^2$ test rejects the null hypothesis if

$$\bar{\chi}_k^2 = \sum_{i=1}^{k} w_i(\bar{y}_i^* - \bar{y})^2 \geq C,$$

for some positive constant C, or equivalently if

$$\sum_{i=1}^{k} v_i(z_i^* - \bar{z})^2 \geq \frac{C}{ab},$$

where $z = b^{-\frac{1}{2}}\bar{y}_i$, $i = 1, 2, \ldots, k$, z_i^* is the isotonic regression of z_i, with weights v_i, and $\bar{z}$ is the weighted mean of the zs. Now, for each value of i, z_i has the $N(\theta_i, (abv_i)^{-1})$ distribution. It follows that, as $ab \to \infty$, z_i converges in probability to θ_i, and therefore that z_i^* converges in probability to θ_i^* and $\bar{z}$ to $\bar{\theta}$, the weighted mean of the θs. Thus

$$\sum_{i=1}^{k} v_i(z_i^* - z)^2$$

converges in probability to

$$\sum_{i=1}^{k} v_i(\theta_i^* - \bar{\theta})^2,$$

which is positive if $l_\mu \geq 2$. The observation that

$$\frac{C}{ab} \to 0$$

as $ab \to \infty$ concludes the proof of the first part of the theorem.

In order to prove the second part, the following notation is introduced:

$$\mathrm{Av}_\mu A = \sum_{i \in A} w_i \mu_i \bigg/ \sum_{i \in A} w_i,$$

where A is any subset of $K = \{1, 2, \ldots, k\}$; $\mathrm{Av}_\theta A$, $\mathrm{Av}_z A$, and $\mathrm{Av}_y A$ are similarly defined. It is known from the Minimum Lower Sets algorithm that a necessary and sufficient set of conditions for $l_\mu = 1$ is that, for every lower set L which is a proper subset of K,

$$\mathrm{Av}_\mu L \geq \mathrm{Av}_\mu K = \bar{\mu}.$$

Thus, if $(\mu_1, \mu_2, \ldots, \mu_k)$ lies in the interior of the region such that $l_\mu = 1$, the above conditions must hold with every weak inequality made strict. For any L, if

$$\mathrm{Av}_\mu L > \bar{\mu}$$

it follows that

$$\mathrm{Av}_\theta L > \bar{\theta}$$

Now, as $ab \to \infty$, $\mathrm{Av}_z L$ and z converge in probability to $\mathrm{Av}_\theta L$ and $\bar{\theta}$ respectively. The condition

$$\mathrm{Av}_\theta L > \bar{\theta}$$

then implies that

$$\Pr\{\mathrm{Av}_z L > z\} \to 1,$$

and therefore that

$$\Pr\{\mathrm{Av}_y L > \bar{y}\} \to 1.$$

It may be now deduced that, if $(\mu_1, \mu_2, \ldots, \mu_k)$ lies in the interior of the region such that $l_\mu = 1$, the probability that, for every lower set L which is a proper subset of K,

$$\mathrm{Av}_y L > \bar{y},$$

tends to unity as $ab \to \infty$. It follows that the statistic $\bar{\chi}_k^2$ converges in probability to zero, and hence the power of the test tends to zero, as $ab \to \infty$. ∎

The most important application of the above theorem is to situations where the values of μ_i, σ_i^2, $i = 1, 2, \ldots, k$, are fixed and the sample size becomes large. Then, if $l_\mu \geq 2$, the test is consistent. However, though the knowledge that the test is consistent against a given alternative is reassuring, the test will not be regarded as satisfactory against that alternative unless its power approaches unity reasonably quickly as sample size increases. Table 3.9 therefore gives values of the power function of the $\bar{\chi}_4^2$ test at the 5% significance level against the simple order alternative when the weights are equal. The power of the χ^2 test has been included for comparison.

Table 3.9 Power Functions of the $\bar{\chi}_4^2$ Test Against Simple Order, and of the χ^2 Test

Test	Spacing	\multicolumn{5}{c}{Δ}				
		0	1	2	3	4
$\bar{\chi}_4^2$	1, 2, 3, 4	0.050	0.238	0.594	0.885	0.985
	1, 3, 2, 4	0 050	0.203	0.514	0.824	0.968
	2, 1, 4, 3	0.050	0.152	0.400	0.723	0.929
	3, 2, 1, 4	0.050	0.111	0.292	0,580	0.834
	3, 2, 1, 3	0.050	0.066	0.142	0.288	0.484
	3, 2, 1, 2	0.050	0.021	0.017	0.017	0.017
χ^2	any	0.050	0.115	0.350	0.710	0.945

The first mean configuration in the table is that for which, it is conjectured, the power of the $\bar{\chi}_4^2$ test is greatest for any value of Δ^2. The next four configurations lie outside the alternative hypothesis region, but within the region of consistency of the test. It is seen that, despite the reversal in the order (1, 2, 3, 4), the power of the $\bar{\chi}_4^2$ test against that mean configuration remains good, while against (2, 1, 3, 4) its power is still greater than that of χ^2 for moderate values of Δ. However, against the next two configurations, which differ more markedly from an increasing set, its power is lower than that of the χ^2 test. It is concluded that minor departures from the restrictions of the alternative hypothesis have little adverse effect on the power of $\bar{\chi}^2$, but that in the presence of more pronounced departures, while $\bar{\chi}^2$ may be a consistent test, it cannot be said to perform satisfactorily.

The results for the final configuration (3, 2, 1, 2) illustrate the behaviour of the $\bar{\chi}^2$ test when the population means lie on the boundary of the region such that $l_\mu = 1$, a situation which was not dealt with in Theorem 3.4. Then the asymptotic distribution of $\bar{\chi}_k^2$ is a central $\bar{\chi}_j^2$ distribution for some $j \leq k$ (in this case $j = 2$), with strict inequality except at points corresponding to the null hypothesis, and the power of the $\bar{\chi}^2$ test tends to a positive value not exceeding its significance level.

The question of optimality

The foregoing power calculations have demonstrated that the $\bar{\chi}^2$ test is often a useful improvement over the χ^2 test in terms of power. Attention will now be turned to the question of whether the $\bar{\chi}^2$ test is, in any sense, the best test for use against the restricted alternatives being considered. Though it is sometimes possible to show that likelihood ratio tests are asymptotically most powerful, the general methods used to establish such results involve the assumption that the null hypothesis corresponds to an interior point of the parameter space. As the null hypothesis in our problems corresponds to a boundary point of the parameter space, it follows that use cannot be made of these results here. Instead some results will be presented which, while not conclusive, suggest that $\bar{\chi}^2$ cannot be greatly improved on. In Chapter 4 the power function of $\bar{\chi}^2$ will be compared with those of other tests which have been devised for use against ordered alternatives, and it will be seen that none of these tests is uniformly better than $\bar{\chi}^2$.

The results relate to the question of whether any improvement over the $\bar{\chi}^2$ test is possible using a test based on the bivariate statistic $(l, \bar{\chi}^2)$. Such a test is considered on the grounds that a large value of l is itself evidence of a departure from the null hypothesis; and, while large values of l tend to be associated with large values of $\bar{\chi}^2$, the association is not perfect and some further information may be gained by using both statistics.

A test at level α based on the bivariate statistic will have the form:

$$\text{Reject } H_0 \text{ if } \bar{\chi}^2_k \geq C_l(\alpha),$$

where the Cs are chosen to satisfy

$$\sum_{l=2}^{k} P(l, k) \Pr\{\chi^2_{l-1} \geq C_l(\alpha)\} = \alpha. \qquad (3.64)$$

We must consider whether there is any choice of the Cs which may be described as optimum. Details of an investigation of this question, for $k = 3$ and the simple order alternative, are given in Bartholomew (1961a). It was demonstrated there that the optimum choice of the Cs depends on both Δ and β. There is therefore no satisfactory answer to the question of whether $\bar{\chi}^2_3$ may be improved upon by a bivariate test; but numerical comparisons (Bartholomew, 1961a, Table 8) do not reveal any bivariate test which is clearly better than $\bar{\chi}^2_3$.

One of the above bivariate tests deserves special mention. The Cs may be chosen so that $\Pr\{\bar{\chi}^2_k \geq C_{l_0} | l = l_0\}$ takes the same value, α' say, for $l_0 = 2$ ... k; C_{l_0} must be the upper $100\alpha'\%$ point of the $\chi^2_{l_0-1}$ distribution, where $\alpha' = \alpha/\{1 - P(1, k)\}$. The resulting test, which will be referred to as the conditional test, has obvious practical advantages, particularly if we are prepared to make the approximation (good for large k, and exact if the alternative hypothesis is decomposable) that $P(1, k)$ is zero. Then standard tables of critical values of χ^2 are all that is required to carry out the test.

Some values of the power of the conditional test have been calculated. Those for $k = 3$ and the simple order alternative with equal weights are given in Table 3.10; for other results see Bartholomew (1961a, Table 11). These show that, for the cases considered, the conditional test is more powerful than the χ^2 test but rarely as powerful as $\bar{\chi}^2$. As they involve small values of k, however, they do not relate to the type of situation in which it is most likely that the conditional test might be used.

Table 3.10 *Power Functions of the $\bar{\chi}_3^2$ and Conditional Tests Against the Simple Order Alternative*

Test	Spacing	Δ				
		0	1	2	3	4
$\bar{\chi}_3^2$	1, 2, 3	0 050	0.244	0.605	0.892	0.987
	1, 2, 2	0.050	0.221	0.569	0.872	0.983
Conditional	1, 2, 3	0.050	0.214	0.533	0.839	0.973
	1, 2, 2	0.050	0.206	0.541	0.856	0.979
χ^2	any	0.050	0.130	0.402	0.776	0.959

3.5 COMPLEMENTS

As so often happens in scientific research, several people began to work independently at about the same time on the problems discussed in this chapter. The first paper to be published on the $\bar{\chi}^2$ test was Bartholomew (1959a) but Chacko had independently obtained many of the same results although his work did not appear until 1963. Also, van Eeden (1958) had investigated the problem of estimating ordered parameters and had discussed a hypothesis testing problem. In her case, however, the null hypothesis was $\mu_1 \leq \mu_2 \leq \ldots \leq \mu_k$, the alternative being that at least one inequality was reversed. Kudô's (1963) test for the multivariate mean which is to be described in Chapter 4 is also closely related to $\bar{\chi}^2$ and it represents a further independent derivation of essentially the same result. In addition to this work on tests, the estimation theory which is needed for the likelihood ratio test had been developed in Brunk's work described in Chapter 2.

The original test against the alternative of simple order when the variance is unknown, given in Bartholomew (1959a), was based on $\bar{F}$ (see Section 3.2). An asymptotic investigation of the power of this test (in Bartholomew (1961b)) showed it to be considerably less powerful than the likelihood ratio test based on $\bar{E}^2$.

There have been several attempts to obtain the null hypothesis distribution of $\bar{\chi}^2$ either for the case of simple order or more generally. Theorem 3.1 is easily proved for $k = 2$ since $|\bar{y}_1 - \bar{y}_2|$ is distributed independently of the sign of $(\bar{y}_1 - \bar{y}_2)$. Bartholomew's original extension of this argument for the case of simple order aimed to show that the distribution of $\bar{\chi}^2$, conditional on the restraints imposed on the $\bar{y}_i$s by the averaging process, was that of χ^2. This was achieved, as in the proof of Theorem 3.1, by dividing the restraints into two classes. The first class includes all those restraints involving inequalities between means which were subsequently averaged; the second is the set of inequalities which the final set of means satisfy, for example $Y_1 \leq Y_2 \leq \ldots \leq Y_l$ in the case of simple order. The individual means were shown to be independent of the first class and a heuristic argument was used to show their independence of the second.

An alternative method of proof was given in Bartholomew (1961a) and later used by Shorack (1967). This was an inductive proof based on the number of essential restrictions defining the alternative hypothesis (see Section 2.3 for definition of essential restrictions). In the case of simple order there are $k - 1$ essential restrictions $\mu_i \leq \mu_{i+1}$ ($i = 1, 2, \ldots, k - 1$) and the proof then simplifies considerably.

Proof of Theorem 3.1 for the case of simple order

The theorem requires that the conditional distribution of $\bar{\chi}_k^2$, given l, be shown to be that of χ^2 with $l - 1$ degrees of freedom. Assume this result to be true when $k = m - 1$. Consider the case when $k = m$. Either the m means $\bar{y}_1, \bar{y}_2, \ldots, \bar{y}_m$ will be in ascending order of magnitude or there will be at least one inversion (ties have probability zero). In the former case the theorem is true by Lemma C. In the latter case suppose that $\bar{y}_i > \bar{y}_{i+1}$; then according to the Pool-Adjacent-Violators algorithm the ith and $(i + 1)$th groups must be amalgamated and will never be separated subsequently. The joint distribution of the $m - 1$ means left after this amalgamation, given $\bar{y}_i > \bar{y}_{i+1}$, is the same as the unconditional distribution since $\bar{y}_i - \bar{y}_{i+1}$ is distributed independently of $(w_i \bar{y}_i + w_{i+1} \bar{y}_{i+1})/(w_i + w_{i+1})$. Hence Theorem 3.1 holds by hypothesis. The theorem certainly holds when $k = 1$ and hence it holds generally by induction on k. ■

It is an essential part of this argument that two means, once pooled, are never separated. While this is true of the simple order and some partial orders it is not true in general. This is the point which is not adequately covered in the general inductive proofs of Bartholomew and Shorack.

Shorack (1967) showed that Theorem 3.1 also holds if the $\bar{y}_i$s are dependent with covariances

$$\sigma_{ij} = \delta_{ij}\, w_j^{-1} - \left(\sum_{h=1}^{k} w_h\right)^{-1} \quad (i, j = 1, 2, \ldots, k).$$

This result is obvious on noting that these are the correlation coefficients of the set of variables $z_i = \bar{y}_i - \bar{y}$. It is clear that the subtraction of $\bar{y}$ from each mean has no effect on the calculation of $\bar{\chi}_k^2$ and hence its distribution is the same. The problem arises in this form in one of the applications in Chapter 4 when $\bar{\chi}^2$ is calculated from the ranks of the observations.

The probabilities, $P(l, k)$, for equal weights have arisen in many contexts. It has already been noted that they are distribution-free and their study is essentially a combinatorial exercise. The first derivation of our main result appears to be due to Sparre Andersen (1954). The recurrence formula

$$P(l, k) = \frac{1}{l} \sum_{j=l-1}^{k-1} P(1, k-j) P(l-1, j) \tag{3.65}$$

was conjectured by Bartholomew (1959a) and proved by Miles (1959). In the same paper Miles gave the recurrence relation of Corollary B to Theorem 3.3 and several other properties. Barton and Mallows (1961) considered the problem afresh and re-derived Miles' result by a simpler method. The following discussion is based on their work and provides an alternative treatment to that given in the body of the chapter.

The probabilities $P(l, k)$ to be considered now arise in a situation rather different from that discussed so far in this chapter. However, it will be shown that these probabilities must be equal to those in Section 3.3.

Consider an array of numbers

$$\begin{pmatrix} x_1, x_2, \ldots, x_k \\ v_1, v_2, \ldots, v_k \end{pmatrix}, \tag{3.66}$$

where the vs are positive weights and the xs and vs are such that no pair of weighted means of the xs are equal. Now submit the columns of this array to a random permutation: that is, give equal probability to each of the $k!$ possible permutations of the integers $1, 2, \ldots, k$. Having obtained a random array by this method, calculate the isotonic regression of the xs in this array, using the appropriate weights, with respect to simple order; l and $P(l, k)$ are now defined as before. It will be shown that the values of the probabilities

$P(l, k)$ so obtained do not depend on the xs or the vs provided that the condition of no ties is satisfied. If this is so, the probabilities defined here must be the same as those arising in the situation described in the body of the chapter. To see this note that the distribution of $\bar{y}_1, \bar{y}_2, \ldots, \bar{y}_k$ can be regarded as arising in two stages, as follows: $\bar{y}_{(1)}, \bar{y}_{(2)}, \ldots, \bar{y}_{(k)}$ arise as an ordered sample from a normal distribution and are then subject to a random permutation. Thus, provided $\bar{y}_{(1)}, \bar{y}_{(2)}, \ldots, \bar{y}_{(k)}$ satisfy the condition of no ties, the conditional distribution of l may be obtained by considering permutations of the columns of the array

$$\begin{pmatrix} \bar{y}_{(1)}, \bar{y}_{(2)}, \ldots, \bar{y}_{(k)} \\ 1, \quad 1, \quad \ldots, 1 \end{pmatrix},$$

and is thus given by the probabilities $P(l, k)$ as defined above. Finally, as ties have zero probability, this is also the marginal distribution of l.

The main result concerning the $P(l, k)$s will now be presented. It is important to note that, though for the present purposes it would be sufficient to prove the theorem for the particular case when the weights $v_1, v_2, \ldots, v_k$ in the array (3.66) are all equal, the proof given must be for the more general case, as it is inductive and, even if the weights are equal initially, involves the application of the result to arrays where the weights are not equal.

Theorem 3.5

The probabilities $P(l, k)$ are given by

$$P(l, k) = |S_k^l|/k! \qquad (l = 1, 2, \ldots, k)$$

where $|S_k^l|$ is the coefficient of t^l in $t(t + 1) \ldots (t + k - 1)$.

Proof. Denote the probability generating function of l, for given k, by $\pi_k(t)$. We must prove that $\pi_k(t) = t(t + 1) \ldots (t + k - 1)/k!$, and proceed by induction, assuming the result true for $k = 1, 2, \ldots, m - 1$. If it can then be proved true for $k = m$, then it must be true for all k, as it is obviously true for $k = 1$.

We may, without loss of generality, suppose that x_m is the largest of $x_1, x_2, \ldots, x_m$. Now consider all $m!$ permutations of the columns of the array, and note the position of x_m in each. In $(m - 1)!$ permutations, x_m is in the last column; thus the amalgamation process must leave the last column unchanged. The conditional distribution of l is thus that of $l' + 1$, where l' results from random permutation of the remaining columns, and the conditional probability generating function of l is thus $t\pi_{m-1}(t)$. The remaining $(m - 1)(m - 1)!$ permutations may be enumerated as follows. For each i,

$i = 1, 2, \ldots, m - 1$, let x_m be in the column which precedes x_i. Then, treating these two columns as a unit, consider all $(m - 1)!$ permutations of this unit and the remaining $(m - 2)$ columns. In all these cases, the two tied columns may be amalgamated as the first stage of the isotonic regression calculation, which then proceeds to operate on the resulting $m - 1$ columns in the same way. Thus the conditional probability generating function of l is $\pi_{m-1}(t)$. Combining these results,

$$\pi_m(t) = \frac{t}{m} \pi_{m-1}(t) + \frac{m-1}{m} \pi_{m-1}(t). \tag{3.67}$$

By the induction hypothesis, $\pi_{m-1}(t) = t(t + 1) \ldots (t + m - 2)/(m - 1)!$. Substituting this result in (3.67) gives

$$\pi_m(t) = t(t + 1) \ldots (t + m - 1)/m! \quad \blacksquare \tag{3.68}$$

Miles' approach was somewhat similar, the main difference being that he considered the $(k!)^2$ permutations produced by permuting each row of the array (3.66) independently.

Chacko (1963) derived some theory for the case of equal weights though he did not obtain the explicit representation in terms of Stirling numbers. His approach was first to show that

$$P(1, k) = \frac{1}{k}$$

by an appeal to Sparre Andersen's results and then to obtain the $P(l, k)$s in the form given in Theorem 3.3 with $P(1, b_i) = 1/b_i$. Chacko also gave the result that

$$\sum_{l=1}^{k} P(l, k)(-1)^l = 0$$

which implies that the sum of the $P(l, k)$s for odd l is equal to the sum for even l. The result is easily obtained by putting $t = -1$ in (3.68). The same result appears to hold for the case of unequal weights and for any partial order when the ys are normally distributed. This remark is based on extensive computation for $k \leq 4$ but a proof for general k is lacking. If true in general the result would be useful for computing the $P(l, k)$s.

CHAPTER 4

Testing the Equality of Ordered Means: Extensions and Generalizations

4.1 A ONE-SIDED TEST FOR THE MULTIVARIATE NORMAL MEAN

Introduction

The following problem, which has received considerable attention, is closely related to the work of the previous chapter. A p-dimensional random variable has a multivariate normal distribution with mean vector $\boldsymbol{\tau} = (\tau_1, \tau_2, \ldots, \tau_p)'$ and covariance matrix $\boldsymbol{\Sigma}$. A test is required of the hypothesis

$$H_0 : \boldsymbol{\tau} = \mathbf{0}$$

against the alternative

$$H_1 : \boldsymbol{\tau} \neq \mathbf{0}, \quad \tau_i \geq 0, \quad i = 1, 2, \ldots, r$$

(for some $r \leq p$). The problem is thus a natural generalization of the one-tail test for a single normal mean. The situation when $\boldsymbol{\Sigma}$ is known will be considered first; later this restriction will be relaxed.

The connection between this problem and those of the previous chapter may be illustrated by an example. Suppose we are testing against the simple order alternative. If the transformations

$$z_i = \bar{y}_{i+1} - \bar{y}_i$$

and

$$\tau_i = \mu_{i+1} - \mu_i, \quad i = 1, 2, \ldots, k-1 \tag{4.1}$$

are made, the null and alternative hypotheses are seen to be equivalent to H_0 and H_1 respectively (with $r = p = k - 1$), and the test is to be based on the vector $\mathbf{z} = (z_1, z_2, \ldots, z_{k-1})'$, which is sufficient for $\boldsymbol{\tau}$. If the population variances are known, it follows that the covariance matrix of $\mathbf{z}$ is known and that we are in the sort of situation described in the previous paragraph.

The transformation (4.1) has appeared already, in Section 3.3, where it was used in finding $P(k, k; \mathbf{w})$. Similar transformations enable us to express many of the hypothesis testing problems of the previous chapter in the above multivariate form. There is not, however, an exact correspondence between the two types of problem. In Chapter 3 alternative hypotheses were dealt with which could be defined by restricting certain differences between group means to take non-negative values. The methods of this chapter enable us to develop tests against alternatives similarly defined in terms of more complex contrasts among the means. They also make it possible to extend the results of Chapter 3 to situations where the ys are correlated. On the other hand, some of the problems of Chapter 3 cannot be transformed in such a way that the methods of this section apply. For example, if $k = 4$, and the alternative hypothesis is that of simple loop order, the transformation (4.1) leads to a multivariate problem in which the alternative hypothesis region is restricted by $\tau_1 \geq 0$, $\tau_1 + \tau_2 \geq 0$, $\tau_2 + \tau_3 \geq 0$ and $\tau_3 \geq 0$. Problems such as this, where the number of restrictions defining the alternative hypothesis exceeds p, may be dealt with by an extension of the methods of this section; however, details will not be given here.

Theory of the one-sided test

It will be supposed that the data are in the form of a random sample of size n, $z_1, z_2, \ldots, z_n$, with mean $\bar{z}$. Then it may be shown (see Kudô (1963) or Nüesch (1966) for details when $r = p$) that the likelihood ratio test rejects H_0 for large values of

$$n\{\bar{z}'\Sigma^{-1}\bar{z} - (\bar{z} - \tau^*)'\Sigma^{-1}(\bar{z} - \tau^*)\}, \tag{4.2}$$

where τ^* is the maximum likelihood estimator of τ in the region $H_1 : \tau_i \geq 0$, $i = 1, 2, \ldots, r$ (as in Chapter 3, the same notation is used for the region in the parameter space corresponding to the null and alternative hypotheses together, as is used to denote the alternative hypothesis).

The determination of τ^* is thus central to the theory of this test. We seek the vector in H_1 which minimizes the quadratic form $(\bar{z} - \tau)'\Sigma^{-1}(\bar{z} - \tau)$ with respect to τ. Obviously, if $\bar{z}$ lies in H_1 estimation poses no problem, for then $\tau^* = \bar{z}$. Otherwise τ^* must lie on the boundary of H_1 which we shall find convenient to split into $2^r - 1$ "faces"—each of these being defined by restricting some of $\tau_1, \tau_2, \ldots, \tau_r$ to be zero, and the others to be positive. In accordance with this definition the interior of H_1 will also be referred to as a face. H_1 has thus been partitioned into 2^r faces. The problem of finding τ^* is then solved if it is known to which face it belongs: for it is simply the projection, with respect to the inner product $(\mathbf{x}, \mathbf{y}) = \mathbf{x}'\Sigma^{-1}\mathbf{y}$, of $\bar{z}$ onto the linear subspace generated by that face. Thus τ^* could be found by finding all 2^r such projections and choosing that which gives the smallest value of $(\bar{z} - \tau)'\Sigma^{-1}(\bar{z} - \tau)$. This method may be feasible for small r, but for larger values of r a method would be preferred which cuts down the amount of computation needed. Kudô (1963) described such a method (for the case $r = p$) which avoids the examination of many of the above projections. His method is not, strictly speaking, an algorithm, as it involves intuitive judgement in the selection of subspaces onto which to project. It is therefore most suitable for use with a desk calculator. For computer applications we have developed an algorithm based on Kudô's method (Bremner, 1967). We have, however, no proof that this algorithm always leads to a solution, although it has worked on all occasions on which we have used it.

The expression for the likelihood ratio test statistic (4.2) may be simplified. As τ^* is the projection of $\bar{z}$ onto a linear subspace, it follows that

$$\bar{z}'\Sigma^{-1}\bar{z} = (\bar{z} - \tau^*)'\Sigma^{-1}(\bar{z} - \tau^*) + \tau^{*'}\Sigma^{-1}\tau^*;$$

thus the test statistic may be written as

$$\bar{\chi}^2 = n\tau^{*'}\Sigma^{-1}\tau^*. \tag{4.3}$$

One of the reasons underlying the choice of the notation $\bar{\chi}^2$ for this statistic is that, if it results from the application of a transformation such as (4.1) to one of the problems of Chapter 3, it has the same value as the $\bar{\chi}^2$ statistic calculated from the untransformed data. A more important reason is that the null hypothesis distribution of the statistic (4.3) is a mixture of χ^2 distributions such as we have seen before. This result may be stated as the following theorem, proofs of which (for $r = p$) appear in Kudô (1963) and Nüesch (1964, 1966).

Theorem 4.1

If H_0 is true, then

$$\Pr\{\bar{\chi}^2 \geq C\} = \sum_{j=1}^{p} Q(j,p) \Pr\{\chi_j^2 \geq C\}, \quad C > 0$$

$$\Pr\{\bar{\chi}^2 = 0\} = Q(0,p),$$

where $Q(j,p)$ is the probability that τ^* has exactly j nonzero elements, and χ_j^2 denotes a random variable having the χ^2 distribution with j degrees of freedom.

The similarity between the roles played by the $P(l,k)$s in Theorem 3.1 and the $Q(j,p)$s in Theorem 4.1 is at once evident. This similarity extends to the problem of their determination: the $Q(j,p)$s, like the $P(l,k)$s, may be expressed in terms of orthant probabilities of certain multivariate normal distributions (see Kudô (1963) for details), and no general solution in closed form exists for

$r > 3$. Thus the range of solutions in which the test based on the statistic (4.3) can be used is rather severely restricted.

In many practical applications the covariance matrix $\boldsymbol{\Sigma}$ is unknown, but is known to be of the form $\sigma^2\boldsymbol{\Lambda}$, where $\boldsymbol{\Lambda}$ is completely known but σ^2 is unknown. If this is so, the likelihood ratio test is similar to the $\bar{E}^2$ test of the previous chapter. It is based on the statistic

$$\bar{E}^2 = \frac{n\boldsymbol{\tau}^{*\prime}\boldsymbol{\Lambda}^{-1}\boldsymbol{\tau}^*}{\sum_{i=1}^{n}\mathbf{z}_i'\boldsymbol{\Lambda}^{-1}\mathbf{z}_i}, \tag{4.4}$$

which has its null hypothesis distribution given by the following theorem.

Theorem 4.2

If H_0 is true, then

$$\Pr\{\bar{E}^2 \geq C\} = \sum_{j=1}^{p} Q(j, p) \Pr\{B_{\frac{1}{2}j,\,\frac{1}{2}(np-j)} \geq C\}, \qquad C > 0$$

$$\Pr\{\bar{E}^2 = 0\} = Q(0, p),$$

where $Q(j, p)$ is defined in Theorem 4.1, and $B(a, b)$ denotes a random variable having the Beta distribution with parameters a and b.

It has been remarked above that the $\bar{\chi}_k^2$ statistic calculated for the testing problems of the previous chapter is the same as the $\bar{\chi}^2$ statistic calculated for the corresponding $(k-1)$-variate multivariate problem. This is not so, however, for the $\bar{E}^2$ statistics. To see this, consider the one-way analysis of variance situation, with $n_1 = n_2 = \ldots = n_k = n > 1$. The transformation to vectors $\mathbf{z}_1, \mathbf{z}_2, \ldots, \mathbf{z}_n$ where the elements of $\mathbf{z}_j$ are given by

$$z_{ji} = y_{i+1,j} - y_{ij}, \qquad i = 1, 2, \ldots, k-1 \tag{4.5}$$

allows the use of the methods of this chapter, but the $\bar{E}^2$ statistic obtained is not the same as the $\bar{E}_k^2$ statistic: nor does it have the same distribution. The reason for this can be seen in the fact that the use of transformation (4.5) implies an association between $y_{1j}, y_{2j}, \ldots, y_{kj}$ which is not inherent in the one-way analysis of variance model. Such an association is present if the ys are considered to be data from a randomized blocks experiment, and analysis under the assumption of that model leads to an $\bar{E}_k^2$ statistic identical with the $\bar{E}^2$ statistic for the transformed data. The denominators of the two $\bar{E}^2$ statistics differ by an amount equal to the "blocks" sum of squares.

Ordered tests for the general linear hypothesis

In Section 3.2 it was shown how the $\bar{E}^2$ test could be used in the analysis of orthogonal designs, to provide a test of the equality of main effects against the alternative that they were ordered. The method used there breaks down if the design is not orthogonal. A more general approach, related to the multivariate problem of this section, removes this restriction and is more widely applicable. This approach will be described in the context of the general linear model and, as an example, we shall indicate how it may be applied to the two-way classification with unequal cell frequencies.

The model used is well-known. A vector **y** denotes our observations $y_1, y_2, \ldots, y_N$, and the joint distribution of the observations is given by

$$\mathbf{y} = \mathbf{X}\boldsymbol{\theta} + \mathbf{e}, \tag{4.6}$$

where **X** is a known matrix of order $N \times p$ and rank p, $\boldsymbol{\theta} = (\theta_1, \theta_2, \ldots, \theta_p)'$ is a vector of unknown parameters, and **e** is a vector of independent random variables, each having a normal distribution with mean zero and variance σ^2. On the basis of the data **y** it is desired to test the null hypothesis that the first q ($1 \leq q \leq p$) of the θs are zero against the alternative that at least one of these parameters is nonzero and the first r of them ($1 \leq r \leq q$) are nonnegative. It should be noted that the majority of problems likely to be encountered involving ordered hypotheses in linear models will not appear in this form when expressed in terms of the natural parametrization of the model. However, nearly all such problems, whether the model is of full rank or not, may be expressed in this form after suitable reparametrization. The only exceptions occur when the number of inequality restrictions imposed by the alternative hypothesis exceeds the number of independent constraints imposed by the null hypothesis. Then a transformation to the above form cannot be made; but, as has already been remarked, the methods of this section readily extend to such situations.

In what follows it will be found convenient to partition $\boldsymbol{\theta}$ into two components, $\boldsymbol{\theta}_1 = (\theta_1, \theta_2, \ldots, \theta_q)'$ and $\boldsymbol{\theta}_2 = (\theta_{q+1}, \theta_{q+2}, \ldots, \theta_p)'$. Thus the argument assumes that $q < p$; a simplified version of this argument shows that the same results are valid for $q = p$. The null hypothesis is now

$$H_0: \boldsymbol{\theta}_1 = \mathbf{0},$$

and the alternative hypothesis is

$$H_1: \boldsymbol{\theta}_1 \neq \mathbf{0}, \quad \theta_i \geq 0, \quad i = 1, 2, \ldots, r.$$

Following the earlier conventions the notation H_0 will also be used to denote the region $\boldsymbol{\theta}_1 = \mathbf{0}$, and the notation H_1 to denote the region $\theta_i \geq 0$, $i = 1, 2, \ldots, r$.

First suppose that σ^2 is known. The likelihood ratio test then rejects H_0 for large values of

$$\min_{\theta \in H_0} S - \min_{\theta \in H_1} S, \qquad (4.7)$$

where $S = (\mathbf{y} - \mathbf{X}\theta)'(\mathbf{y} - \mathbf{X}\theta)$. Now it is a familiar result of least squares theory that

$$S = (\mathbf{y} - \mathbf{X}\hat{\theta})'(\mathbf{y} - \mathbf{X}\hat{\theta}) + (\hat{\theta} - \theta)'\Gamma(\hat{\theta} - \theta), \qquad (4.8)$$

where $\Gamma = \mathbf{X}'\mathbf{X}$ and $\hat{\theta} = \Gamma^{-1}\mathbf{X}'\mathbf{y}$ is the vector of (unrestricted) least squares estimators of the parameters. Further, if Γ is partitioned as

$$\Gamma = \begin{bmatrix} \Gamma_{11} & \Gamma_{12} \\ \Gamma_{21} & \Gamma_{22} \end{bmatrix},$$

where Γ_{11} is a $q \times q$ matrix, the second term of the right hand side of (4.8) may be rewritten:

$$S = (\mathbf{y} - \mathbf{X}\hat{\theta})'(\mathbf{y} - \mathbf{X}\hat{\theta}) + (\hat{\theta}_1 - \theta_1)'(\Gamma_{11} - \Gamma_{12}\Gamma_{22}^{-1}\Gamma_{21})(\hat{\theta}_1 - \theta_1)$$
$$+ [\Gamma_{22}^{-1}\Gamma_{21}(\hat{\theta}_1 - \theta_1) + \hat{\theta}_2 - \theta_2]'\Gamma_{22}[\Gamma_{22}^{-1}\Gamma_{21}(\hat{\theta}_1 - \theta_1) + \hat{\theta}_2 - \theta_2]. \qquad (4.9)$$

Of the three terms in this expression, the first does not involve θ and the minimum of the third with respect to θ_2 is zero irrespective of the value of θ_1. Thus the test statistic (4.7) becomes

$$\min_{\theta \in H_0} (\hat{\theta}_1 - \theta_1)'\Lambda_{11}^{-1}(\hat{\theta}_1 - \theta_1) - \min_{\theta \in H_1} (\hat{\theta}_1 - \theta_1)'\Lambda_{11}^{-1}(\hat{\theta}_1 - \theta_1),$$
$$= \hat{\theta}_1'\Lambda_{11}^{-1}\hat{\theta}_1 - (\hat{\theta}_1 - \theta_1^*)'\Lambda_{11}^{-1}(\hat{\theta}_1 - \theta_1^*). \qquad (4.10)$$

where $\Lambda_{11} = (\Gamma_{11} - \Gamma_{12}\Gamma_{22}^{-1}\Gamma_{21})^{-1}$ and θ_1^* is the value of θ_1 achieving the second minimization above. As $\hat{\theta}_1$ is distributed with covariance matrix $\sigma^2\Lambda_{11}$, comparison of (4.10) and (4.2) shows that the test procedure amounts to ignoring $\hat{\theta}_2$ and performing a test of the type discussed in the earlier part of this section, based on the single observation $\hat{\theta}_1$. Thus the test rejects H_0 for large values of

$$\bar{\chi}^2 = \theta_1^{*'}\Lambda_{11}^{-1}\theta_1^*/\sigma^2, \qquad (4.11)$$

the null hypothesis distribution of this statistic being as in Theorem 4.1 with q substituted for p.

When σ^2 is unknown the likelihood ratio test of H_0 against H_1 rejects the null hypothesis for large values of

$$\bar{E}^2 = \frac{\theta_1^{*'}\Lambda_{11}^{-1}\theta_1^*}{\hat{\theta}_1'\Lambda_{11}^{-1}\hat{\theta}_1 + (\mathbf{y} - \mathbf{X}\hat{\theta})'(\mathbf{y} - \mathbf{X}\hat{\theta})}. \qquad (4.12)$$

The null hypothesis distribution of this statistic is given by

$$\Pr\{\bar{E}^2 \geq C\} = \sum_{j=1}^{q} Q(j,q) \Pr\{B_{\frac{1}{2}j,\frac{1}{2}(N-p+q-j)} \geq C\}, \quad C > 0 \\ \Pr\{\bar{E}^2 = 0\} = Q(0,q).$$ (4.13)

The above methods will now be illustrated by applying them to the problem of testing the equality of main effects in the two-way cross-classification against the alternative of an increasing trend. The problem has already been examined in Section 3.2 under the assumption of equal cell frequencies; that restriction is now relaxed. The model is

$$y_{tij} = \mu + \alpha_t + \beta_i + \gamma_{ti} + e_{tij}$$ (4.14)

$$(t = 1, 2, \ldots, k; i = 1, 2, \ldots, m; j = 1, 2, \ldots, n_{ti})$$

where

$$\sum_{t=1}^{k} \alpha_t = \sum_{i=1}^{m} \beta_i = \sum_{t=1}^{k} \gamma_{ti} = \sum_{i=1}^{m} \gamma_{ti} = 0,$$ (4.15)

and the es are uncorrelated random variables with zero means and the same (unknown) variance. A brief description will be given of three tests of the null hypothesis

$$\alpha_1 = \alpha_2 = \ldots = \alpha_k$$

against the simple order alternative

$$\alpha_1 \leq \alpha_2 \leq \ldots \leq \alpha_k,$$

with at least one inequality strict. Which (if any) of the tests is appropriate in any given situation depends on our objectives and state of knowledge concerning the interaction parameters in (4.14). Some of the tests illustrate simplifications and modifications of the general procedure described above.

First suppose that we are prepared to postulate an additive model i.e. to delete the γs from (4.14). If the reparameterization given by

$$\theta_i = \alpha_{i+1} - \alpha_i, \quad i = 1, 2, \ldots, k-1 \\ \theta_{k-1+i} = \beta_{i+1} - \beta_i, \quad i = 1, 2, \ldots, m-1 \\ \theta_{k+m-1} = \mu$$ (4.16)

is made, the problem is exactly in the form that has been described, with $p = k + m - 1, q = r = k - 1$. The appropriate test statistic is of the form (4.12). Its null hypothesis distribution is as given by (4.13), with $q = k - 1$ and the second parameter of the Beta distribution replaced by $\frac{1}{2}(N - m - j)$.

A second situation which may be encountered is one where we are not prepared to assume an additive model, but wish to draw inferences about the αs only if a preliminary test fails to reject the hypothesis of no interaction (for a warning concerning the validity of such a procedure, see Scheffé (1959), p. 94). In such circumstances the practice frequently adopted is that the statistics on which tests of hypotheses concerning the αs are based involve the residual mean square for the model with interactions present rather than that for the additive model. For the problem of testing the equality of the αs against the simple order alternative, it seems natural to modify the $\bar{E}^2$ statistic calculated as described in the previous paragraph by replacing the residual sum of squares $(\mathbf{y} - \mathbf{X}\hat{\boldsymbol{\theta}})'(\mathbf{y} - \mathbf{X}\hat{\boldsymbol{\theta}})$ in the denominator by the "within cells" sum of squares. The null hypothesis distribution of this statistic is as in (4.13), but with $q = k - 1$ and the second parameter of the Beta distribution replaced by $\frac{1}{2}(N - km + k - j - 1)$. It should be noted that when all cell frequencies are equal this test reduces to that based on the statistic (3.14).

A further possibility when we have data which arise in accordance with the model (4.14) is that we may wish to test that the αs are equal without making the assumption that the γs are zero. In such situations the choice of the constraints imposed on the parameters of the model is important, because the interpretation to be placed on the parameters depends on these constraints; see Scheffé (1959, Chapter 4), for a discussion of this question. We shall suppose that the constraints (4.15) are appropriate. Then a test could be carried out by the general method described above, by adding $(k - 1)(m - 1)$ independent interaction parameters to (4.16). It is not, however, necessary to introduce this degree of complexity into the analysis. If $\mu_t = \mu + \alpha_t$, $t = 1, 2, \ldots, k$, the null and alternative hypotheses take the same form when expressed in terms of the μs as they do when expressed in terms of the αs. The least squares estimators of the μs are given by

$$\hat{\mu}_t = \frac{1}{m} \sum_{i=1}^{m} \bar{y}_{ti\cdot}, \qquad t = 1, 2, \ldots k$$

with the usual notation for means. Now as these are uncorrelated the methods of isotonic regression may be used to find constrained estimates of the μs, and the expressions for sums of squares are considerably simplified. An argument similar to that used in Section 3.2 to derive a test for the case when all cell frequencies are equal leads to a test statistic which resembles (3.14), differing in that the expressions for sums of squares are modified to take account of the unequal variances of the $\hat{\mu}$s and in that the isotonic regression calculations are for general weights. The test statistic becomes equal to (3.14) when cell frequencies are all the same. Thus the two statistics discussed in this and the previous paragraph are identical in this situation; this is not, however, true in general.

4.2 TESTS BASED ON CONTRASTS AMONG THE MEANS

he test statistics

The use of likelihood ratio tests up to this point has been prompted by the fact that these statistics have a strong intuitive appeal. We are aware that there are cases where the adoption of the method gives absurd results but, in the great majority of cases where it has been used, the resulting tests have been satisfactory. Often it has been possible to show that such tests have optimum properties—in an asymptotic sense at least. Also there are obvious advantages in having a general method which is capable of coping with the great variety of order restrictions which have been considered. It has been shown in Section 3.4 that the power of the likelihood ratio statistic for ordered alternatives shows a satisfactory increase over the standard χ^2-test. Nevertheless it remains an open question as to whether the $\bar{\chi}^2$ and $\bar{E}^2$ tests are optimum in any sense or whether better tests can be found. In this section a partial answer will be given to such questions by considering an alternative class of statistics. As far as the authors are aware all known tests for ordered alternatives, except $\bar{\chi}^2$, $\bar{E}^2$, and the distribution-free statistics described in Section 4.4, are included within this class.

We have noted that $\bar{\chi}^2$ could be regarded as a generalization of the one-tail test. The class of statistics to which we now turn is obtained by a different generalization. To be specific, consider the case where there are k independent random variables $y_1, y_2, \ldots, y_k$ with y_i drawn from a normal population with mean μ_i and variance σ^2 ($i = 1, 2, \ldots, k$). As before, we wish to test the null hypothesis

$$H_0: \mu_1 = \mu_2 = \ldots = \mu_k.$$

To begin with, consider the alternative hypothesis of simple order. If $k = 2$ the standard procedure is to form the difference $y_2 - y_1$ and reject H_0 if the difference is significantly large. A natural extension of this test when $k > 2$ is to construct a statistic based on the $\binom{k}{2}$ differences $(y_i - y_j)$ $(i > j)$. Only positive values of each difference count against H_0 and perhaps the simplest way of combining the differences is

$$T = \sum_{i=2}^{k} \sum_{j=1}^{i-1} (y_i - y_j) = \sum_{i=1}^{k} (2i - k + 1) y_i. \qquad (4.17)$$

If the ys conform closely to the order predicted by the alternative hypothesis most of the differences $(y_i - y_j)$ will be positive and so T will tend to be large. H_0 would thus be rejected if T were significantly larger than zero.

The sampling distribution and the power of T are easily obtained. Under both the null and alternative hypotheses T is normally distributed. The mean and variance are given by

$$E(T) = \sum_{i=1}^{k} (2i - k + 1)\mu_i,$$

which is zero when $\mu_1 = \mu_2 = \ldots = \mu_k$, and

$$\text{var}(T) = \sum_{i=1}^{k} (2i - k + 1)^2 \sigma^2.$$

If, instead of one, there are n observations from each population with means $\bar{y}_1, \bar{y}_2, \ldots, \bar{y}_k$ and if the common population variance is unknown, the appropriate statistic is

$$T' = \sqrt{n} \sum_{i=1}^{k} (2i - k + 1)\bar{y}_i \Big/ \left\{ \sum_{i=1}^{k} (2i - k + 1)^2 \right\}^{\frac{1}{2}} s \qquad (4.18)$$

where s is the "within groups" estimate of σ. The null hypothesis distribution of T' now has the Student form. This test is easy to calculate and its size and power are readily found. It relates only to the case of simple order and it assumes equal standard deviations in the populations. Later in this section its power function will be compared with that of $\bar{\chi}^2$.

The statistic introduced in (4.17) is a special case of the more general contrast

$$T = \sum_{i=1}^{k} c_i y_i \quad \text{with} \quad \sum_{i=1}^{k} c_i = 0. \qquad (4.19)$$

It is not obvious that it is best to give all the differences $(y_i - y_j)$ equal weight when forming T. The statistic given by (4.19) allows the consideration of any set of cs and so we can investigate which ever one gives the greatest power. The extension needed if n_i observations are available from each population and if σ^2 is unknown is immediate. The simpler case will be treated for which

$$T \sim N\left(\sum_{i=1}^{k} c_i \mu_i, \sigma^2 \sum_{i=1}^{k} c_i^2 \right). \qquad (4.20)$$

The statistic T has sometimes been arrived at (e.g. by Armitage (1955)) using regression arguments. Suppose that the μs are represented as functions of an independent variable x as in Chapter 1. The problem is then to estimate the regression of y on x. In the isotonic regression problem we fit an isotonic function by least squares and the solution does not depend on our being able to assign numerical values to the xs. Suppose, however, that a set of scores $s_1, s_2, \ldots, s_k$ are introduced in place of the xs satisfying the same order restrictions viz. $s_1 < s_2 < \ldots < s_k$. It would then be possible to carry out a standard least squares analysis fitting any monotonic function to the points (y_i, s_i). In particular, if a straight line were to be fitted, the test of significance would be based essentially on T with $c_i = s_i - \bar{s}$. Any set of scores will meet the case providing that they satisfy the condition of simple order. This leads immediately to the question of whether or not any particular set of scores can be said to be "best".

The optimum choice of scores

The obvious criterion of optimality is that of power. Thus a set of scores might be sought such that the power of T was at least as great at any point in $\mu_1 \leq \mu_2 \leq \ldots \leq \mu_k$ as could be obtained using any other set. This requirement is easily formulated as follows. The upper $100\alpha\%$ significance point of T is easily obtained from (4.20) as

$$u_\alpha \sigma \left\{ \sum_{i=1}^{k} c_i^2 \right\}^{\frac{1}{2}}$$

where u_α is the upper $100\alpha\%$ point of the standard normal distribution. The power at a given point μ is therefore

$$\text{Power} = \Phi\left(\sum_{i=1}^{k} c_i \mu_i \Big/ \left\{ \sum_{i=1}^{k} c_i^2 \right\}^{\frac{1}{2}} \sigma - u_\alpha \right)$$

$$= \Phi(r\Delta - u_\alpha) \qquad (4.21)$$

where

$$\Delta^2 = \sum_{i=1}^{k} (\mu_i - \bar{\mu})^2 / \sigma^2,$$

$$\bar{\mu} = \sum_{i=1}^{k} \mu_i / k,$$

$$r = \sum_{i=1}^{k} (\mu_i - \bar{\mu}) c_i \Big/ \left\{ \sum_{i=1}^{k} c_i^2 \right\}^{\frac{1}{2}} \Delta.$$

From here on and without loss of generality we shall assume $\sigma = 1$ as in Section 3.4.

The parameter Δ measures the "distance" of μ from the null hypothesis and r is the product moment correlation coefficient between the μs and the corresponding cs. The power is thus greatest for a given Δ if r is as large as possible, that is, equal to 1. Thus if the scores $\{c_i\}$ are guessed correctly the power is as great as it would have been if the μs had been known. However, it is of the essence of the problem that only the ordering of the μs is known.

Since the optimum cs are not known the next best thing would seem to be to use estimates of them obtained from the data. We wish to estimate $c_i = \mu_i - \bar{\mu}$ and this can be easily done using the isotonic estimators of the μs. The form of the resulting statistic is given by:

Theorem 4.3

The statistic

$$T = \sum_{i=1}^{k} (\hat{\mu}_i^* - \bar{y})y_i = \bar{\chi}_k^2.$$

Proof.

$$\sum_{i=1}^{k} (\hat{\mu}_i^* - \bar{y})y_i = \sum_{i=1}^{k} \hat{\mu}_i^{*2} - k\bar{y}^2 = \sum_{i=1}^{k} (\hat{\mu}_i^* - \bar{y})^2 = \bar{\chi}_k^2. \blacksquare$$

(This result follows directly from Theorem 1.4, of course.) It should be noted that this result holds for all partial orderings as well as simple order. The theorem provides an interesting and important alternative justification of the $\bar{\chi}^2$ test statistic. It also suggests that the class of statistics, T, is not likely to yield a test which is significantly superior to $\bar{\chi}^2$. Another approach to the problem of choosing the cs was proposed by Abelson and Tukey (1963) and their proposal will now be considered and compared with $\bar{\chi}^2$.

Abelson and Tukey sought to choose the cs such that they would be as highly correlated with the μs as possible. This is a natural criterion to adopt and (4.21) shows how the power depends on the value of r. For any given $\{c_i\}$, r has an upper limit of 1 and a lower limit which will be called $r_{\min}$. We should expect this value to be positive and reasonably large because the *rank* correlation between the cs and μs is necessarily unity. Abelson and Tukey suggested choosing the set $\{c_i\}$ so as to maximize the value of $r_{\min}$ and they called this value maximin r. It is clear from (4.21) that maximin r also maximizes the minimum power for given Δ. It was seen in Section 3.4 that there was a power "band" associated with the $\bar{\chi}^2$ test. The effect of choosing the cs in the way just described can therefore be thought of as raising the lower limit of the corresponding band for the test based on scores as high as possible.

We shall not go into the details of the determination of maximin r, but the principle of the method is easily described in geometric terms. Consider a k-dimensional Euclidean space of points $\boldsymbol{\mu}$. The null hypothesis $\mu_1 = \mu_2 = \ldots = \mu_k$ is then represented by a line in this space; each inequality $\mu_i < \mu_j$ defines a region on one side of the $(k-1)$-dimensional subspace $\mu_i = \mu_j$. The intersection of the half spaces defined by the set of inequalities is the region in which $\boldsymbol{\mu}_0 = (\mu_{01}, \mu_{02}, \ldots, \mu_{0k})'$, the true value, is constrained to lie. Let

$$\bar{\mu}_0 = \sum_{i=1}^{k} \mu_{0i}/k;$$

then $\boldsymbol{\mu}_0$ is a point on the subspace

$$\sum_{i=1}^{k}(\mu_i - \bar{\mu}_0) = 0.$$

Any set of cs satisfying

$$\sum_{i=1}^{k} c_i = 0$$

can be thought of as another point in this subspace and r is then the cosine of the angle between the lines joining the point $\mu_i = \bar{\mu}_0$ ($i = 1, 2, \ldots, k$) to the points $\boldsymbol{\mu}_0$ and $\mathbf{c}$. Thus if the two lines are coincident the angle is zero and $r = 1$. For any fixed point $\mathbf{c}$, r will vary as $\boldsymbol{\mu}_0$ moves within the restricted parameter space and there will be a certain $\boldsymbol{\mu}_0$ for which r is a minimum. This minimum value will vary with $\mathbf{c}$ and the problem is to find that $\mathbf{c}$ for which the minimum r is greatest.

For the case of simple order Abelson and Tukey showed that maximin r was obtained by taking

$$c_i \propto \left\{(i-1)\left(1 - \frac{i-1}{k}\right)\right\}^{\frac{1}{2}} - \left\{i\left(1 - \frac{i}{k}\right)\right\}^{\frac{1}{2}}, \qquad (i = 1, 2, \ldots, k). \quad (4.22)$$

It is interesting to note that, except when $k = 2$ and 3, the scores given by this formula are not equally spaced. The original T statistic of (4.17) is therefore not optimum in this sense if $k > 3$. The value of maximin r as a function of k decreases from 1 when $k = 2$ to zero as $k \to \infty$. For large k

$$\text{maximin } r \approx \sqrt{2}\{2 + \log(k-1)\}^{-\frac{1}{2}}$$

indicating that the approach to zero with k is extremely slow. The minimum value is attained when all but one of the differences $\{\mu_{i+1} - \mu_i\}$ are zero. A table of optimum scores is given in Abelson and Tukey (1963); the generalization to the case of unequal weights is given in Section 4.5.

The same method can be used for partial ordering. For example, if the alternative is $\mu_1 \leq [\mu_2, \mu_3, \ldots, \mu_k]$ the optimum scores are

$$\mathbf{c}' \propto (-(k-1), 1, 1, \ldots, 1). \tag{4.23}$$

In this case maximin $r = (k-1)^{-1}$ which is attained when

$$(\boldsymbol{\mu} - \bar{\boldsymbol{\mu}})' \propto (-1, -1, \ldots, -1, k-1),$$

where $\bar{\boldsymbol{\mu}}$ is a vector in which each element is $\bar{\mu}$.
The lower limit of the power function is then

$$\Phi\left(\frac{\Delta}{k-1} - u_\alpha\right).$$

It is clear from this expression that the lower bound for the power falls away much faster, with increasing k, than was the case for simple order. This result is typical of what may be expected when the μs are subject only to partial ordering. This point will be elaborated on further after looking at some computations of the power functions.

Power comparisons with $\bar{\chi}^2$

The power band for T is obtained by substituting $r = 1$ and $r = $ maximin r in

$$\Phi(r\Delta - u_\alpha).$$

The values of the power function for $\bar{\chi}^2$ given in the following tables are taken from Chapter 3. Table 4.1 compares the power bands of $\bar{\chi}_k^2$ and T for the case of simple order. The band for $\bar{\chi}_k^2$ lies inside the band for T so, for ease of reading, the power of the two tests, for given Δ, have been listed in decreasing order. There is very little to choose between $\bar{\chi}_k^2$ and T on grounds of power if k is small. The guaranteed power with $\bar{\chi}_k^2$ is greater than with T but this has to be offset against a lower maximum power. As k gets larger, the gap between the lower bound of $\bar{\chi}_k^2$ and that of T increases as the table given below indicates. The risk of having low power is now more serious and the advantage seems to lie with $\bar{\chi}_k^2$.

Table 4.2 makes a more limited comparison for the case of the simple tree alternative $\mu_1 \leq [\mu_2, \mu_3, \ldots, \mu_k]$ for $k = 3$ and 4 where the difference between the tests is more clear cut.

For this alternative the power band is very much wider for T than for $\bar{\chi}^2$. For $\Delta = 3$ the lower limit of the power of $\bar{\chi}^2$ is greater than the lower limit for T by a factor of nearly 2 and the position of T will deteriorate rapidly as k increases. In the absence of any prior knowledge of the μs other than that conveyed by the order restrictions the use of T for this alternative has little

Table 4.1 Comparison of the Power of $\bar{\chi}_k^2$ with the Optimum T-test for the Case of Simple Order

k		Δ				
		0	1	2	3	4
3	T (max)	0.050	0.260	0.639	0.912	0.991
	$\bar{\chi}^2$	0.050	0.244	0.605	0.892	0.987
		0.050	0.221	0.569	0.872	0.983
	T(min)	0.050	0.218	0.535	0.830	0.966
4	T(max)	0.050	0.260	0.639	0.912	0.991
	$\bar{\chi}^2$	0.050	0.238	0.594	0.885	0.985
		0.050	0.202	0.531	0.849	0.977
	T(min)	0.050	0.201	0.488	0.781	0.943
12	T(max)	0.050	0.260	0.639	0.912	0.991
	approx. $\bar{\chi}^2$(min)	0.067	0.159	0.414	0.766	0.963
	T(min)	0.050	0.166	0.385	0.649	0.855

Table 4.2 Comparison of the Power of $\bar{\chi}_k^2$ with the Optimum T-test for the Simple Tree Alternative $\mu_1 \leq [\mu_2, \ldots \mu_k]$

k		Δ				
		0	1	2	3	4
3	T(max)	0.050	0.260	0.639	0.912	0.991
	$\bar{\chi}^2$	0.050	0.210	0.543	0.857	0.979
		0.050	0.173	0.494	0.831	0.974
	T(min)	0.050	0.126	0.260	0.442	0.639
4	T(max)	0.050	0.260	0.639	0.912	0.991
	$\bar{\chi}^2$	0.050	0.187	0.489	0.813	0.967
		0.050	0.140	0.416	0.767	0.995
	T(min)	0.050	0.092	0.159	0.254	0.370

to commend it. Of course, if additional prior information were available which led us to suppose that the configuration of the μs was such as to make r near to 1 then the claims of T would be much stronger. However, in such a situation it would presumably be possible to produce an improved version of $\bar{\chi}_k^2$.

The T statistic is even more unsatisfactory if there are some μs which are not ordered with respect to the others. For example, if the alternative is $\mu_1 \leq \mu_2 \leq \ldots \leq \mu_{k-1}$ with the last mean μ_k not subject to any order restriction then it is easy to show that maximin $r = 0$.

It may be useful to summarize the advantages and disadvantages of the optimal T-test as compared with $\bar{\chi}^2$ as follows.

Advantages

(a) T is easier to calculate if the weights are equal.
(b) The distribution of T under both the null and alternative hypotheses is normal with easily calculated mean and variance.
(c) It is easy to do a two-sided version of the test if the direction of the ordering is not specified.

Disadvantages

(a) If the alternative involves a partial ordering the minimum power of T may be very much lower than that of $\bar{\chi}_k^2$. In some cases this lower bound may not even rise above the significance level.
(b) If the true alternative lies *outside* the alternative hypothesis region the power of T may be very low indeed. It was seen in Section 3.4 that $\bar{\chi}^2$ does not suffer from this disadvantage to the same extent.

Our general conclusion is that $\bar{\chi}^2$ provides the most satisfactory *general* way of incorporating prior information about ordering into the test procedure. However in the case of simple order and small k, T is a suitable alternative statistic. If additional prior information about the *spacing* as well as the ordering is available a T-statistic, with scores chosen according to this spacing, should give a power near to the values tabulated for $T(\max)$. In these circumstances T may be preferred to $\bar{\chi}^2$.

The one-sided test for the multivariate normal mean

The approach of Abelson and Tukey (1963) can be extended to deal with the problem of the multivariate normal mean discussed in Section 4.1. This extension has been made by Schaafsma (1966) and Schaafsma and Smid (1966). In the notation of Section 4.1 the test is based on a statistic T_s, say, of the form

$$T_s = \sum_{i=1}^{p} a_i \bar{z}_i$$

where $a_1, a_2, \ldots, a_p$ is a set of positive numbers. (These correspond to differences between the cs in the previous problem.) Any such statistic will be the uniformly most powerful test of the hypothesis $\tau = 0$ against the alternative $\tau > 0$ where $\boldsymbol{\tau}$ is restricted to lie on the line $\tau_i = a_i \tau$ ($i = 1, 2, \ldots, p; \tau \geq 0$). The object is then to choose the as so that the loss of power sustained at other points in $\boldsymbol{\tau} \geq \mathbf{0}$, at an equal distance from the origin, is minimized. We shall elaborate on this approach in Section 4.5.

4.3 ORDERED TESTS FOR NON-NORMAL DISTRIBUTIONS

Two limiting results

The theory of tests of significance when the observations are not normally distributed is much less well developed than in the normal case. Almost all of the existing work relates to the case of simple order and this is the only alternative considered in detail here. We shall begin by stating some general results which serve to suggest approximations for a wide class of distributions. Then we shall consider some important special cases in greater detail.

The set-up assumed in the following discussion is one in which there are independent random variables y_{ij} ($i = 1, 2, \ldots, k; j = 1, 2, \ldots, n$) where y_{ij} has the distribution function $F(y - \mu_i)$. It is assumed that the form of the distribution is known but that the means, μ_i, are unknown. The object is to find tests of the null hypothesis $\mu_1 = \mu_2 = \ldots = \mu_k$ against the simple order alternative $\mu_1 \leq \mu_2 \leq \ldots \leq \mu_k$. Most of our results could easily be extended to the case when j runs from 1 to n_i and we leave to the reader to make the necessary changes.

The first limiting result relates to the case when k is fixed and n is allowed to tend to infinity. Provided that the distribution F has finite variance the central limit theorem can be invoked to infer the approximate normality of the sample means $\bar{y}_i$ ($i = 1, 2, \ldots, k$). It is then possible to use the theory for normal distributions given in Chapter 3: this will be possible, of course, whether or not the alternative is that of simple order. If F belongs to the exponential family, the set of sample means $\bar{y}_i$ will be jointly sufficient for the population means and no information is lost by the condensation. In other cases it might be more efficient to use some other estimator of the population mean although it would be necessary for it to have an asymptotically normal distribution.

The second limiting result is due to Boswell and Brunk (1969). It relates to the case $n = 1$, $k \to \infty$, when F is a member of the exponential family and when the test is based on the likelihood ratio statistic. We first state the main theorem and then discuss its implications.

Theorem 4.4

For each fixed $l_0 = 1, 2, \ldots$

$$\lim_{k \to \infty} \Pr\{-2 \log \lambda \leq K | l = l_0\} = \Pr\{\chi^2_{l_0-1} \leq K\}.$$

where λ is the likelihood ratio statistic.

It is important to be clear about what this theorem says. In particular, it is not a statement about the limiting form of the distribution of λ as k goes to infinity but only about the conditional distribution given l. However, it is reasonable to expect that when k is large

$$\Pr\{-2\log\lambda \leq K | l = l_0\}$$

will be closely approximated by its limiting value and hence that the unconditional distribution will be approximately that of $\bar{\chi}_k^2$. The probabilities $P(l, k)$ are the same as in the normal case because we have assumed equal weights. A numerical comparison of the approximations with some exact results is given under the discussion of the exponential distribution below.

A proof of Theorem 4.4 will not be given but the following heuristic argument may give some insight into its nature. It follows from the fact that the distribution F belongs to the exponential family that the maximum likelihood estimators of the means will be the averages of the groups formed by amalgamation (see Section 2.4). Consequently λ will be a function only of these averages. It is also known (Miles (1959)) that l, the number of groups after amalgamation, has mean and variance which both tend to $\log k$ as k tends to infinity. Hence the size of almost all groups will tend to infinity as k tends to infinity. The central limit theorem then ensures that the group averages will be asymptotically normal. $-2 \log \lambda$ is now expanded in the standard way so that its leading term is a quadratic form in the group averages. This leads us to expect that the distribution, given l, will be that of χ^2 with $l - 1$ degrees of freedom.

The binomial distribution

If $\Pr\{y_{ij} = 1\} = p_i$ and $\Pr\{y_{ij} = 0\} = 1 - p_i$ $(i = 1, 2, \ldots, k; j = 1, 2, \ldots, n)$ the problem becomes one of testing the equality of k proportions against the alternative of a trend. The data in this case have the following form.

	Sample					Total
	1	2	3	...	k	
1	a_1	a_2	a_3	...	a_k	A
0	$n - a_1$	$n - a_2$	$n - a_3$	...	$n - a_k$	$N - A$
	n	n	n	...	n	N

The likelihood ratio statistic is easily calculated but its exact distribution is unknown. Since the binomial distribution belongs to the exponential family

Theorem 4.3 may be used to deduce a suitable approximation when k is large. There are, however, a number of other exact and approximate tests which will now be described.

The proportions a_i/n will be approximately normal by the central limit theorem but the application of the $\bar{\chi}^2$ test to them is not straightforward. This is because the variances of the proportions under H_0 depend on the common value of the p_i, which is unknown. There are two ways of dealing with this situation. The first is to transform the observations by

$$z_i = \sin^{-1}(a_i/n) \qquad (i = 1, 2, \ldots, k).$$

It is well known that

$$z_i \stackrel{.}{\sim} N\left(\sin^{-1} p_i, \frac{1}{4n}\right).$$

Since the transformation is monotonic it will suffice to test the hypothesis $E(z_i) = $ constant for all i against the alternative $E(z_1) \leq E(z_2) \leq \ldots \leq E(z_k)$ and this can be done using normal theory. The second method was proposed by Bartholomew (1959a) and Shorack (1967). It consists of applying the standard χ^2 test of association to the table after the amalgamation procedure has been carried out on the proportions. This amounts to the same thing as estimating the common p by A/N and computing $\bar{\chi}^2$ on the assumption that

$$\frac{a_i}{n} \stackrel{.}{\sim} N\left(\frac{A}{N}, \frac{A}{N}\left(1 - \frac{A}{N}\right)\Big/n\right)$$

when the null hypothesis is true. The justification of this procedure requires us to find the distribution of the statistic conditional upon both A and N. This principle has been much debated but is commonly accepted in the analysis of contingency tables. Shorack (1967) showed that the distribution of the $\bar{\chi}^2$-statistic was unaffected by the fact that the proportions are subject to the restraint

$$\sum_{i=1}^{k} a_i/n = A/n$$

(see the discussion of this point in Section 3.5).

Another class of approximate tests can be obtained by considering linear functions of the form

$$T = \sum_{i=1}^{k} c_i a_i/n, \qquad \sum_{i=1}^{k} c_i = 0.$$

These were discussed in detail in Section 4.2 and the results given there will apply approximately if the normal variates are replaced by proportions.

Closely related to the T statistics are tests based on rank correlation coefficients. The $2 \times k$ table given above can be regarded as a pair of rankings with ties. The ordering of the k populations defines one ranking—the n members in each column are tied in this ranking. The categories 0 and 1 define the second ranking—all members of any row are ranked equal on this criterion. Armitage (1955) proposed Kendall's τ and gave the theory necessary for its application. Stuart (1963) gave the corresponding theory for Spearman's rho. Nothing directly appears to be known about the power of these tests but insofar as they depend on normal approximations we expect the conclusions of Chapter 3 to apply in general terms here. These tests will be met again in Section 4.4.

One exact test for this problem, not mentioned so far, was proposed by Bennett (1962). His proposal was to determine a set $\mathscr{A}$ of vectors $\mathbf{a} = (a_1, a_2, \ldots, a_k)'$ such that

$$\Pr\{\mathbf{a} \in \mathscr{A} | H_0\} \leq \alpha.$$

Since the test was to be sensitive to ordered alternatives all points included in $\mathscr{A}$ had to satisfy the inequalities $a_1 \leq a_2 \leq \ldots \leq a_k$. This test is similar to that of Chassan (1960 and 1962) mentioned in the complements to this chapter and it suffers from the same disadvantages. In particular it precludes H_0 from ever being rejected when there is an inversion in the ordering of the as, however insignificant it may be.

The exponential distribution

In this subsection we shall describe tests suitable for testing for trend when the distribution F is exponential and when $n = 1$. The case $n > 1$ is covered by the tests for the gamma distribution in the following subsection.

If the density function of the exponential distribution is written in the form

$$f(y, \mu) = \mu^{-1} e^{-y/\mu}, \quad (\mu > 0, y \geq 0)$$

then the likelihood ratio statistic is

$$-2 \log \lambda = 2 \sum_{i=1}^{k} \log (\bar{y}/\hat{\mu}_i^*) \qquad (4.24)$$

where $\bar{y}$ is the mean y and the $\hat{\mu}_i^*$ are the isotonic estimators of the μs calculated in the usual way. Boswell and Brunk (1969) were able to compute the exact distribution of $-2 \log \lambda$. They provided tables of percentage points for λ, an extract of which is given in Table 4.3.

In addition to providing the means of carrying out the likelihood ratio

Table 4.3 *Percentage Points of the Likelihood Ratio Statistic* λ, *for* $2 \leqslant k \leqslant 10$

100α \ k	2	3	4	5	6	7	8	9	10
0.5	0.020	0.010	0.004	0.004	0.002	0.002	0.001	0.001	0.001
1.0	0.040	0.020	0.011	0.009	0.006	0.005	0.004	0.004	0.003
2.0	0.076	0.040	0.025	0.020	0.016	0.014	0.010	0.010	0.009
5.0	0.175	0.103	0.070	0.057	0.046	0.039	0.035	0.030	0.026
10.0	0.390	0.214	0.153	0.123	0.101	0.089	0.077	0.070	0.064

test these exact results enable the adequacy of the approximation suggested by Theorem 4.3 to be judged. The calculations in Table 4.4 relate to the lower tail of the distribution of λ (upper tail of $-2 \log \lambda$); this is the only part of the distribution which is required for significance tests. In every case the approximate value is smaller than the exact value. This means that if we used the approximation for making significance tests we should err by thinking that the size of the test was smaller than it actually was. The magnitude of this error can be gauged from the table; it is moderate but not so large as to make the approximation valueless. The "approximate" result is, of course, exact for normal variables. The exponential distribution exhibits a very high degree of non-normality, so with distributions showing smaller departures from normality, a better degree of approximation might be expected.

Another class of tests for trend with exponential variables is made possible

Table 4.4 *Comparison of the Exact and Approximate Distribution Functions of the Likelihood Ratio Statistic* λ, *for the Exponential Distribution*

k	2		5		10	
λ	Exact	Approx.	Exact	Approx.	Exact	Approx.
0.00	0.000	0.000	0.000	0.000	0.000	0.000
0.05	0.013	0.007	0.046	0.032	0.083	0.063
0.10	0.026	0.016	0.082	0.062	0.143	0.116
0.20	0.053	0.036	0.153	0.123	0.243	0.209
0.30	0.082	0.060	0.218	0.184	0.328	0.295
0.40	0.113	0.088	0.282	0.247	0.408	0.375
0.50	0.146	0.120	0.345	0.312	0.482	0.452

by the following well-known result. If $y_1, y_2, \ldots, y_k$ are independent exponential variables with mean μ then the joint distribution of

$$u_i = \sum_{j=1}^{i} y_j / \sum_{j=1}^{k} y_j \quad (i = 1, 2, \ldots, k-1)$$

is that of the order statistics of a sample of size $k - 1$ from the rectangular distribution on the interval $(0, 1)$. Since

$$E(u_i) = i/k$$

(whatever the distribution of the ys providing they are non-negative), evidence of an increasing trend in the means of the ys would be revealed by a tendency for the us to be smaller than their expectation on the null hypothesis. This suggests constructing statistics as functions of the differences $(u_i - i/k)$ which preserve their signs. One such statistic is

$$D_k^- = \min_i (u_i - i/k).$$

It is known (see, for example, Birnbaum and Tingey (1951)) that

$$\lim_{k \to \infty} \Pr\{D_k^- < D\} = 1 - e^{-2nD^2}$$

and tables of percentage points for finite k are given in Owen (1962). A second statistic is the arithmetic mean of the us. It is easy to show (see Bartholomew (1956)) that this has the following representation:

$$\bar{u} = -\left[\int_0^1 (F_k(u) - u) du - \frac{1}{2}\right] \quad (4.25)$$

where $F_k(u)$ is the empirical distribution function computed from the u_is. Under the null hypothesis $\bar{u}$ has a distribution which is very close to normal, even for small k. The result required for tests of significance is

$$\bar{u} \sim N\left(\frac{1}{2}, \frac{1}{12k}\right). \quad (4.26)$$

The average, or sum, of any monotonic function of the us will also be sensitive to a trend in the means of the ys. One such statistic

$$Q = -2 \sum_{i=1}^{k-1} \log u_i$$

has the useful property that it is distributed like χ^2 with $2(k-1)$ degrees of freedom.

The results which are available on the power of the statistics based on the

us all relate to the case where the alternative hypothesis specifies a distribution for u other than the rectangle. This kind of alternative is not relevant for the application here. Our alternative is that the ys continue to have exponential distributions but with unequal parameters in a monotonic sequence. To the best of the authors' knowledge nothing is known about the power under such alternatives.

The foregoing discussion does not exhaust the possibilities for testing for trend under the exponential distribution. Before leaving the subject one further class of tests is mentioned which is easily generalized to the gamma distribution discussed in the following subsection. Suppose the sequence $\{y_i\}$ is divided arbitrarily into two halves. Let

$$Y_1 = \sum_{i=1}^{d} y_i, \qquad Y_2 = \sum_{i=d+1}^{k} y_i$$

where the division takes place between y_d and y_{d+1}. Then, on the null hypothesis, $dY_2/(k-d)Y_1$ is distributed like an F random variable with $2d$ and $2(k-d)$ degrees of freedom. In the presence of an increasing trend the ratio will tend to be larger so large values of F would be significant. After reading the next section the reader should be able to devise other tests by breaking up the sequence at more than one point.

The gamma distribution

The problem of testing for trend in gamma variables may arise in several different ways. For example, if instead of taking $n = 1$ in the previous subsection we had taken $n > 1$ then gamma variables would have arisen as follows. The sample sums

$$\sum_{j=1}^{n} y_{ij} = y_i \qquad (i = 1, 2, \ldots, k)$$

are sufficient statistics for the population means μ_i and so a test can be based on them without loss of information. But these sums are gamma variables. A second important example arises when testing the homogeneity of normal variances against ordered alternatives because the sample variances have gamma distributions. The problem may be expressed formally as follows. Observations y_i are available on independent random variables the ith of which has the gamma density

$$f(y_i; \mu, p) = \frac{1}{\mu_i^p \Gamma(p)} y_i^{p-1} e^{-y_i/\mu_i}$$

$$(y \geq 0, \mu > 0) \qquad (i = 1, 2, \ldots, k).$$

In the example involving exponential variables, we put $p = n$. For testing the equality of k normal variances set $p = m - 1$, $\mu_i = \sigma_i^2$, $y_i = \frac{1}{2}(m - 1)s_i^2$, where s_i^2 is the unbiased sample estimator based on $m - 1$ observations.

The likelihood ratio statistic in this case is

$$-2 \log \lambda = 2 \sum_{i=1}^{k} p \log (\bar{y}/p\hat{\mu}_i^*) \tag{4.27}$$

where the $\hat{\mu}_i^*$ are the isotonic estimators obtained from the unrestrained estimators of the μ_i, namely $\hat{\mu}_i = y_i/p$. In the sense of Theorem 4.4 the statistic of (4.27) will have an asymptotic $\bar{\chi}^2$ distribution. As p gets larger the ys will have a distribution approaching normality and so the approximation may be expected to be better than when $p = 1$.

The u-transformation which was used in the exponential case can be applied here also but its use is more limited. If p is an integer the variable

$$u_i = \sum_{j=1}^{i} y_j / \sum_{j=1}^{k} y_j$$

will now have the same distribution as the ipth order statistic in a sample of size $kp - 1$ from a rectangular distribution. The statistics $\bar{D}_k$ and $\bar{u}$ will still be suitable for testing for trend but their distribution will be more difficult to determine since the joint distribution of the us is that of a particular subset of the order statistics of the rectangle instead of all of them.

The F-ratio test described for the exponential is still available. The statistic in this case is $dY_2/(k - d)Y_1$ but the degrees of freedom are now $2pd$ and $2p(k - d)$.

A test for inequality of variances against ordered alternatives was proposed by Vincent (1961). This test was derived on the supposition that σ_i^2 ($i = 1, 2, \ldots, k$) was a linear function of i. On this assumption a reasonable statistic to use is the regression coefficient of $\hat{\sigma}_i^2$ on i. This leads to a statistic of the family discussed under the designation of "T-statistics" in Section 4.2.

4.4 DISTRIBUTION-FREE TESTS FOR ORDERED ALTERNATIVES

Robustness of $\bar{\chi}^2$, $\bar{E}^2$ and T

The greater part of the distribution theory which has been derived for tests against ordered alternatives has been based on the assumption of normally distributed observations. In this section we shall consider the tests which are available for use when nothing is known about the form of the distribution. Before doing this it is pertinent to ask how sensitive the distributions of the normal theory statistics are to departures from the assumption of normality.

If these tests prove to be robust then the need for an extensive theory of distribution-free tests is diminished.

Such information as is available suggests that the normal theory tests are robust. First consider statistics of the class T when σ is known; these are linear functions of the observations. If the observations are exactly normal then T is also normal. In the contrary case we may appeal to the central limit theorem, to infer the approximate normality of T. The adequacy of the approximation will depend both on k (the central limit theorem requires $k \to \infty$) and on the degree of the departure from normality of the observations. The approximation is of the same kind as used for some of the distribution-free tests considered later. On this ground, therefore, there would seem little to be gained by abandoning T in favour of a distribution-free test. Other functions to be considered will be discussed later. When σ is unknown, and has to be estimated, the T-statistic will be affected by the fact that the variance estimator is more sensitive to non-normality than is a linear combination. However, the effect on Student's t-test is not serious and we anticipate that the same will be true for T which is essentially the same.

The position in the case of $\bar{\chi}^2$ and $\bar{E}^2$ is a little more complicated but, here again, the evidence suggests that they are robust. This conclusion is based partly upon known results about the robustness of the ordinary χ^2 and F tests and partly on results obtained in earlier sections. In Section 3.3 it was noted that for simple order the probabilities $P(l, k)$ were distribution-free in the case of equal weights. To that extent, therefore, the null hypothesis distribution does not depend on the form of the distribution of the observations. The robustness of the χ^2 and F tests has been extensively studied (see, for a summary of results, Scheffé (1959)) leading to the conclusion that moderate departures from normality are not serious. These results apply also to the conditional distribution of $\bar{\chi}^2$ and $\bar{E}^2$ given l. Further support for this conclusion is provided by Theorem 4.4. There it was shown that as $k \to \infty$ the conditional distribution of $\bar{\chi}_k^2$ given l approached the normal form if the observations were drawn from any member of the exponential family. This family includes highly non-normal distributions so this adds weight to the evidence deduced from other sources. We conclude that $\bar{\chi}^2$ and $\bar{E}^2$ are likely to be robust, particularly when the weights are equal and k is large.

It might appear from the foregoing discussion that little is to be gained by devising distribution-free tests. However, in addition to safeguarding the user against faulty assumptions they sometimes offer other worthwhile advantages. A distribution-free test is often easier to carry out and it may have greater power under non-normality than the normal theory test. In the following sections some of the tests which are available will be described.

Tests for the case of one observation per group

Suppose that there are k independent random variables y_i having the distribution function $F_i(y)$ ($i = 1, 2, \ldots, k$). Our object is to test the null hypothesis

$$H_0 : F_i = F \qquad (i = 1, 2, \ldots, k)$$

against the alternative

$$H_1 : F_1 \geq F_2 \geq \ldots \geq F_k$$

where at least one of the inequalities is strict. Our discussion here is confined mainly to the case of simple order since this is the area in which most work has been done. Also the hypotheses have been expressed in terms of the distribution functions rather than in terms of a parameter such as the mean. The distribution-free approach allows this extra degree of generality and so we take advantage of it.

It would not be possible to construct a sensible test of H_0 with only one observation from each distribution if H_1 merely said that the Fs were not the same. Any amount of dispersion in the ys could be explained wholly in terms of the spread of the unknown F under H_0. It is the order information which makes a test possible.

If the null hypothesis is true, all permutations of the ys are equally likely. If H_1 is true there will be a tendency for the ys to occur in the order predicted by H_1. H_0 can therefore be tested by any criterion based on a rank correlation coefficient. However, unless k is reasonably large it will not be possible to obtain convincing evidence against H_0. Another test, suggested by Brunk (1960) though implicit in Sparre Andersen (1954), is based on the fact that the probabilities $P(l, k)$ are distribution-free (on the null hypothesis). The means of the distributions F_i are estimated by the technique of Chapter 2 and the number of distinct values which result is recorded. This is the quantity denoted by l. When H_0 is true its distribution is $P(l, k)$ ($l = 1, 2, \ldots, k$). Large values of l would indicate a significant departure from H_0. Once more, k needs to be fairly large before the probabilities in the upper tail become small enough to establish significance. Nothing appears to be known about the relative power of the tests but the l-method has the important advantage of also providing estimates of the distribution means in the event that H_0 is rejected.

Tests for the case of several observations per group

When there are several observations in each group all of the tests in the last section can be extended in a natural way. Historically, the subject has been approached from many angles; different tests have been proposed only for it to be discovered later that they were equivalent. All of them can, in fact, be derived using the ideas underlying $\bar{\chi}^2$ and T. We shall adopt this approach giving alternative forms and historical references where appropriate. Some further interesting developments are described in Section 4.5.

The first of these tests is a distribution-free version of $\bar{\chi}^2$ introduced by Chacko (1963) and extended by Shorack (1967). The method of calculating the statistic is as follows. As before let the jth observation in the ith group be y_{ij} ($i = 1, 2, \ldots, k; j = 1, 2, \ldots, n_i$); then all the observations are ranked from the smallest to the largest. Let r_{ij} denote the rank of y_{ij}, then compute $\bar{\chi}_k^2$ using the *ranks* instead of the ys and taking the weights to be

$$w_i = \frac{12n_i}{N(N+1)} \quad (i = 1, 2, \ldots, k).$$

Since the mean rank is $(N+1)/2$ the test statistic may be written

$$\bar{\chi}_{\text{rank}}^2 = \frac{12}{N(N+1)} \sum_{i=1}^{k} n_i \left(\hat{\mu}_i^* - \frac{N+1}{2} \right)^2 \quad (4.28)$$

where the $\hat{\mu}^*$s are now calculated from the ranks. The statistic defined in (4.28) is, in fact, the basis of the Kruskal–Wallis test for the one-way classification after amalgamation. It is interesting to observe that

$$w_i = n_i / \sigma_{\text{rank}}^2$$

where

$$\sigma_{\text{rank}}^2 = \frac{1}{N-1} \sum_{i=1}^{k} \sum_{j=1}^{n_i} (r_{ij} - \tfrac{1}{2}(N+1))^2.$$

This result makes the definition (4.28) a natural extension of the normal theory $\bar{\chi}^2$-statistic. It is worth noting that the definition covers $n_i = 1$ ($i = 1, 2, \ldots, k$) as a special case. However, under these circumstances $\bar{\chi}_{\text{rank}}^2$ turns out to be almost the same as the statistic l when l is calculated from ranks rather than variate values. Thus $\bar{\chi}_{\text{rank}}^2$ takes its largest value if and only if $l = k$. Its next largest value occurs when there is exactly one inversion in which case $l = k - 1$. There are two distinct values of $\bar{\chi}_{\text{rank}}^2$ arising when $l = k - 2$—one when three adjacent groups are amalgamated and one when two pairs are combined. Proceeding in this way it is clear that the distribution of l, using ranks, is simply a grouped form of the distribution of $\bar{\chi}_{\text{rank}}^2$. It is thus a coarser, but essentially equivalent, statistic. Since Brunk's test uses actual values rather than ranks and has a known distribution it seems generally preferable to $\bar{\chi}^2_{\text{rank}}$ in this case.

The limiting distribution of $\bar{\chi}^2_{\text{rank}}$ on the null hypothesis is given by the following theorem.

Theorem 4.5

In the limit as $N \to \infty$ *with*

$$\lim_{N \to \infty} \frac{n_i}{N} = \nu_i,$$

$\bar{\chi}^2_{\text{rank}}$ *has the same null hypothesis distribution as* $\bar{\chi}^2$ *for the case of simple order.*

Proof. It was shown by Kruskal (1952) that under the conditions stated the vector of mean ranks $\mathbf{r} = (\bar{r}_{1.}, \bar{r}_{2.}, \ldots, \bar{r}_{k.})'$ has a limiting multivariate normal distribution with means $\frac{1}{2}(N+1)$ and covariance matrix with (i,j)th element equal to

$$\frac{N}{12}\left(\frac{\delta_{ij}}{\nu_i} - 1\right) \tag{4.29}$$

where δ_{ij} is the Kronecker delta function. This is identical with the form of the joint distribution of the random variables

$$\{\bar{y}_{i.} - \bar{y}_{..}\}, \qquad (i = 1, 2, \ldots, k)$$

except that the factor $N/12$ in the covariance matrix is now replaced by σ^2/N. Hence a $\bar{\chi}^2$ statistic calculated from the $\bar{r}_i$ with weights $12n_i/N^2$ will have the same distribution, in the limit, as the statistic calculated using the $y_i - \bar{y}$ with weights n_i/σ^2. The latter statistic is obviously the normal theory $\bar{\chi}^2$. The former, apart from the asymptotically negligible factor $(N+1)/N$, is $\bar{\chi}^2_{\text{rank}}$. Hence the result follows. ∎

Nothing appears to be known about the accuracy of the asymptotic approximation in the finite case. Instead of using ranks it would be possible and, perhaps, desirable to use normal scores. The same asymptotic distribution theory would apply.

The asymptotic power of $\bar{\chi}^2_{\text{rank}}$ can be simply related to the power of $\bar{\chi}^2$ as follows. Assume that y_i has a continuous distribution function $F_i(y)$ under the alternative hypothesis having the form $F(y - (\theta_i/\sqrt{N}))$. The ordering on the F_is implies that $\theta_1 \leq \theta_2 \leq \ldots \leq \theta_k$. A result of Andrews (1954), quoted by Chacko (1963), shows that the mean ranks $\bar{r}_i$ $(i = 1, 2, \ldots, k)$ have a limiting multivariate normal distribution with the same covariance matrix as in the null case. Their expectations are easily calculated from the fact that

$$E(\bar{r}_i) = E(r_{ij})$$
$$= \sum_{g=1}^{k} \sum_{h=1}^{n_g} \Pr\{X_{ij} > X_{gh}\} + 1$$
$$= \sum_{g=1}^{k} \sum_{h=1}^{n_g} \int_{-\infty}^{\infty} (1 - F_i(x)) f_g(x)\, dx - \tfrac{1}{2} + 1$$
$$= \sum_{g=1}^{k} n_g \int_{-\infty}^{\infty} (1 - F_i(x)) f_g(x)\, dx + \tfrac{1}{2}.$$

Expanding F_i and f_g about $\theta_i = 0$, $\theta_g = 0$

$$E(\bar{r}_i) = \frac{1}{2} + \sum_{g=1}^{k} n_g \left[\int_{-\infty}^{\infty} (1 - F(x)) f(x)\, dx + \frac{\theta_i}{\sqrt{N}} \int_{-\infty}^{\infty} f^2(x)\, dx \right.$$
$$\left. - \frac{\theta_g}{\sqrt{N}} \int_{-\infty}^{\infty} f'(x)(1 - F(x))\, dx + O(N^{-1}) \right]$$
$$= \frac{N+1}{2} + \sqrt{N}(\theta_i - \bar{\theta}) \int_{-\infty}^{\infty} f^2(x)\, dx + O(1)$$

where

$$\bar{\theta} = \sum_{g=1}^{k} v_g \theta_g$$

with $v_g = n_g/N$. Hence

$$E\left(\frac{12}{N}\right)^{\frac{1}{2}} \left(\bar{r}_i - \frac{1}{2}(N+1)\right) = \sqrt{12}(\theta_i - \bar{\theta}) \int_{-\infty}^{\infty} f^2(x)\, dx + O(N^{-\frac{1}{2}}). \quad (4.30)$$

In the normal case the corresponding quantity is

$$\frac{\sqrt{N}}{\sigma}(\bar{y}_{i.} - \bar{y}_{..}) \quad (4.31)$$

which has expectation

$$(\theta_i - \bar{\theta})/\sigma. \quad (4.32)$$

It follows, by comparing (4.30) and (4.31), that the power of $\bar{\chi}^2$ when the location parameters have the values

$$\frac{\sqrt{12}\sigma\theta_i \int_{-\infty}^{\infty} f^2(x)\, dx}{\sqrt{N}},$$

will be the same as that of $\bar{\chi}^2_{\text{rank}}$ when the parameters are $\theta_i/\sqrt{N}$. The total sample size may therefore be adjusted in the first case (holding the νs constant) to make the powers of the two tests equal against the alternative $\{\theta_i/\sqrt{N}\}$. If N and N_{rank} denote the sample sizes for $\bar{\chi}^2$ and $\bar{\chi}^2_{\text{rank}}$ respectively then for equal power their ratio must be given by

$$\frac{N}{N_{\text{rank}}} = 12\sigma^2 \left[\int_{-\infty}^{\infty} f^2(x)\,dx\right]^2. \tag{4.33}$$

Thus we have

Theorem 4.6

The asymptotic relative efficiency of $\bar{\chi}^2_{\text{rank}}$ relative to $\bar{\chi}^2$ is

$$12\sigma^2 \left[\int_{-\infty}^{\infty} f^2(x)\,dx\right]^2.$$

Hodges and Lehmann (1956) have shown that

$$0.864 \le 12\sigma^2 \left\{\int_{-\infty}^{\infty} f^2(x)\,dx\right\}^2 \le \infty. \tag{4.34}$$

If $f(x)$ is the normal density, the efficiency is $3/\pi = 0.955$. Hence very little is lost, asymptotically, in terms of power by using ranks instead of variate values. The value of the asymptotic relative efficiency found here is exactly the same as in the unordered case. In both cases it can be deduced that the asymptotic relative efficiency using normal scores for the calculation of $\bar{\chi}^2_{\text{rank}}$ instead of ranks would be no less than unity.

Nothing that has been done when deriving the asymptotic distribution theory of $\bar{\chi}^2_{\text{rank}}$ has depended upon the ordering being simple. We therefore conclude that all results hold if the ordering is partial.

The other tests which have been proposed for the case of several observations in each group are essentially rank versions of the T-statistics of Section 4.2. Like their normal theory counterparts their main usefulness is in the case of simple order so the discussion will be limited to that case only.

The Jonckheere–Terpstra test

The statistic suggested independently by Terpstra (1952) and Jonckheere (1954a) may be defined as follows.

$$S = \sum_{i=1}^{k-1} \sum_{j=i+1}^{k} \sum_{r=1}^{n_i} \sum_{s=1}^{n_j} \text{sgn}\,(y_{ir} - y_{js}) \tag{4.35}$$

where

$$\operatorname{sgn}(x) = 1 \quad \text{if} \quad x > 0$$
$$= -1 \quad \text{if} \quad x < 0$$
$$= 0 \quad \text{if} \quad x = 0.$$

If there is an increasing trend in the means then one would expect y_{js} to be greater than y_{ir} if $j > i$. Consequently large negative values of S are regarded as significant. Terpstra (1955, 1956) found that the set of alternatives against which the test was consistent depended on the ratios of the n_is. He showed that this feature could be removed by re-defining the statistic so as to divide sgn $(y_{jr} - y_{is})$ in (4.35) by $n_i n_j$. The theory of S is somewhat simpler and so only that case will be discussed.

The statistic S is also proportional to Kendall's rank correlation coefficient with ties in one ranking. The N observations produce one of the rankings and the ordered categories constitute the other. In the latter case all the observations in the same group are assumed to be ranked equal on the "category" ranking. The test is thus a natural generalization of the rank correlation coefficient which we proposed for the case $n_i = 1$, for all i, earlier in this section.

There is a third representation of S which brings out more clearly its relation with the T-statistics discussed in Section 4.2. If $k = 2$, S is equivalent to the well-known Mann–Whitney statistic for testing the equality of two distributions. This statistic, for the ith and jth groups considered alone, may be written in the form

$$\frac{n_i n_j}{n_i + n_j} (\bar{r}_i - \bar{r}_j) \tag{4.36}$$

where $\bar{r}_i$ is the mean rank of the ith group when groups i and j are ranked together. This follows easily from the fact that the rank of the rth member of group i is

$$\tfrac{1}{2} \sum_{s=1}^{n_j} \operatorname{sgn}(y_{ir} - y_{js}) + \tfrac{1}{2}(n_i + n_j + 1).$$

Summing over i and j with $j > i$ then gives S. This procedure is exactly the same as we followed when generalizing the one-sided test for two normal means to the k-sample case in Section 4.2. Here also we could introduce arbitrary scores to form a weighted sum but this has not been explored. Similarly, instead of using ranks, order statistics from some suitable distribution could be used. Puri (1965) discussed the general properties of such statistics.

There is a further way of arriving at a rank version of T which does not yield the S statistic. The calculation of T involves a separate ranking operation for each pair of samples. Suppose instead that the process used when calculating $\bar{\chi}^2_{\text{rank}}$ is followed and *all* N observations are ranked and a T-statistic computed from the ranks. Such a statistic would have the form

$$R = \sum_{i=1}^{k} c_i \bar{r}_i, \quad \left(\sum_{i=1}^{k} c_i = 0 \right) \tag{4.37}$$

where $\bar{r}_i$ is now the average rank in the ith group for this method of ranking. Suppose we assign to each member of each group a score equal to the average rank implied by the alternative hypothesis. Thus each member of group i has the score

$$s_i = \sum_{j=1}^{i-1} n_j + \frac{n_i + 1}{2};$$

then it follows at once that R with $c_i = n_i(s_i - \bar{s})$ is proportional to Spearman's rank correlation coefficient. It will be seen below that the optimum scores of (4.22) give a test which is asymptotically optimum.

The asymptotic null hypothesis distributions of both S and R are easy to obtain. Since the asymptotic distribution of the Mann–Whitney statistic is normal with zero mean it follows at once that any linear combination of such statistics, and hence S, has a normal distribution with the same mean. Similarly the multivariate normality of the rank means $\{\bar{r}_i\}$ ensures the normality of R. The variances are as follows

$$\left. \begin{aligned} \text{Var}(S) &= \frac{1}{18} \left\{ N^2(2N + 3) - \sum_{i=1}^{k} n_i^2(2n_i + 3) \right\}, \\ \text{Var}(R) &= \frac{N}{12} \left\{ \sum_{i=1}^{k} c_i^2 - \left(\sum_{i=1}^{k} c_i \sqrt{v_i} \right)^2 \right\}. \end{aligned} \right\} \tag{4.38}$$

It follows, by an immediate application of the results on the asymptotic power of $\bar{\chi}^2_{\text{rank}}$, that R has an asymptotic relative efficiency, with respect to the normal theory T test having the same cs, of

$$12\sigma^2 \left[\int_{-\infty}^{+\infty} f^2(x)\, dx \right]^2,$$

The optimum choice of cs for the normal case is thus also optimum in the rank case. The same method yields the same result for S (see Bartholomew (1961b) and Puri (1965)).

All the results about the asymptotic power of rank tests for ordered alternatives may thus be summed up in the following theorem.

Theorem 4.7

The asymptotic relative efficiencies of $\bar{\chi}^2_{\text{rank}}$, S and R with respect to their normal theory counterparts are

$$12\sigma^2 \left[\int_{-\infty}^{+\infty} f^2(x)\, dx \right]^2.$$

This theorem greatly extends the applicability of the power comparisons which have been made in the normal case. The conclusions reached there apply asymptotically also to the rank tests with an appropriate adjustment of the sample size.

Tests for the randomized block design

There has been extensive discussion of distribution-free tests for the randomized block design. Suppose that observations are available on random variables y_{ij} ($i = 1, 2, \ldots, n$; $j = 1, 2, \ldots, k$) with continuous distribution function $F_j(y - b_i)$. The b_is represent "block" effects which are of no direct interest. We wish to test the null hypothesis

$$H_0: F_1 = F_2 = \ldots = F_k$$

against the alternative

$$H_1: F_1 \geq F_2 \geq \ldots \geq F_k$$

as in the one-way classification, after the block effects have been eliminated. An obvious way to do this is to rank the y_{ij}s separately within each block. Thus, for example, a typical result with $k = 4$ and $n = 3$ might be as follows:

$$\text{Blocks} \begin{cases} \text{I} & 1 \quad 3 \quad 2 \quad 4 \\ \text{II} & 1 \quad 2 \quad 4 \quad 3 \\ \text{III} & 2 \quad 1 \quad 3 \quad 4 \end{cases}$$

Any statistic based on these ranks makes use only of the rankings within blocks and so is independent of block differences.

Five different statistics have been proposed for testing H_0 against H_1. In each case the statistic will be described and its asymptotic null hypothesis distribution and power given. The first three are natural extensions of tests already discussed for the one-way classification.

The test proposed by Shorack (1967) is $\bar{\chi}^2_{\text{rank}}$ calculated from the column means of the ranks. This may be looked upon as an ordered version of Friedman's (1937) χ^2 test for the concordance of a set of rankings. Asymptotically, this statistic has the $\bar{\chi}^2$ distribution because the rank means are asymptotically normal (by the central limit theorem) subject to the restraint that their sum is fixed. As has already been seen this does not affect the distribution theory. To investigate the asymptotic power a sequence of alternatives, $H_1^{(n)}$ is considered:

$$F\left(y - b_i - \frac{\theta_j}{\sqrt{n}}\right) \quad (i = 1, 2, \ldots, n; j = 1, 2, \ldots, k)$$

where $\theta_1 \leq \theta_2 \leq \ldots \leq \theta_k$. Our result therefore relates to the case where the number of blocks is large. The asymptotic power of the test can be obtained by an extension of the method used for $\tilde{\chi}^2_{\text{rank}}$. Let R_{ij} denote the rank of the jth member of the ith row (block) and let $\bar{R}_{.j}$ denote the mean of the jth column. The expected value of R_{ij} is found by putting $n_g = 1$ for all g and replacing $\theta_i/\sqrt{N}$ by $\theta_i/\sqrt{n}$ in the calculation leading up to (4.30). The result i

$$E(R_{ij}) = \frac{1}{2}(k + 1) + \frac{k}{\sqrt{n}}(\theta_i - \bar{\theta}) \int_{-\infty}^{+\infty} f^2(x)\, dx + O\left(\frac{1}{n}\right). \quad (4.39)$$

This is also the expectation of $\bar{R}_{.j}$. The covariance matrix of the $\bar{R}_{.j}$s is asymptotically the same as under the null hypothesis, namely

$$\text{Cov}(\bar{R}_{.h}, \bar{R}_{.j}) = \frac{k+1}{12n}(\delta_{hj}k - 1), \quad (h, j = 1, 2, \ldots, k) \quad (4.40)$$

and the joint distribution is normal.

The foregoing results are now compared with those for the joint distribution of $(\bar{y}_{.j} - \bar{y}_{..})$ $(j = 1, 2, \ldots, k)$. In this case

$$E(\bar{y}_{.j} - \bar{y}_{..}) = (\theta_j - \bar{\theta})/n$$

and

$$\text{Cov}(\bar{y}_{.h} - \bar{y}_{..}, \bar{y}_{.j} - \bar{y}_{..}) = \frac{\sigma^2}{kn}(\delta_{hj}k - 1).$$

Hence,

$$\sigma\left[\frac{12}{k(k+1)}\right]^{\frac{1}{2}}\left(\bar{R}_{.j} - \frac{k+1}{2}\right) \quad \text{and} \quad \bar{y}_{.j} - \bar{y}_{..}$$

$$(j = 1, 2, \ldots, k)$$

have the same joint multinormal distribution except that in the first case the mean is

$$\sigma\left[\frac{12}{k+1}\right]^{\frac{1}{2}} \frac{(\theta_j - \bar{\theta})}{\sqrt{n}} \int_{-\infty}^{+\infty} f^2(x)\, dx \quad (j = 1, 2, \ldots, k)$$

and in the second it is

$$\frac{\theta_j - \bar{\theta}}{\sqrt{n}}.$$

Now, if $\bar{\chi}^2$ is computed using the variates

$$\sigma \left[\frac{12}{k(k+1)}\right]^{\frac{1}{2}} \left(\bar{R}_{.j} - \frac{k+1}{2}\right) \quad (j = 1, 2, \ldots, k)$$

and weights n/σ^2 Shorack's test is obtained. Hence this test will have the same asymptotic power as the normal theory $\bar{\chi}^2$ test if the sample sizes are chosen such that

$$\frac{N}{N_{\text{rank}}} = \frac{n}{n_{\text{rank}}} = \frac{12k\sigma^2}{k+1} \left[\int_{-\infty}^{+\infty} f^2(x)\,dx\right]^2. \quad (4.41)$$

Note that this is always somewhat less than the corresponding efficiency for the test in the one-way classification because $k/(k+1) < 1$ but the difference is small if k is large.

Tests based upon rank correlation coefficients

It was noted earlier in this section that the agreement of a single ranking with a postulated ranking could be tested by any of the standard rank correlation coefficients. This idea readily generalizes to the present case where there are n independent rankings. Evidence for a trend in any one of these rankings would be reflected in the value of its rank correlation with the postulated order. A simple statistic for testing for trend in all the rankings is the sum or average of these coefficients. Jonckheere (1954b) proposed the statistic

$$\tau = \sum_{i=1}^{n} \tau_i \quad (4.42)$$

where τ_i is Kendall's rank correlation coefficient between the ranks in the ith row and the group ordering. Page (1963) proposed

$$\rho = \sum_{i=1}^{n} \rho_i \quad (4.43)$$

where ρ_i is the corresponding Spearman coefficient. Page also gave some exact significance levels for his statistic but an approximation is easily obtained for both tests. For example, since the ρs in (4.43) are independent,

$$E(\rho) = \sum_{i=1}^{n} E(\rho_i) \quad \text{and} \quad \text{Var}(\rho) = \sum_{i=1}^{n} \text{Var}(\rho_i).$$

On the null hypothesis, $E(\rho_i) = 0$, Var $(\rho_i) = 1/(k-1)$ and the distribution is approximately normal so

$$\rho \sim N(0, n/(k-1)).$$

A similar result holds for τ. Hollander (1967) investigated the asymptotic power of both tests and found conditions under which they were consistent. The question of consistency has already been raised in Chapter 3 in relation to the normal theory $\bar{\chi}^2$ test. Within the region of the parameter space defined by the alternative hypothesis any reasonable test would be expected to be consistent. However, other things being equal, it would be preferable for a test to be consistent even if the true parameter point lay outside the alternative region. This point will be taken up again in the following section but we note in passing that the regions of consistency for ρ and τ are not identical.

Hollander found the asymptotic relative efficiencies of each test compared with the normal theory T-test with equally spaced scores and for the two tests compared with one another. The methods used to derive these results are essentially the same as those used previously so the details are omitted and the conclusions stated in the following two theorems.

Theorem 4.8

The A.R.E. of ρ compared with τ, assuming normally distributed variables, is

$$k(2k + 5)/2(k + 1)^2.$$

This ratio is 1 when $k = 2$ (in which case the tests are equivalent). As k increases it reaches a maximum value of 1.042 at $k = 5$ and then decreases to its limiting value of 1. The ρ-test thus has a very slight advantage over τ for all values of k.

Theorem 4.9

The A.R.E. of ρ compared with the normal theory statistic

$$T = \sum_{j=1}^{k} c_j \bar{y}_{.j}$$

is

$$12\rho^2 \frac{k}{(k+1)} \left\{ \int_{-\infty}^{+\infty} f^2(x) \, dx \right\}^2 \left[\frac{\sum_{j=1}^{k}(2j-k-1)\theta_j}{\sum_{j=1}^{k} c_j \theta_j} \right]^2$$

where the cs are scaled so that

$$\sum_{j=1}^{k} c_j^2 = \sum_{j=1}^{k} (2j - k - 1)^2 = \tfrac{1}{3}k(k^2 - 1).$$

This result is a generalization of Hollander's Theorem 4.1 which serves to clarify his conclusions. If we choose $c_j = (2j - k + 1)$ the A.R.E. is exactly that which was found for $\bar{\chi}^2_{\text{rank}}$ and $\bar{\chi}^2$ in (4.41). However, the T-test with equally spaced scores is not optimum in the sense of Abelson and Tukey (1963) if $k > 3$. If the optimum scores (4.22) given in Section 4.2 were to be chosen the factor in square brackets could be either greater or less than unity. It is essentially a ratio of correlation coefficients between the θ_js and the two sets of scores. If the θs are equally spaced then the numerator will be greater than the denominator unless the cs are also equally spaced. Conversely, if the spacing of the θs is very irregular the denominator will be larger since the optimum scores are chosen to maximize the minimum value of the correlation coefficient.

Similar conclusions apply to τ by combining the conclusions of Theorems 4.8 and 4.9.

The tests of Hollander and Doksum

Hollander (1967) and Doksum (1967a) have devised closely related tests which are asymptotically more powerful than either ρ or τ. To provide some motivation for their proposals we first observe that all of our statistics, distribution-free or otherwise, depend essentially on a function of the form

$$\sum_{i=1}^{k-1} \sum_{j=i+1}^{k} s_{ij} \qquad (4.44)$$

where s_{ij} is some measure of the difference between group i and group j. If we score 1 if the gth member of group j is greater than the hth member of group i and zero otherwise, and sum over g and h, we get τ. If, instead, s_{ij} is taken as the difference in mean ranks, ρ is obtained. That there exists a possibility of improvement can be seen by considering τ. In this case there is a contribution of 1 to the total score if $y_{hi} < y_{hj}$ regardless of how big the difference is. However, "big" differences are more "significant" than small ones so some means of incorporating the size of the differences should improve the test. Hollander and Doksum proposed to do this by using scores.

$$s_{ij} = \sum_{h=1}^{n} \psi_{ij}^{(h)} R_{ij}^{(h)}$$

where

$$\psi_{ij}^{(h)} = 1 \quad \text{if } y_{hi} < y_{hj}$$
$$= 0 \quad \text{otherwise}$$

and $R_{ij}^{(h)}$ is the rank of $|y_{hi} - y_{hj}|$ in the ranking from the least to the greatest. Hollander's statistic is then

$$Y = \sum_{i=1}^{k-1} \sum_{j=i+1}^{k} s_{ij}. \qquad (4.45)$$

Doksum's statistic is

$$Y' = \sum_{i=1}^{k-1} \sum_{j=i+1}^{k} (s_{i.} - s_{.j}). \qquad (4.46)$$

Although the scores s_{ij} are distribution-free, the statistics Y and Y' are not because the correlation coefficient between any pair with a common i depends on the form of $F(x)$. In order to circumvent this difficulty the correlation coefficients can be estimated from the data using a device first proposed by Lehmann (1964). The resulting test is asymptotically distribution-free. The execution of these tests is more tedious than the others which have been described and the adequacy of the asymptotic distribution for finite samples remains unexplored. The reader is referred to the original papers for further details but in order to show the advantages of the tests the following results are quoted on their asymptotic efficiency. Doksum showed that Y' is always asymptotically more powerful the Y for all F and k. However, the difference is negligible; values of the maximum value of the A.R.E. given by Hollander are 1.004 for F normal, 1.002 for F rectangular and 1.008 when F is exponential. The asymptotic gain using Y (or Y') as compared with ρ can be seen from Table 4.5.

Table 4.5 *The A.R.E.s of Y and ρ Compared with the Statistic*

$$T = \sum_{j=1}^{k} (2j - k - 1)\bar{y}_{.j}$$

when the θs are Equally Spaced Assuming Normal Distributions

k	2	3	4	6	10	20	50	∞
Y with T	0.955	0.963	0.968	0.974	0.980	0.984	0.987	0.989
ρ with T	0.637	0.716	0.764	0.819	0.868	0.909	0.936	0.955

It is clear that the tests which have been discussed for the two-way classification could be extended to orthogonal designs in higher dimensions. Whether or not distribution-free tests can be devised for nonorthogonal designs remains a subject for further research. In view of the great variety of tests available for the simple cases discussed here it seems likely that further work in this direction will lead to an even greater proliferation of statistics. It seems to the authors that striving after increased power, often leading to only marginal improvements, should give way to the search for a unified, simple approach to a large class of designs. The rank analogues of the $\bar{\chi}^2$ and T statistics seem, at this stage, to offer the greatest hope of progress in this direction.

4.5 COMPLEMENTS

The tests for the multivariate normal mean described in Section 4.1 were originally developed by Kudô (1963) and Nüesch (1964, 1966). These authors assumed that (in the notation of Section 4.1) Σ was known, and developed tests appropriate when $r = p$, though Kudô pointed out the possibility of making the extensions we have described. For the extension to the case $\Sigma = \sigma^2 \Lambda$, he suggested a test based on the statistic

$$\frac{n\tau^{*\prime}\Lambda^{-1}\tau^*}{\sum_{j=1}^{n}(\mathbf{z}_i - \bar{\mathbf{z}})'\Lambda^{-1}(\mathbf{z}_i - \bar{\mathbf{z}})}$$

Such a test would be asymptotically equivalent to the test based on $\bar{E}^2$.

The case when Σ is completely unknown has also received some attention. An attempt by Nüesch (1964, 1966) to find a test for this situation was invalidated by an error, pointed out by Shorack (1967). More recently Perlman (1969) has dealt with a variety of problems involving tests against restricted alternatives when Σ is unknown. His results for the problem may be summarized as follows. We assume that $n > p$, so that a nondegenerate estimator

$$\mathbf{S} = \frac{1}{n-1}\sum_{i=1}^{n}(\mathbf{z}_i - \bar{\mathbf{z}})(\mathbf{z}_i - \bar{\mathbf{z}})'$$

of Σ may be obtained. Then, if τ_s^* is the vector in H_1 which minimizes $(\bar{\mathbf{z}} - \tau)'\mathbf{S}^{-1}(\bar{\mathbf{z}} - \tau)$ with respect to τ, the likelihood ratio test rejects H_0 in favour of H_1 for large values of

$$U = \frac{\tau_s^{*\prime}\mathbf{S}^{-1}\tau_s^*}{n - 1 + (\bar{\mathbf{z}} - \tau_s^*)'\mathbf{S}^{-1}(\bar{\mathbf{z}} - \tau_s^*)}.$$

The null hypothesis distribution of this statistic is given by:

$$\Pr\{U \geq C\} = \sum_{j=1}^{p} Q(j,p) \Pr\{\chi_j^2/\chi_{n-p}^2 \geq C\}, \quad C > 0$$

$$\Pr\{U = 0\} = Q(0,p),$$

where $Q(j,p)$ is defined as in Theorem 4.1 and χ_j^2, χ_{n-p}^2 denote independent χ^2 random variables with j and $n-p$ degrees of freedom respectively. It follows that the probability of Type I error for the likelihood ratio test depends on the unknown Σ through the probabilities $Q(j,p)$. A test with size α may be obtained by rejecting H_0 for $U \geq C_0$, where C_0 satisfies

$$\Pr\{\chi_{p-1}^2/\chi_{n-p}^2 \geq C_0\} + \Pr\{\chi_p^2/\chi_{n-p}^2 \geq C_0\} = 2\alpha.$$

However, the probability of a Type I error for this test may vary within a wide range, with lower limit $\frac{1}{2}\Pr\{\chi_1^2/\chi_{n-p}^2 \geq C_0\}$. For this reason, and for others discussed by Perlman, it is doubtful whether this test is of much practical value.

An earlier paper which has some connection with the problems of Section 4.1 is that of Chernoff (1954). He derived the $\bar{\chi}^2$ test for some particular cases when $p = 2$, and showed that, for a wide class of testing problems with parameters restricted by linear constraints, $-2 \log \lambda$ (where λ denotes the likelihood ratio) has asymptotically a distribution of the $\bar{\chi}^2$ type.

An alternative approach to many of the problems discussed in this chapter was given by Schaafsma (1966). His treatment is in the spirit of Section 4.2 although his motivation was somewhat different from that given there. Schaafsma's work was referred briefly to at the end of Section 4.2 and here we summarize the basic ideas and results which he has obtained.

The general class of problems to which Schaafsma addresses himself is the following. The n-dimensional random vector $\mathbf{y} = (y_1 \ldots y_n)'$ has a multivariate normal distribution with mean vector $\xi = (\xi_1, \xi_2, \ldots, \xi_n)'$ and variance-covariance matrix $\Sigma = \sigma^2 \mathbf{A}^{-1}$ where $\mathbf{A}$ is a known matrix and σ^2 may be known or unknown. The problem posed is (in Schaafsma's notation) to test the null hypothesis

$$H_0: \eta_h = b_{0h} + \sum_{i=1}^{n} b_{ih}\xi_i = 0, \quad (h = n-s+1, \ldots, n-s+r)$$

against the one-sided alternative $\eta_h \geq 0$ for all h listed above, with at least one inequality strict, or the corresponding two-sided alternative. It will be observed that this formulation covers all of the normal theory problems treated in this and the previous chapter.

Schaafsma's approach is as follows. Since it is impossible to find a test which is uniformly most powerful against all members of the chosen alternative he decides to concentrate on the class of tests which are "somewhere most powerful" (SMP). Such tests attain the greatest possible power against at least one member of the alternative class. In fact if a test is most powerful when $\eta_h = \eta_h^{(0)}$ it is also most powerful against $\eta_h = a\eta_h^{(0)}$ for all $a > 0$. It is very easy to show that the most powerful test should depend on the statistic

$$\sum_h \eta_h^{(0)} Y_h$$

where

$$Y_h = b_{0h} + \sum_{i=1}^{n} b_{ih} y_i.$$

Any test based on scores will therefore be most powerful somewhere in the alternative parameter space. To choose among these tests Schaafsma considers the amount by which the power of any such test falls below the maximum attainable. This is termed the shortcoming at that point. The configuration of scores is then sought which makes the maximum shortcoming a minimum. This is described as the *most stringent somewhere most powerful* (MSSMP) test. It will be clear at once that Abelson and Tukey's method described in Section 4.2 is exactly of this form. Schaafsma's formulation of the problem is more general and covers the whole range of problems which we have tackled by the likelihood ratio technique. The techniques which Schaafsma develops for finding the optimum test are based on essentially the same geometrical ideas as were described in Section 4.2.

Schaafsma has obtained explicit results for a number of practically important cases. In particular he gives optimum scores for the case of simple order. If there are k groups with weights $w_1, w_2, \ldots, w_k$ then the optimum statistic of the form

$$T = \sum_{i=1}^{k} c_i y_i$$

has

$$c_i = \{W_{i-1}(W_k - W_{i-1})\}^{\frac{1}{2}} - \{W_i(W_k - W_i)\}^{\frac{1}{2}} \qquad (i = 1, 2, \ldots, k)$$

where

$$W_i = \sum_{j=1}^{i} w_j, \qquad W_0 = 0.$$

In the special case of equal weights these become equivalent to those given in Section 4.2. If the ys are means based on numbers of observations $n_1, n_2, \ldots, n_k$ having the same variance then $w_i \propto n_i$.

The foregoing result enables the equality of means to be tested in, for example, a one-way analysis of variance. Schaafsma extends this test to include the combination of tests on several sets of data. He does this by assuming $y_{tij} \sim N(\mu_{ti}, \sigma^2)$ $(t = 1, 2, \ldots, k;\ i = 1, 2, \ldots, l;\ j = 1, 2, \ldots, n_{ti})$ and testing the null hypothesis

$$H_0: \mu_{1i} = \mu_{2i} = \ldots = \mu_{ki} \qquad (i = 1, 2, \ldots, l)$$

against the alternative

$$H_1: \mu_{1i} \leq \mu_{2i} \leq \ldots \leq \mu_{ki} \qquad (i = 1, 2, \ldots, l)$$

with at least one inequality strict. The resulting test statistic is, as one would expect, a weighted average of the test statistics for each i separately. This test is an example of the use of the general approach for combining independent tests of significance. The simplest problem of this kind discussed by Schaafsma is that of combining k one-sided tests for a normal mean. The two hypotheses here are

$$H_0: \mu_i = 0$$
$$H_1: \mu_i \geq 0 \qquad (i = 1, 2, \ldots, l)$$

where at least one inequality in H_1 is strict. The most stringent somewhere most powerful test here depends on the statistic

$$\sum_{i=1}^{l} n_i^{\frac{1}{2}} \bar{y}_i \tag{4.47}$$

where $\bar{y}_i$ is the ith sample mean and n_i is the corresponding sample size. The likelihood ratio method can also be used for this problem and is, in fact, a special case of Kudô's problem discussed in Section 4.1. The statistic in this case is

$$\sum_{i=1}^{l} n_i \{\max (\bar{y}_i, 0)\}^2. \tag{4.48}$$

Schaafsma shows that (4.47) provides a better test than (4.48) in the sense that what he calls the averaged maximum shortcoming is smaller for the former than the latter. It would be relatively easy to make exact comparisons of power for this problem.

The model given at the beginning of the last paragraph can also be applied to the two-way cross classification. This problem has been discussed in Sections 3.2 and 4.1 where the representation

$$\mu_{ti} = \mu + \alpha_t + \beta_i + \gamma_{ti} \qquad (t = 1, 2, \ldots, k; i = 1, 2, \ldots, l)$$

was assumed with

$$\sum_t \alpha_t = \sum_i \beta_i = \sum_t \gamma_{ti} = \sum_i \gamma_{ti} = 0.$$

The α_ts are described as the "row" effects, the β_is as the column effects and the γ_{ti}s as the interactions. Schaafsma provides two tests associated with this model. For the first it is assumed that the interaction term is zero. The null and alternative hypotheses are

$$\left.\begin{array}{l} H_0: \alpha_t = 0 \quad (t = 1, 2, \ldots, k) \\ H_1: \alpha_1 \leq \alpha_2 \leq \ldots \leq \alpha_k. \end{array}\right\} \quad (4.49)$$

The likelihood ratio test was given for the case $n_{ti} = n$ (all t and i) in Section 3.2 and the general case in Section 4.1. (It was not assumed in Section 3.2 that the interaction term was zero but the necessary modifications are easily made.) Schaafsma's result is contained in the following thorem and corollary.

Theorem 4.10

The MSSMP test for H_0 given in (4.49) is based on the statistic

$$\sum_{t=1}^{k} \sum_{i=1}^{l} n_{ti} w_{ti} \bar{y}_{ti\cdot},$$

where the weights w_{ti} are determined from

$$w_{ti} = b_i - n_{t\cdot}^{-1} \sum_{i=1}^{l} n_{ti} b_i,$$

$$n_{\cdot g} b_g - \sum_{i=1}^{l-1} b_i \sum_{t=1}^{k} n_{t\cdot}^{-1} n_{tg} n_{ti} = -d_g + d_{g-1}, \quad (g = 1, 2, \ldots, k-1)$$

$$b_k = 0$$

and

$$d_g = \left\{ \sum_{t=1}^{k} n_{t\cdot}^{-1} S_{tg}(n_{t\cdot} - S_{tg}) \right\}^{\frac{1}{2}} \quad (g = 0, 1, \ldots, k-1)$$

provided that $b_1, b_2, \ldots, b_l$ so determined satisfy

$$b_1 \leq b_2 \leq \ldots \leq b_k = 0.$$

In the above expressions

$$n_{t\cdot} = \sum_i n_{ti},$$
$$n_{\cdot i} = \sum_t n_{ti},$$
$$S_{tg} = \sum_{i=1}^{g} n_{ti} \quad (g = 1, 2, \ldots, l),$$
$$S_{t0} = 0 \quad (t = 1, 2, \ldots, k).$$

If the bs do not satisfy these inequalities recourse must be had to an approach from first principles. In the special case when the cell frequencies are proportional to the row totals these inequalities are automatically satisfied. The result is contained in the following corollary.

Corollary

If $n_{ti} = n_{t.}p_i$, $p_i \geq 0$, ($t = 1, 2, \ldots, k$; $i = 1, 2, \ldots, l$; $p_1 + p_2 + \ldots + p_k = 1$) the scores are given by

$$w_{ti} = n^{-\frac{1}{2}} p_i^{-1} \left\{ -\left(\sum_{j=1}^{i} p_j \sum_{j=i+1}^{l} p_j \right)^{\frac{1}{2}} + \left(\sum_{j=1}^{i-1} p_j \sum_{j=i}^{l} p_j \right)^{\frac{1}{2}} \right\},$$

where

$$n = \sum_t \sum_i n_{ti}.$$

This expression simplifies even further if p_i is constant which covers the case of equal cell frequencies.

A test for additivity can be obtained by testing the hypothesis

$$H_0: \mu_{t+1,i+1} - \mu_{t+1,i} - \mu_{t,i+1} + \mu_{t,i} = 0$$

$$(t = 1, 2, \ldots, k-1; i = 1, 2, \ldots, l-1).$$

Schaafsma shows how to find a test of this hypothesis against the alternative that all of these interaction terms are non-negative. This implies that increasing the level of both the row and column factors produces a bigger increase in response than would be predicted from the effect of either taken separately. This is a fairly complicated kind of ordering which implies a relatively large amount of prior information about the way the factors interact. The same problem could be tackled using the general likelihood ratio method of Section 4.1.

A number of other common statistical problems can give rise to order restrictions on the alternative. For example, in a goodness of fit problem the alternative may postulate a distribution stochastically larger than the distribution being fitted. Schaafsma considers this problem as it arises with grouped data. He supposes that the groups are such that the proportions falling into them on the null hypothesis are expected to be equal. His alternative is then that the proportions form a monotonic sequence. The maximum likelihood estimators of the proportions for this case were given in Section 2.1 (Example 2.1) and these could be made the basis of a test. A discussion of other goodness of fit problems, with special reference to the exponential distribution, is given in Chapter 6.

Tests of symmetry and tests for positive dependence (in the sense of Lehmann (1959)) can also give rise to ordered alternatives and the interested reader should consult Schaafsma (1966) for further details.

A disadvantage of the optimal T-test is that the scores lack the simplicity of the "equally-spaced" version given in (4.17). In an attempt to combine simplicity with near optimality, Abelson and Tukey (1963) proposed the use of "linear-2-4" scores. These are obtained from equally spaced scores by doubling the scores next to the end and quadrupling those at the end. Thus for $k = 11$ the vector $\mathbf{c}$ would be $(-20, -8, -3, -2, -1, 0, 1, 2, 3, 8, 20)'$. For $k \leq 20$ these contrasts have values of minimum r which are at least 97.5% of the corresponding value for maximin r.

The relationship between the likelihood ratio test and tests based on scores expressed by Theorem 4.3 was first pointed out in Bartholomew (1961b) and again by Hogg (1965). This last paper also raised the important question of what happens if the prior information on which the order restrictions is based proves to be wrong. As has been remarked elsewhere it is desirable that an ordered test should still have reasonable power if the true alternative lies outside the restricted parameter space. One way of investigating this point is to determine the region of consistency for each test—that is, the region of the parameter space within which the power tends to 1 with increasing Δ. This question is easily answered for $\bar{\chi}^2$ and T. It has been shown in Section 3.4 that $\bar{\chi}^2$ is consistent for any alternative such that the isotonic regression of the μs with respect to simple order takes more than one value. The "proportion" of the sample space for which this is so is $1 - P(1, k) = (k - 1)/k$ in the case of equal weights. T is consistent for all points in the half-space

$$\sum_{i=1}^{k} c_i \mu_i > 0$$

and it is easy to see that this region is wholly contained within the domain of consistency of $\bar{\chi}^2$. The $\bar{\chi}^2$ test is thus less vulnerable to error in the specification of the alternative. The scale of the difference can be gauged from Table 4.6 which compares the powers of the optimum T test with $\bar{\chi}^2$ against simple order at points outside the alternative region.

For the single inversion in the ordering considered in the second comparison T and $\bar{\chi}_4^2$ retain their advantage over χ^2 with T having the edge over $\bar{\chi}_4^2$. The third and fourth comparisons show that this advantage is lost for the more extreme violations of simple order but that $\bar{\chi}_4^2$ is much less severely affected than T. This is the basis of the disadvantage of T mentioned in Section 4.2. Hollander (1967) found the domain of consistency for the τ and ρ tests of

Table 4.6 *Power of $\bar{\chi}_4^2$ and T for Various Spacings of the μs Lying Outside the Region of Simple Order. The Maximum Power of Both Tests and of the χ^2 Test are Included for Comparison*

Statistic	Spacing					Δ			
					0	1	2	3	4
T	−0.866,	−0.134,	0.134,	0.866	0.050	0.260	0.639	0.912	0.991
$\bar{\chi}_4^2$	1,	2,	3,	4	0.050	0.238	0.594	0.885	0.985
T	1,	3,	2,	4	0.050	0.225	0.505	0.847	0.972
$\bar{\chi}_4^2$					0.050	0.203	0.514	0.824	0.968
T	2,	1,	3,	2	0.050	0.068	0.090	0.118	0.151
$\bar{\chi}_4^2$					0.050	0.103	0.251	0.502	0.759
T	3,	2,	1,	3	0.050	0.043	0.036	0.030	0.026
$\bar{\chi}_4^2$					0.050	0.066	0.142	0.288	0.484
χ^2	any				0.050	0.115	0.350	0.710	0.945

Section 4.4. He showed that a necessary and sufficient condition for consistency of τ was that

$$\sum_{i<j} \int F_i \, dF_j / k(k-1) \geq \tfrac{1}{4}. \tag{4.50}$$

In the case of ρ the condition is

$$\sum_{j=1}^{k} (2j - k - 1) \sum_{i=1}^{k} \int F_i \, dF_j \geq 0. \tag{4.51}$$

It is not easy to see from these conditions what they imply about the ordering of the Fs but Hollander gives a simple example which helps to clarify the position. Roughly speaking, the boundary of the domain of consistency occurs where the rank correlation between the ordering of the Fs and the natural order is near zero. Points at which τ is consistent and ρ is not can be found by specifying Fs whose ordering is such that τ is positive and ρ is negative. Hollander also gives results on the consistency of his Y statistic.

Tests have been proposed by Chassan (1960 and 1962) and Bennett (1962) which are not even consistent on the boundary of the alternative parameter space. Both relate to the problem of testing the equality of a set of proportions against the alternative of a monotonic trend. In both cases the critical region adopted for the test contains only points for which the observed proportions are in the required order. Such a test will be consistent only if the alternative lies wholly within the alternative region. If it lies *on* the boundary the test will not be consistent. Thus even if only two proportions are equal the chance of an inversion occurring for that pair is approximately $\frac{1}{2}$ and so the power cannot rise above this figure.

The result of Theorem 4.6 was first obtained by Chacko (1963) in the case when the n_is are equal. His proof depended on the assertion that the asymptotic distributions of both $\bar{\chi}^2$ and $\bar{\chi}^2_{\text{rank}}$ were weighted sums of noncentral χ^2 distributions. There is no justification for this assertion so we have re-derived the result from first principles.

CHAPTER 5

Estimation of Distributions

5.1 INTRODUCTION

The empirical distribution is a convenient and natural estimator of a distribution function, F, in the absence of additional information concerning the shape of F. If, on the other hand, we felt that F was a member of some specified parametric class such as the normal this estimator would not be used. In this chapter several intermediate cases are examined. In each case an assumption is made concerning the geometrical shape of the distribution. However, the class of distributions considered cannot be described by a finite number of parameters as was the case in Chapter 2.

The problem of nonparametric density estimation is much harder than the problem of nonparametric distribution estimation. This is related to the fact that the density may be changed considerably over a short range without affecting the probability substantially. The histogram is one of the oldest and most useful solutions to this problem. The histogram estimates densities without making any parametric assumptions just as the empirical distribution estimates distributions nonparametrically. However, there is some arbitrariness in the choice of intervals over which the histogram is constant. This defect can be overcome by assuming additional smoothness conditions such as unimodality or decreasing density and obtaining the maximum likelihood estimate. The maximum likelihood estimator (MLE) in each case turns out to be an isotonic estimator with respect to a "naive" estimator based on the empirical distribution. Properties of isotonic estimators provide useful tools for proving consistency of the MLE in each case. In Section 5.2 the maximum likelihood estimation problem is considered for unimodal and decreasing densities.

In Section 5.3 distributions with monotone failure rate are discussed; i.e., distributions for which $\log[1 - F(x)]$ is either convex or concave. The maximum likelihood estimator for the failure rate function again turns out to be an isotonic estimator with respect to the natural actuarial estimator of the failure rate which in turn is based on the empirical distribution. Strong consistency of the maximum likelihood estimator follows from the strong consistency of the empirical distribution.

In Section 5.4 an attempt is made to improve upon the maximum likelihood estimators considered in Sections 5.2 and 5.3. The MLE suggests a promising class of so-called isotonized "window estimators". By proper choice of the windows the MLE can be improved upon asymptotically.

In Section 5.5 the problem of estimating starshaped and star-ordered classes of distributions is considered. *Ad hoc* isotonic estimators based on the empirical distribution are proposed. The usefulness of the isotonic regression approach to estimation is dramatized by the fact that the MLE in these cases is inconsistent and actually converges to the wrong distribution! The isotonic estimators are, of course, consistent.

5.2 ESTIMATION OF UNIMODAL DENSITIES

Suppose that the distribution F is absolutely continuous with density f which is unimodal with mode M. At first it will be assumed that the mode, M, is known.

Definition 5.1

A function f is *unimodal* at M if and only if f is nondecreasing at x for $x < M$ and f is nonincreasing at x for $x > M$.

If M is the mode and $F = \alpha F_+ + (1 - \alpha)F_-$ where F_+ is the conditional distribution on $[M, \infty)$ and F_- is the conditional distribution on $(-\infty, M)$, then it can easily be seen that the maximum likelihood estimate (MLE) of F is $\hat{\alpha}\hat{F}_+ + (1 - \hat{\alpha})\hat{F}_-$ where $\hat{\alpha}$ is the sample proportion on $[M, \infty)$ and $\hat{F}_+, \hat{F}_-$ are the MLEs of F_+ and F_- respectively.

Let f_+ and f_- be the densities of F_+ and F_- respectively and let $\hat{f}_+, \hat{f}_-$ be their MLEs. It will be sufficient to consider only the estimator $\hat{f}_+$ since the problem of estimating f_- is similar. Without loss of generality, we may assume that $M = 0$ and that $F(x) = 0$ for $x < 0$. Since f is unimodal, f is nonincreasing for $x \geq 0$.

MLE of a decreasing density

Suppose $0 < X_{1:n} \leq X_{2:n} \leq \ldots \leq X_{n:n}$ are n ordered observations from a sample of size n from a distribution F. Let $\mathscr{F}$ be the class of distributions with nonincreasing left continuous densities defined on $[0, \infty)$. Let $X_{0:n} \equiv 0$ and define

$$L(F) = \sum_{i=1}^{n} \log f(X_{i:n})$$

to be the logarithm of the likelihood for $F \in \mathscr{F}$. We now wish to show that the MLE is a step function. For any $F \in \mathscr{F}$, let F^* be the distribution with density

$$f^*(x) = \begin{cases} 0 & \text{for } x \leq 0 \\ Cf(X_{i:n}) & \text{for } X_{i-1:n} < x \leq X_{i:n} \quad 1 \leq i \leq n \\ 0 & \text{for } x > X_{n:n} \end{cases}$$

where C is a normalizing constant. By construction

$$1 = \int_0^\infty f^*(x)\, dx$$

$$= C \sum_{i=1}^n f(X_{i:n})[X_{i:n} - X_{i-1:n}].$$

Since f is nonincreasing and a density on $[0, \infty)$,

$$\sum_{i=1}^n f(X_{i:n})[X_{i:n} - X_{i-1:n}] \leq 1,$$

which implies $C \geq 1$. Hence

$$L(F^*) = n \log C + L(F) \geq L(F).$$

Thus, to each $F \in \mathscr{F}$ corresponds a distribution $F^* \in \mathscr{F}$ whose density is a step function and whose likelihood is at least as large. Therefore, the MLE $\hat{f}_n$ of f will be a step function with steps at the order statistics $X_{i:n}$, $1 \leq i \leq n$.

Hence the problem of maximizing $L(F)$ for $F \in \mathscr{F}$ reduces to the problem:

$$\left.\begin{array}{l} \text{Maximize } \prod_{i=1}^n f_i \\ \text{subject to (i) } f_1 \geq f_2 \geq \ldots \geq f_n \\ \qquad\qquad \text{(ii) } X_{1:n} f_1 + (X_{2:n} - X_{1:n}) f_2 \\ \qquad\qquad\qquad + \ldots + (X_{n:n} - X_{n-1:n}) f_n = 1 \end{array}\right\} \quad (5.1)$$

where $f_i = f(X_{i:n})$.

In order to understand the nature of the solution, let F_n be the empirical distribution where

$$F_n(x) = \begin{cases} 0 & \text{for } x < X_{1:n} \\ \dfrac{i}{n} & \text{for } X_{i:n} \leq x < X_{i+1:n} \quad 1 \leq i \leq n-1 \quad (5.2) \\ 1 & \text{for } x \geq X_{n:n}. \end{cases}$$

It will be shown that

$$f_i = \min_{s \leq i-1} \max_{t \geq i} \frac{F_n(X_{t:n}) - F_n(X_{s:n})}{X_{t:n} - X_{s:n}} \quad (5.3)$$

solves problem (5.1). It will follow that the MLE for the density of $F \in \mathscr{F}$ is

$$\hat{f}_n(x) = \begin{cases} 0 & \text{for} \quad x \leq 0 \\ \hat{f}_i & \text{for} \quad X_{i-1:n} < x \leq X_{i:n} \\ 0 & \text{for} \quad x > X_{n:n}. \end{cases}$$

Note that $\hat{f}_n$ is *left* continuous. Geometrically, (5.3), is the slope of the least concave majorant to F_n at the point $X_{i:n}^-$. Figure 5.1 illustrates the solution.

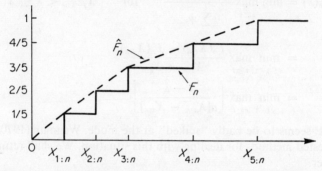

Figure 5.1 Least Concave Majorant of an empirical distribution

From Figure 5.1, we can read off the slopes from the dotted line and obtain the density estimate

$$\hat{f}_n(x) = \begin{cases} 0 & \text{for} \quad x \leq 0 \\ \dfrac{3}{5 X_{3:n}} & \text{for} \quad 0 < x \leq X_{3:n} \\ \dfrac{2}{5[X_{5:n} - X_{3:n}]} & \text{for} \quad X_{3:n} < x \leq X_{5:n} \end{cases}$$

for the particular example illustrated here.

The solution can also be interpreted as the isotonic regression of the following basic estimator:

$$f_n(x) = \begin{cases} 0 & \text{for} \quad x \leq 0 \\ \dfrac{F_n(X_{i:n}) - F_n(X_{i-1:n})}{X_{i:n} - X_{i-1:n}} & \text{for} \quad X_{i-1:n} < x \leq X_{i:n} \quad 1 \leq i \leq n \\ 0 & \text{for} \quad x > X_{n:n}. \end{cases} \quad (5.4)$$

The associated partial ordering is, of course, the simple ordering $1 \succ 2 \succ \ldots \succ n$ on the index set. The estimator f_n is sometimes called the "naive" estimator. (Unfortunately, it is not a very good estimator since it is not even consistent. This is easily verified; for example, take $F(x) = 1 - e^{-x}$ for $x \geq 0$.)

If we choose the weights

$$w_i = [X_{i:n} - X_{i-1:n}],$$

then it is easily seen that

$$\begin{aligned} f_n(x) &= \min_{s \leq i} \max_{t \geq i} \frac{\sum_{j=s}^{t} f_n(X_{j:n}) w_j}{\sum_{j=s}^{t} w_j} \quad \text{for} \quad X_{i-1:n} < x \leq X_{i:n} \\ &= \min_{s \leq i-1} \max_{t \geq i} \frac{F_n(X_{t:n}) - F_n(X_{s:n})}{X_{t:n} - X_{s:n}} \\ &= \min_{s \leq i-1} \max_{t \geq i} \left[\frac{t-s}{n[X_{t:n} - X_{s:n}]} \right]. \end{aligned} \quad (5.5)$$

The MLE seems to be badly "spiked" at the mode. Wegman (1970a, 1970b) has suggested methods for dealing with this situation. We shall return to this point later.

Derivation of the MLE for a decreasing density

The isotonic regression solution (5.5) of the maximum likelihood estimation problem (5.1) follows from Example 1.10 (Chapter 1) on maximizing a product. For convenience, let

$$w_i = (X_{i:n} - X_{i-1:n}) \quad \text{and} \quad g_i = \frac{1}{n(X_{i:n} - X_{i-1:n})} = f_n(X_{i:n}).$$

Problem (5.1) can be restated as:

$$\text{Maximize} \quad \sum_{i=1}^{n} g_i w_i \log f_i$$

$$\text{subject to} \quad f_1 \geq f_2 \geq \ldots \geq f_n$$

$$\sum_{i=1}^{n} [g_i - f_i] w_i = 0.$$

The solution, according to Example 1.10 of Chapter 1, is the isotonic regression, g^*. Equation (5.5) is one representation of this solution.

Consistency of the MLE

The strong consistency of the MLE for a decreasing density is a consequence of the strong consistency of the cumulative sum function (see the Lemma of Section 2.2, Chapter 2). In this case the cumulative sum function is the empirical distribution. To see this the naive estimator (5.4) is taken for the basic estimator with weights

$$w_i = X_{i:n} - X_{i-1:n}.$$

The cumulative sum function at the ith order statistic is

$$F_n(X_{i:n}) = \sum_{j=1}^{i} f_n(X_{j:n})w_j = \frac{i}{n} \quad 1 \leq i \leq n.$$

The cumulative sum function on $[0, \infty)$ has thus been defined so that it coincides with the empirical distribution, F_n. By the Glivenko–Cantelli theorem

$$\sup_x |F_n(x) - F(x)| \to 0$$

with probability one as $n \to \infty$. Clearly

$$\sup_x |F_n(x) - F(x)| = \sup_{1 \leq i \leq n} |F_n(X_{i:n}) - F(X_{i:n})|.$$

We proved in the lemma of Section 2.2 (Chapter 2) that

Lemma A

If F is concave on $[0, \infty)$, F_n is the empirical distribution and $\hat{F}_n$ is the concave majorant to F_n; i.e., at order statistics

$$\hat{F}_n(X_{i:n}) = \sum_{j=1}^{i} \hat{f}_n(X_{j:n})w_j$$

is the cumulative isotonic regression corresponding to the MLE for a decreasing density on $[0, \infty)$, then

$$\sup_x |\hat{F}_n(x) - F(x)| \leq \sup_x |F_n(x) - F(x)|.$$

It follows from the strong consistency of F_n, that

$$\sup_x |\hat{F}_n(x) - F(x)| \to 0$$

with probability one as $n \to \infty$.

From (5.3) it is seen that $\hat{f}_n(x)$ is the (left continuous) slope of $\hat{F}_n$ at the point x. From the following lemma it is seen that the strong uniform consistency of $\hat{F}_n$ implies the strong consistency of its derivative, $\hat{f}_n$.

Lemma B

If h and h_n, $n = 1, 2, \ldots$, are defined and concave on an open interval I, and if
$$\lim_{n \to \infty} h_n(x) = h(x)$$
uniformly in $x \in I$, then
$$h^-(x) \geq \overline{\lim} \, h_n^-(x) \geq \underline{\lim} \, h_n^+(x) \geq h^+(x) \qquad x \in I$$
where $g^-(g^+)$ is the left (right) hand derivative of g, $g = h, h_n$.

Proof. Since h_n is concave, we have for all $\Delta > 0$
$$\frac{h_n(x) - h_n(x - \Delta)}{\Delta} \geq h_n^-(x) \geq h_n^+(x) \geq \frac{h_n(x + \Delta) - h_n(x)}{\Delta}.$$

Letting $n \to \infty$ and then Δ decrease to zero we have
$$h^-(x) \geq \overline{\lim} \, h_n^-(x) \geq \underline{\lim} \, h_n^+(x) \geq h^+(x). \quad \blacksquare$$

Theorem 5.1

If F has decreasing density on $[0, \infty)$ and $\hat{f}_n(x)$ is the maximum likelihood estimate, then
$$f(x^-) \geq \overline{\lim} \, \hat{f}_n(x^-) \geq \underline{\lim} \, \hat{f}_n(x^+) \geq f(x^+) \tag{5.6}$$
with probability one.

Proof. Let $h_n(x) = \hat{F}_n(x)$ in Lemma B and recall that $\hat{F}_n$ converges to F uniformly in x with probability one by Lemma A. Since $\hat{f}_n(x^-)$ ($\hat{f}_n(x^+)$) is the left (right) hand derivative of $\hat{F}_n(x)$, (5.6) follows from Lemma B. $\blacksquare$

"Window" estimators for decreasing densities

Prakasa Rao (1969) has shown, under regularity conditions, that
$$n^{\frac{1}{3}}[\tfrac{1}{2} f(x) f'(x)]^{-\frac{1}{3}}[\hat{f}_n(x) - f(x)]$$
has a limiting distribution related to the heat equation. This is disappointing in that the convergence rate is like $n^{-\frac{1}{3}}$ rather than like $n^{-\frac{1}{2}}$ which is the case for the sample mean, for example. This result has motivated work on isotonic window estimators (Barlow and van Zwet (1970)). A grid $\{t_{i:n}\}_{i=1}^{\infty}$ on $[0, \infty)$ is chosen in advance and the basic estimator is taken to be
$$f_n(x) = \frac{F_n(t_{i:n}) - F_n(t_{i-1:n})}{t_{i:n} - t_{i-1:n}} \qquad \text{for} \quad t_{i-1:n} < x \leq t_{i:n}.$$

The isotonic regression on f_n with respect to weights $w_i = t_{i:n} - t_{i-1:n}$ replaces the MLE. By proper choice of the "windows" (i.e., the interval widths, $t_{i:n} - t_{i-1:n}$) the MLE can be improved. This is discussed in a generalized setting in Section 5.4.

Estimation of unimodal densities with unknown mode

The MLE has been given for a unimodal density when the mode is known. Wegman (1970a) gave a maximum likelihood estimate when the mode is unknown. A peculiar characteristic of the MLE when the mode is known is a "peaking" near the mode. To eliminate this peaking, at least partially, Wegman requires his estimate to have a *modal interval* of length ε, where ε is some fixed positive number. This also has the effect of uniformly bounding the estimate for all sample sizes by $1/\varepsilon$.

Definition 5.2

If f is unimodal at every point of an interval, I, (i.e., if each point of I is a mode of f) then I is called the *modal interval* of f. The centre of I will be called the *centre mode*.

The estimate

Let $X_{1:n} \leq X_{2:n} \leq \ldots \leq X_{n:n}$ be an ordered sample selected according to a unimodal density f with mode M. Let L and R be points such that $R - L = \varepsilon$. Let $l(n)$ and $r(n)$ be the subscripts of the largest observation less than or equal to L and the smallest observation greater than or equal to R respectively.

First it is shown that the MLE is a step function. Suppose h is any estimate which has $[L, R]$ for its modal interval. Define $g(x)$ by

$$g(x) = \begin{cases} h(X_{l(n):n}) & X_{l(n):n} \leq x < L \\ h(X_{r(n):n}) & R < x \leq X_{r(n):n} \\ h(x) & \text{otherwise.} \end{cases}$$

It follows that

$$\hat{f} = \left[\int_{-\infty}^{\infty} g(x)\, dx \right]^{-1} g$$

has modal interval $[L, R]$ and has a likelihood product no smaller than that of h. Similarly, any estimate, which is to maximize the likelihood product, must be constant on every open interval joining consecutive observations in the complement of $[L, R]$. The problem then is to find an estimate which is constant on

$$A_1 = [X_{1:n}, X_{2:n}),$$
$$A_2 = [X_{2:n}, X_{3:n}), \ldots, A_{l(n)} = [X_{l(n):n}, L),$$
$$A_{l(n)+1} = [L, R],$$
$$A_{l(n)+2} = (R, X_{r(n):n}], \ldots, A_k = (X_{n-1:n}, X_{n:n}].$$

Call $m = (R + L)/2$ the centre mode (not necessarily the true mode). Let

$$\hat{g}_{nm} = \sum_{i=1}^{k} n_i [n\lambda(A_i)]^{-1} 1_{A_i} \tag{5.7}$$

where $\lambda(A_i)$ is the length of interval A_i, 1_{A_i} is the indicator function for A_i and n_i is the number of observations in A_i. The maximum likelihood estimate f_{nm} given the modal interval $[L, R]$ can be computed in terms of $\hat{g}_{nm}$. Let

$$G_{nm}(x) = \int_{(-\infty, x]} g_{nm}(u) \, du.$$

Now find the interval $[L, R]$ such that $[G_{nm}(b) - G_{nm}(a)]/(b - a)$ is maximized. On $[a, b]$, the MLE is given by $[G_{nm}(b) - G_{nm}(a)]/(b - a)$. To the left of a, it is the slope of the greatest convex minorant of G_{nm} and to the right of b, it is the slope of the least concave majorant of G_{nm}. A min–max type formula can also be written down for the estimate as in (5.5).

$$\hat{f}_{nm}(x) = \begin{cases} 0 & x < X_{1:n} \\ \hat{f}_{nm}(X_{i:n}) & X_{i:n} \leq x < X_{i+1:n} \leq a \\ \dfrac{G_{nm}(b) - G_{nm}(a)}{b - a} & a \leq x \leq b \\ \hat{f}_{nm}(X_{i+1:n}) & X_{i:n} < x \leq X_{i+1:n} \quad \text{for} \quad b \leq X_{i:n} \\ 0 & x > X_{n:n} \end{cases}$$

(extending the steps, not in the model interval, to complete the definition) where

$$\hat{f}_{nm}(X_{i:n}) = \min_{a \geq X_{t:n} \geq X_{i:n}} \max_{X_{s:n} \leq X_{i:n}} \frac{\sum_{j=s}^{t} n_j}{n(X_{t:n} - X_{s:n})} \quad \text{for} \quad X_{i:n} \leq a$$

$$\hat{f}_{nm}(X_{i:n}) = \min_{b \leq X_{s:n} \leq X_{i:n}} \max_{X_{t:n} \geq X_{i:n}} \frac{\sum_{j=s}^{t} n_j}{n(X_{t:n} - X_{s:n})} \quad \text{for} \quad X_{i:n} \geq b.$$

The form of the maximum likelihood estimate given the location of the modal interval is known. We wish to determine the location of the modal interval. Wegman (1970a, 1970b) proves that if $\hat{f}_{nm}$ is the maximum likelihood estimate, then it may be assumed that one of the endpoints of the modal interval lies in the set $\{X_1, X_2, \ldots, X_n\}$. There are at most $2n$ intervals of the form $[L, R]$ where one of L or R belongs to $\{X_1, X_2, \ldots, X_n\}$. Since the modal interval is unspecified, compute $2n$ estimates $\hat{f}_{nm_i}$, $i = 1, 2, \ldots, 2n$, corresponding to these $2n$ intervals. Calculate the $2n$ likelihood products,

$$L_{nm_i} = \prod_{j=1}^{n} \hat{f}_{nm_i}(X_{j:n}).$$

Select the maximum of these $2n$ numbers and let $\hat{f}_n$ be the corresponding estimate. Let m_n be the centre mode. (If two or more of the L_{nm_i} are equal we adopt the convention of choosing $\hat{f}_n$ to be the estimate with the smallest centre mode.)

Consistency

The maximum likelihood estimate of $f(x)$ corresponding to a modal interval of length $\varepsilon > 0$ is a strongly consistent estimate except on an interval of length ε. To be precise, let m be the mode of

$$\int_{-\infty}^{\infty} [\log f_m(x)] f(x) \, dx$$

assumed finite for every m. Then m_n, the maximum likelihood estimate of the mode, converges to m with probability one. For $x \notin [m - \varepsilon/2, m + \varepsilon/2]$

$$\hat{f}_n(x) \to f(x)$$

with probability one, while for $x \in [m - \varepsilon/2, m + \varepsilon/2]$

$$\hat{f}_n(x) \to \int_{m-\varepsilon/2}^{m+\varepsilon/2} f(u) \, du/\varepsilon$$

with probability one. These results were obtained by Wegman (1970a, 1970b).

If $\varepsilon = 0$, the same method yields a maximum likelihood estimate. The lack of a uniform bound, however, causes the consistency arguments used by Wegman to fail.

5.3 MAXIMUM LIKELIHOOD ESTIMATION FOR DISTRIBUTIONS WITH MONOTONE FAILURE RATE

In many statistical studies involving failure data, biometric mortality data, and actuarial mortality data, the failure rate $r(x) = f(x)/[1 - F(x)]$ for

$F(x) < 1$, (corresponding to a lifetime distribution with density f and distribution F), is of prime importance. If F is the life distribution of a person, an aircraft engine, a light bulb, a machine tool, etc., then the failure rate function, $r(x)$, describes the way in which the item in question wears out. If the item is ageing very rapidly the failure rate function will tend to increase rather steeply. The shape of the failure rate function in the case of equipment often suggests when appropriate maintenance actions should be taken. A problem of considerable interest, therefore, is the estimation of the failure rate function from a sample of n independent identically distributed lifetimes. Many observed failure rate function estimators seem to first decrease and then increase (are U shaped) or simply increase or decrease. The U shaped case is analogous to the case of a unimodal density. For simplicity, attention will be confined to *monotone* failure rate functions. The U shaped case can be estimated by suitably modifying the monotonic estimators, see for example Bray, Crawford and Proschan (1967).

Although most physical applications of the failure rate function correspond to distributions with support on $[0, \infty)$, the failure rate function $r(x) = f(x)/[1 - F(x)]$ (assumed right continuous) is well defined for $F(x) < 1$, if F has a density at x. Note that $r(x)\,dx$ is, heuristically, the conditional probability of failure in $(x, x + dx)$ given survival to time x. The normal distribution, for example, has an increasing failure rate. The reciprocal of its failure rate, called Mills' ratio, is sometimes used in applications. The IFR (for increasing failure rate) class of distributions is now formally defined. (It is to be understood that "increasing" is written for "nondecreasing" and "decreasing" for "nonincreasing".)

Definition 5.3

F is IFR if $-\log[1 - F(x)]$ is convex on the support of F (i.e., the points of increase of F), an interval contained in $[0, \infty)$.

It can be shown that F IFR implies that F is absolutely continuous except for the possibility of a jump at the right hand endpoint of its interval of support. As will be seen, the MLE for the failure rate function always jumps to plus infinity at the largest observation, corresponding to a jump in the estimate of the distribution function.

The estimator

Let $0 \equiv X_{0:n} \leq X_{1:n} \leq X_{2:n} \leq \ldots \leq X_{n:n}$ be the order statistics obtained by ordering a random sample from an unknown distribution $F \in \mathscr{F}$, the class of IFR distributions. Since $f(X_{n:n})$ can be chosen arbitrarily large, it is not possible to obtain a maximum likelihood estimator for $F \in \mathscr{F}$ directly by maximizing

$$\prod_{i=1}^{n} f(X_{i:n}).$$

Consequently, we first consider the subclass $\mathscr{F}^M$ of distributions $F \in \mathscr{F}$ with corresponding failure rates bounded by M, obtaining

$$\sup_{F \in \mathscr{F}^M} \prod_{i=1}^{n} f(X_{i:n}) \leq M^n$$

since $f(x) \leq r(x) \leq M$. As will be seen, there is a unique distribution, $\hat{F}_n^M \in \mathscr{F}^M$ at which the supremum is attained. As $M \to \infty$, $\hat{F}_n^M$ converges in distribution to an estimator $\hat{F}_n \in \mathscr{F}$ which is called the *maximum likelihood estimator* for $F \in \mathscr{F}$.

For convenience, if F is IFR we define $r(x) = \infty$ for all x such that $F(x) = 1$. It is not difficult to see that for any distribution F and any x for which r is finite on $[0, x)$

$$1 - F(x) = \exp\left[-\int_0^x r(u)\, du\right].$$

Hence the log likelihood $L = L(F)$ is given, for $F \in \mathscr{F}^M$ by

$$L = \sum_{i=1}^{n} \log r(X_{i:n}) - \sum_{i=1}^{n} \int_{-0}^{X_{i:n}} r(u)\, du. \tag{5.8}$$

We now show that the maximum likelihood estimator for r is a step function. To do this, let $F \in \mathscr{F}^M$ have failure rate r and let F^* be the distribution with failure rate

$$\left.\begin{aligned} r^*(x) &= 0 & & x < X_{1:n} \\ &= r(X_{i:n}) & & X_{i:n} \leq x < X_{i+1:n} \quad i = 1, 2, \ldots, n-1 \\ &= r(X_{n:n}) & & x \geq X_{n:n} \end{aligned}\right\} \tag{5.9}$$

(Note that r^* is *right continuous*.) Then $F^* \in \mathscr{F}^M$, and $r(x) \geq r^*(x)$ so that

$$-\int_0^{X_{i:n}} r(u)\, du \leq -\int_0^{X_{i:n}} r^*(u)\, du$$

for all i. Hence $L(F) \leq L(F^*)$. Thus, L may be replaced by the function

$$\sum_{i=1}^{n} \log r(X_{i:n}) - \sum_{i=1}^{n-1} (n-i)(X_{i+1:n} - X_{i:n}) r(X_{i:n}). \tag{5.10}$$

Since we wish to maximize (5.10) subject to $r(X_{1:n}) \leq r(X_{2:n}) \leq \ldots \leq r(X_{n:n}) \leq M$, clearly $r(X_{n:n}) = M$ must be chosen. Let $r(X_{i:n}) = r_i$. Then our problem can be stated as:

$$\left.\begin{array}{l} \text{Maximize} \quad \sum_{i=1}^{n-1} [\log r_i - (n-i)(X_{i+1:n} - X_{i:n})r_i] \\ \text{subject to} \quad 0 \leq r_1 \leq r_2 \leq \ldots \leq r_{n-1} \leq M. \end{array}\right\} \quad (5.11)$$

Since M may be chosen arbitrarily large, (5.11) can be identified as the Poisson extremum problem, Example 1.8 considered in Chapter 1. As was seen in Example 1.8, g^* (the isotonic regression) solves the Poisson extremum problem:

$$\left.\begin{array}{l} \text{Maximize} \quad \sum_{x} [g(x) \log f(x) - f(x)] w(x) \\ \text{subject to} \quad f \text{ isotonic.} \end{array}\right\} \quad (5.12)$$

Rewrite (5.11) as:

$$\text{Maximize} \quad \sum_{i=1}^{n-1} \{[(n-i)(X_{i+1:n} - X_{i:n})]^{-1} \log r_i - r_i\} \times (n-i)(X_{i+1:n} - X_{i:n})$$

subject to $0 \leq r_1 \leq r_2 \leq \ldots \leq r_{n-1} \leq M$.

We recognize this as (5.12) where now $X = \{1, 2, \ldots, n-1\}$, $w(i) = (n-i)(X_{i+1:n} - X_{i:n})$, $g(i) = [(n-i)(X_{i+1:n} - X_{i:n})]^{-1}$ and $f(i) = r_i$. The solution can be expressed as

$$\hat{r}_i = \hat{r}_n(X_{i:n}) = \min_{i \leq t \leq n-1} \max_{1 \leq s \leq i} \frac{t - s + 1}{\sum_{j=s}^{t} (n-j)(X_{j+1:n} - X_{j:n})}. \quad (5.13)$$

The resulting estimator is

$$\hat{r}_n^M(x) = \begin{cases} 0 & x < X_{1:n} \\ \hat{r}_i & X_{i-1:n} \leq x \leq X_{i:n} \\ M & x \geq X_{n:n}, \end{cases} \quad (5.14)$$

assuming $M > \max_{1 \leq i \leq n-1} \left[\dfrac{1}{(n-i)(X_{i+1:n} - X_{i:n})} \right]$.

The MLE is obtained by letting $M \to \infty$, so that $\hat{r}_n(x) = +\infty$ for $x \geq X_{n:n}$.

From Theorem 1.3 (Equation 1.17) and Theorem 2.1 it follows that

$$\sum_{i=1}^{n-1}[r_n(X_{i:n}) - r(X_{i:n})]^2 w(X_{i:n})$$
$$\geq \sum_{i=1}^{n-1}[r_n(X_{i:n}) - \hat{r}_n(X_{i:n})]^2 w(X_{i:n}) + \sum_{i=1}^{n-1}[\hat{r}_n(X_{i:n}) - r(X_{i:n})]^2 w(X_{i:n})$$

where r is the true failure rate. Hence, in a least squares sense, $\hat{r}_n$ is closer to r than r_n is to r. Also $\hat{r}_n$ is the closest increasing estimator to r_n in a least squares sense.

The total time on test transformation

The MLE can also be interpreted in terms of the following transform. Let

$$H_F^{-1}(t) = \int_0^{F^{-1}(t)} [1 - F(u)]\, du \qquad 0 \leq t \leq 1. \tag{5.15}$$

For F IFR, F is strictly increasing on its support, an interval. It follows that the inverse function F^{-1} is uniquely defined on $(0, 1)$. $F^{-1}(1)$ is taken to be equal to the right hand endpoint of the support of F (possibly $+\infty$). Note that $H_F^{-1}(0) = 0$ and

$$H_F^{-1}(1) = \int_0^\infty [1 - F(u)]\, du = \mu$$

where μ is the mean of F. (If F is IFR then $\mu < \infty$.) Hence H_F is a distribution with support on $[0, \mu]$ since H_F^{-1} (the inverse of H_F) is strictly increasing on $[0, 1]$. If $G(x) = 1 - e^{-x}$ for $x \geq 0$, then $H_G^{-1}(t) = t$ and H_G is the uniform distribution on $[0, 1]$. When it is clear from the context which distribution is being transformed, H^{-1} will be written for H_F^{-1}.

Note that

$$\frac{d}{dt} H_F^{-1}(t)\Big|_{t=F(x)} = \frac{1 - F(x)}{f(x)} = \frac{1}{r(x)}.$$

If F is IFR, then r is increasing and $H_F^{-1}(t)$ is a concave function of $t \in [0, 1]$ or H_F is a convex distribution on $[F^{-1}(0), \mu]$ where

$$\mu = \int_0^\infty x\, dF(x).$$

If F is replaced by F_n, then we define

$$H_n^{-1}\left(\frac{i}{n}\right) \stackrel{\text{def}}{=} H_{F_n}^{-1}\left(\frac{i}{n}\right) = \int_0^{X_{i:n}} [1 - F_n(u)]\, du \tag{5.16}$$

for $1 \leq i \leq n$. $H_n^{-1}(t)$ for $0 \leq t \leq 1$ is defined by linear interpolation. For example,

$$H_n^{-1}(t) = nX_{1:n}t \quad \text{for} \quad 0 \leq t \leq \frac{1}{n}$$

while

$$H_n^{-1}(t) = \frac{1}{n}[nX_{1:n} + \ldots + (n-i+1)(X_{i:n} - X_{i-1:n})]$$
$$+ \left[t - \frac{i}{n}\right](n-i)(X_{i+1:n} - X_{i:n})$$

for

$$\frac{i}{n} \leq t \leq \frac{i+1}{n} \quad \text{and} \quad 1 \leq i \leq n-1.$$

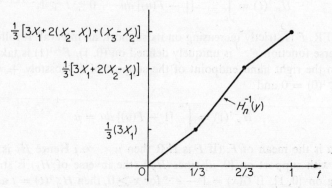

Figure 5.2 Graph of H_n^{-1} as a function of three observations

To clarify ideas, Figure 5.2 is a graph of H_n^{-1} for $n = 3$. Note that H_n^{-1} is continuous and that $H_n^{-1}(1)$ is the sample mean. In general,

$$H_n^{-1}\left(\frac{i}{n}\right) = \frac{1}{n}[nX_{1:n} + (n-1)(X_{2:n} - X_{1:n})$$
$$+ \ldots + (n-i+1)(X_{i:n} - X_{i-1:n})]$$

for $1 \leq i \leq n$. The expression in square brackets can be interpreted as the *total time on test* up to the ith observation. If n items are placed on life test at time 0, then n items survive up to time $X_{1:n}$, $n-1$ items survive through the interval $[X_{1:n}, X_{2:n})$, etc., while $(n-i+1)$ items survive the interval $[X_{i-1:n}, X_{i:n})$. The total time on test statistic is a fundamental tool in solving life testing problems.

It is easy to verify that

$$H_n^{-1}\left(\frac{t}{n}\right) - H_n^{-1}\left(\frac{s}{n}\right) = \frac{1}{n}\sum_{j=s}^{t-1}(n-j)(X_{j+1:n} - X_{j:n}). \quad (5.17)$$

Hence (5.13) can be rewritten as

$$\hat{r}_n(X_{i:n}) = \min_{t \geq i+1} \max_{s \leq i} \frac{F_n(X_{t:n}) - F_n(X_{s:n})}{H_n^{-1}[F_n(X_{t:n})] - H_n^{-1}[F_n(X_{s:n})]} \quad (5.18)$$

for $1 \leq i \leq n-1$. It is seen that $1/\hat{r}_n(x)$ is the *slope* (from the right) of the concave majorant of $H_n^{-1}(y)$ evaluated at the point $y = F_n(x)$.

Consistency

In the proof of consistency of the MLE, H_n^{-1} plays an analogous role to that played by F_n in the proof of consistency of a monotone density in Section 5.2. In fact both results can be proved at once using the generalization in the next section. However, for clarity in exposition and because of the importance of both density and failure rate estimation, both have been treated separately at the expense of conciseness.

Lemma A

If $F(0) = 0$ and

$$\mu = \int_0^\infty x \, dF(x) < \infty,$$

then $H_n^{-1}(t) \to H^{-1}(t)$ uniformly in $t \in [0, 1]$ with probability one.

Proof. By the strong law

$$H_n^{-1}(1) - H_n^{-1}\left(\frac{1}{n}\right) = \frac{1}{n}\sum_{j=1}^{n} X_{j:n} - X_{1:n} \xrightarrow{\text{a.s.}} \mu - F^{-1}(0)$$

$$= H^{-1}[F^{-1}(1)] - H^{-1}[F^{-1}(0)].$$

Strong uniform convergence of H_n^{-1} to H^{-1} is now an easy consequence of the Glivenko–Cantelli theorem (i.e., $F_n(x) \to F(x)$ uniformly in x with probability one as $n \to \infty$). ∎

In proving consistency of the MLE for $r(x)$, it is convenient to consider the estimator for $1/r(x)$. The basic or "naive" estimate for $1/r(x)$ is

$$\frac{1}{r_n(x)} = \frac{[1 - F_n(X_{i:n})]}{f_n(X_{i:n})} \quad \text{for} \quad X_{:n} \leq x < X_{i+1:n} \quad (1 \leq i < n-1)$$

or

$$\frac{1}{r_n(x)} = \frac{[1-(i/n)][X_{i+1:n} - X_{i:n}]}{F_n(X_{i+1:n}) - F_n(X_{i:n})} \quad \text{for} \quad X_{i:n} \leq x < X_{i+1:n}.$$

If we let

$$g_n(i) = \frac{1}{r_n(X_{i:n})} \quad \text{and} \quad w_i = F_n(X_{i+1:n}) - F_n(X_{i:n}),$$

the cumulative sum function (see Chapter 2, Equation (2.5)) is

$$G_n(i) = \sum_{j=1}^{i} g_n(j) w_j$$

$$= \sum_{j=1}^{i} \frac{1}{n}(n-j)(X_{j+1:n} - X_{j:n})$$

$$= H_n^{-1}\left(\frac{i}{n}\right).$$

Hence the transform, H_n^{-1} can be interpreted as the cumulative sum function with respect to the maximum likelihood estimator for $1/r$.

Lemma B

If F is IFR so that

$$\int_0^\infty x \, dF(x) < \infty$$

and $\hat{H}_n^{-1}$ is the least concave majorant of H_n^{-1}, then

$$\hat{H}_n^{-1}(t) \to H^{-1}(t)$$

uniformly in $t \in [0, 1]$ with probability one.

Proof. Clearly

$$\sup_{0 \leq t \leq 1} |\hat{H}_n^{-1}(t) - H^{-1}(t)| = \sup_{1 \leq i \leq n} \left| \hat{H}_n^{-1}\left(\frac{i}{n}\right) - H^{-1}\left(\frac{i}{n}\right) \right|$$

so that

$$\sup_{0 \leq t \leq 1} |\hat{H}_n^{-1}(t) - H^{-1}(t)| \leq \sup_{0 \leq t \leq 1} |H_n^{-1}(t) - H^{-1}(t)|$$

by the lemma in Section 2.2 of Chapter 2. The result follows from this fact and the previous lemma. ∎

The strong consistency theorem can now be stated.

Theorem 5.2

If F is IFR, then for every x_0,

$$r(x_0^-) \leq \varliminf \hat{r}_n(x_0) \leq \varlimsup \hat{r}_n(x_0) \leq r(x_0^+) \qquad (5.19)$$

with probability one.

Proof. By definition, $\hat{H}_n^{-1}(t)$ is concave in t $(0 \leq t \leq 1)$. Hence, for every $\Delta > 0$

$$\left. \begin{array}{c} \dfrac{\hat{H}_n^{-1}[F_n(x_0)] - \hat{H}_n^{-1}[F_n(x_0) - \Delta]}{\Delta} \\ \\ \geq \dfrac{1}{\hat{r}_n(x_0^-)} \geq \dfrac{1}{\hat{r}_n(x_0^+)} \\ \\ \geq \dfrac{\hat{H}_n^{-1}[F_n(x_0) + \Delta] - \hat{H}_n^{-1}[F_n(x_0)]}{\Delta} \end{array} \right\} \qquad (5.20)$$

since

$$\frac{1}{\hat{r}_n(x_0^-)} \quad \text{and} \quad \frac{1}{\hat{r}_n(x_0^+)}$$

are the left and right hand derivatives, respectively, of $\hat{H}_n^{-1}(t)$ evaluated at $t = F_n(x_0)$.

Since F is IFR,

$$F(0) = 0, \qquad \int_0^\infty x \, dF(x) < \infty \quad \text{and} \quad \hat{H}_n^{-1}(t) \to H^{-1}(t)$$

uniformly in t $(0 \leq t \leq 1)$ with probability one by Lemma B. Letting $n \to \infty$ in (5.20) we see that

$$\frac{H^{-1}[F(x_0)] - H^{-1}[F(x_0) - \Delta]}{\Delta} \geq \varlimsup \frac{1}{\hat{r}_n(x_0^-)} \geq \varliminf \frac{1}{\hat{r}_n(x_0^+)}$$

$$\geq \frac{H^{-1}[F(x_0) + \Delta] - H^{-1}[F(x_0)]}{\Delta}$$

with probability one for all Δ such that $0 < \Delta < \min [F(x_0), 1 - F(x_0)]$. Letting $\Delta \to 0$,

$$\frac{1}{r(x_0^-)} \geq \varlimsup \frac{1}{\hat{r}_n(x_0^-)} \geq \varliminf \frac{1}{\hat{r}_n(x_0^+)} \geq \frac{1}{r(x_0^+)}$$

with probability one. This implies (5.19). ∎

Corollary

If r is increasing and continuous on $[a, b]$, then for $t \in [a, b]$

$$\lim_{n \to \infty} |\hat{r}_n(t) - r(t)| = 0$$

with probability one.

Proof. This follows from the same methods as in the usual proof of the Glivenko–Cantelli theorem and Theorem 5.2. ∎

Prakasa Rao (1970) has shown, under regularity conditions, that

$$\left[\frac{2nf(x)}{r'(x)r^2(x)} \right]^{\frac{1}{3}} [\hat{r}_n(x) - r(x)] \tag{5.21}$$

has density $\frac{1}{2}\psi(u/2)$ when ψ is the density of the minimum value of $W(t) + t^2$ and $W(t)$ is a two sided Wiener–Levy process with mean 0, variance 1 per unit t and $W(0) = 0$. Because of the relatively slow rate of convergence indicated by (5.21) isotonic "window" estimators are considered in the next section.

Maximum likelihood estimator for the distribution function

Since

$$1 - F(x) = \exp\left[- \int_{-\infty}^{x} r(u) \, du \right],$$

the maximum likelihood estimator for the distribution function is of course

$$\hat{F}_n(x) = 1 - \exp\left[- \int_{-\infty}^{x} \hat{r}_n(u) \, du \right].$$

Strong consistency of $\hat{F}_n$ is an almost immediate consequence of Theorem 5.2.

Theorem 5.3

If F is IFR and r is continuous, then for all x,

$$\lim_{n \to \infty} \hat{F}_n(x) = F(x)$$

with probability one.

Proof. It is sufficient to prove the theorem for x satisfying $F(x) < 1$, in which case $\hat{F}_n(x) < 1$ for sufficiently large n. By the corollary to Theorem 5.2,

$$\lim_{n \to \infty} \hat{r}_n(u) = r(u)$$

ESTIMATION FOR DISTRIBUTIONS WITH MONOTONE FAILURE RATE   241

uniformly on $[0, x]$ with probability one. For $u \in [0, x]$, $\hat{r}_n(u) < \infty$, and by the Lebesgue dominated convergence theorem,

$$\lim_{n \to \infty} \int_0^x \hat{r}_n(u)\, du = \int_0^x r(u)\, du$$

with probability one. Hence

$$\lim_{n \to \infty} [1 - \hat{F}_n(x)] = 1 - F(x)$$

with probability one. ∎

Monte Carlo experiments indicate that for small sample sizes $\hat{F}_n$ is not a good estimator for F in the tails of the distribution. In fact, the empirical distribution seems to be preferable in the tails. This is perhaps to be expected since, of course, $\hat{r}_n$ will also behave badly in the tails of the distribution.

Distributions with decreasing failure rate

Decreasing failure rate (DFR) distributions frequently arise in applications. They can, for example, arise as mixtures of exponential distributions.

Definition 5.4

A distribution F is said to be DFR if the support of F is of the form $[\alpha, \infty)$, $\alpha > -\infty$, and if $\log[1 - F(x)]$ is convex on $[\alpha, \infty)$.

Note that a DFR distribution may only have a jump at α. It is easy to check that the derivative of the absolutely continuous part must be decreasing on (α, ∞). Thus DFR is, in a sense, a strengthening of the decreasing density assumption. For convenience it is supposed that α is known. Suppose that $\alpha = X_{1:n} = \ldots = X_{k:n} < X_{k+1:n} < \ldots < X_{n:n}$. In case $k = 0$, define $X_{0:n} = \alpha$. It can be shown that the maximum likelihood estimator is

$$\hat{r}_n(\alpha) = \frac{k}{n} = \hat{F}(\alpha^+)$$

and

$$\hat{r}_n(x) = \hat{r}_n(X_{i:n}) \quad \text{for} \quad X_{i-1:n} < x \leq X_{i:n} \quad i = 1, 2, \ldots, n$$

where

$$\hat{r}_n(X_{i:n}) = \max_{t \geq i} \min_{s \leq i-1} \frac{t - s}{[(n-s)(X_{s+1:n} - X_{s:n}) + \ldots + (n-t+1)(X_{t:n} - X_{t-1:n})]}$$

and $X_{0:n} = \alpha$ in case $k = 0$.

Note that the estimator is left continuous in this case. Also, unlike the IFR case, this DFR estimator is not unique. It is determined by the likelihood equation only for $x \leq X_{n:n}$, and may be extended beyond $X_{n:n}$ in any manner that preserves the DFR property.

5.4 ISOTONIC WINDOW ESTIMATORS FOR THE GENERALIZED FAILURE RATE FUNCTION

If we let $G(x) = 1 - e^{-x}$ and $g(x) = e^{-x}$ for $x \geq 0$ and note that $G^{-1}F(x) = -\log[1 - F(x)]$, then the failure rate function can be written as

$$r(x) = \frac{d}{dx} G^{-1}F(x) = \frac{f(x)}{g[G^{-1}F(x)]} = \frac{f(x)}{1 - F(x)}. \tag{5.22}$$

If $G(x) = x$ for $0 \leq x \leq 1$, then

$$r(x) = \frac{d}{dx} G^{-1}F(x) = \frac{f(x)}{g[G^{-1}F(x)]} = f(x) \tag{5.23}$$

is the density of F on the support of F.

We call

$$r(x) = \frac{f(x)}{g[G^{-1}F(x)]} \tag{5.24}$$

the *generalized failure rate function*. If $r(x)$ is nondecreasing, then $G^{-1}F(x)$ is convex on the support of F. This motivates the notion of a partial ordering on the space of distributions called *c*-ordering. We say that $F <_c G$ or F *c*-precedes G if $G^{-1}F$ is convex on the support of F. If $F <_c G$ and G is the uniform distribution, then F has an increasing density. If $F <_c G$ and G is the exponential distribution then F is IFR. As another example, the gamma distributions are *c*-ordered with respect to the shape parameter. Let

$$f_\alpha(x) = \frac{\lambda^\alpha x^{\alpha-1} e^{-\lambda x}}{\Gamma(\alpha)} \quad \text{for} \quad x \geq 0, \quad \alpha, \lambda > 0,$$

and

$$F_\alpha(x) = \int_0^x f_\alpha(u)\, du.$$

Van Zwet (1964) showed that $\alpha_1 > \alpha_2 > 0$ implies

$$F_{\alpha_1} <_c F_{\alpha_2}.$$

Also, $F \underset{c}{<} G$ implies that $F(x)$ crosses $G(\theta x)$ at most once, and from below if at all, for all $\theta > 0$. In this sense, F has "lighter" tails than G. For example, if G were chosen to be the normal distribution then $F \underset{c}{<} G$ implies that F has lighter tails than the normal distribution.

Our objective is to consider a wide class of isotonic estimators for $r(x)$ which will include, as special cases, the maximum likelihood estimators considered in Sections 5.2 and 5.3. It will be shown that, asymptotically, the maximum likelihood estimators can be improved upon. It is *not* assumed as in Section 5.3 that $F(0) = 0$.

It will be convenient to let $\mathscr{F}$ be the class of absolutely continuous distribution functions F on $(0, \infty)$ with positive and right (or left) continuous density f on the *interval* where $0 < F < 1$. It follows that the inverse function F^{-1} is uniquely defined on $(0, 1)$. $F^{-1}(0)$ and $F^{-1}(1)$ are taken to be equal to the left and right hand endpoints of the support of F (possibly $-\infty$ or $+\infty$).

Isotonic estimators

G will be assumed specified, $F \underset{c}{<} G$, F, $G \in \mathscr{F}$ and an ordered sample $X_{1:n} \leq X_{2:n} \leq \ldots \leq X_{n:n}$ from F available. Let F_n be the empirical distribution corresponding to our sample. Let ρ_n be an initial or *basic estimator* for r based on $X_{1:n} \leq X_{2:n} \leq \ldots \leq X_{n:n}$. For each n, a *grid* is defined on $(-\infty, \infty)$, i.e., a finite or infinite sequence $\ldots < t_{n,-2} < t_{n,-1} < t_{n,0} < t_{n,1} < \ldots < t_{n,i} < \ldots$. In each *window* $[t_{n,j}, t_{n,j+1})$ a point $t_{n,j} \leq x_{n,j} < t_{n,j+1}$ is chosen and to each point $x_{n,j}$ is assigned a non-negative weight $w(x_{n,j})$. For convenience, $x_{n,j}$ is abbreviated to x_j. We call

$$r_n(x) = \min_{s \geq i+1} \max_{r \leq i} \left[\frac{\sum_{j=r}^{s-1} \rho_n(x_j) w(x_j)}{\sum_{j=r}^{s-1} w(x_j)} \right] \tag{5.25}$$

for $t_{n,i} \leq x < t_{n,i+1}$, the *monotonic regression* or more generally the *isotonic regression* of ρ_n with respect to the discrete measure w. Note that r_n is a nondecreasing step function and that $r_n = \rho_n$ whenever $\rho_n(x_j)$ happens to be nondecreasing in j.

If $X_{1:n} \leq t_{n,i} < t_{n,i+1} \leq X_{n:n}$, then $gG^{-1}F_n(x_i) > 0$ since $F, G \in \mathscr{F}$ and $r(x)$ may be estimated for $t_{n,i} \leq x < t_{n,i+1}$ by

$$\hat{\rho}_n(x) = \frac{f_n(x_i)}{gG^{-1}F_n(x_i)} = \frac{F_n(t_{n,i+1}) - F_n(t_{n,i})}{gG^{-1}F_n(x_i)(t_{n,i+1} - t_{n,i})}, \tag{5.26}$$

the *naive* estimator for r. With $\hat{\rho}_n$ for the basic estimator and weights

$$w(x_j) = gG^{-1}F_n(x_j)(t_{n,j+1} - t_{n,j}),$$

the isotonic estimator $\hat{r}_n$ is defined for

$$X_{1:n} \leq t_{n,i} \leq x < t_{n,i+1} < X_{n:n}$$

by

$$\hat{r}_n(x) = \min_{\substack{s \geq i+1 \\ t_{n,s} \leq X_{n:n}}} \max_{\substack{r \leq i \\ t_{n,r} \geq X_{1:n}}} \frac{F_n(t_{n,s}) - F_n(t_{n,r})}{\sum_{j=r}^{s-1} gG^{-1}F_n(x_j)(t_{n,j+1} - t_{n,j})}. \qquad (5.27)$$

Note that the conditions on $t_{n,i}$, $t_{n,i+1}$, $t_{n,r}$ and $t_{n,s}$ originate from the fact that $\hat{\rho}_n(x_j)$ is defined only if $X_{1:n} \leq t_{n,j} < t_{n,j+1} \leq X_{n:n}$. An interesting special case arises when we consider the random grid $t_{n,j} = X_{j:n}$, $j = 1, 2, \ldots, n$, determined by the order statistics. For this grid $x_j = X_{j:n}$, (5.27) becomes

$$\hat{r}_n(x) = \min_{i+1 \leq s \leq n} \max_{1 \leq r \leq i} \frac{s - r}{n \sum_{j=r}^{s-1} gG^{-1}(j/n)(X_{j+1:n} - X_{j:n})} \qquad (5.28)$$

for $X_{i:n} \leq x < X_{i+1:n}$, $i = 1, 2, \ldots, n - 1$. This is the MLE in the case when G is the exponential or the uniform distribution. Note that if G is the exponential distribution

$$ngG^{-1}\left(\frac{j}{n}\right)(X_{j+1:n} - X_{j:n}) = (n - j)(X_{j+1:n} - X_{j:n})$$

and hence the weights $w(x_j)$ in (5.25) are proportional to the total time on test between $X_{j:n}$ and $X_{j+1:n}$. Since the total time on test for an interval is a measure of our information over the interval, this choice of weights is intuitively appealing in this case. We shall call

$$gG^{-1}\left(\frac{j}{n}\right)(X_{j+1:n} - X_{j:n})$$

the *total time on test weights* for general G also.

An alternative estimator r_n^* is obtained using the same basic estimator $\rho_n^* = \hat{\rho}_n$ with weights $w(x_j) = t_{n,j+1} - t_{n,j}$, thus

$$r_n^*(x) = \min_{\substack{s \geq i+1 \\ t_{n,s} \leq X_{n:n}}} \max_{\substack{r \leq i \\ t_{n,r} \geq X_{1:n}}} \frac{\sum_{j=r}^{s-1} \dfrac{F_n(t_{n,j+1}) - F_n(t_{n,j})}{gG^{-1}F_n(x_j)}}{t_{n,s} - t_{n,r}} \qquad (5.29)$$

for $X_{1:n} \leq t_{n,i} \leq x < t_{n,i+1} \leq X_{n:n}$.

If the second member of

$$r(x) = \frac{d}{dx} G^{-1}F(x) = f(x)/gG^{-1}F(x) \qquad (5.30)$$

is considered rather than the third one, a logical choice for the basic estimator is the *graphical* estimator

$$\tilde{r}_n(x) = \min_{\substack{s \geq i+1 \\ t_{n,s} \leq X_{n:n}}} \max_{\substack{r \leq i \\ t_{n,r} \geq X_{1:n}}} \frac{G^{-1}F_n(t_{n,s}) - G^{-1}F_n(t_{n,r})}{t_{n,s} - t_{n,r}} \qquad (5.31)$$

for $X_{1:n} \leq t_{n,i} \leq x < t_{n,i+1} \leq X_{n:n}$. Notice that if $G^{-1}(1) = \infty$ and $t_{n,i+1} = X_{n:n}$, then $\tilde{\rho}_n(x) = \tilde{r}_n(x) = \infty$ for $t_{n,i} \leq x < t_{n,i+1}$. When G is the uniform distribution, $\hat{r}_n = r_n^* = \tilde{r}_n$.

Attention will be concentrated on the estimator $\hat{r}_n$ because of its intuitive appeal and because it generalizes the maximum likelihood estimator. We shall attempt to improve upon the MLE through appropriate choice of the grid.

Consistency of nonrandom window estimators

For the grid consisting of order statistics, we note that the basic estimator $\hat{\rho}_n$ is *not* consistent even though the isotonized estimator is consistent. The grid based on order statistics will be called a *narrow* grid. By choosing a fixed grid $\{t_{n,j}\}_{j=1}^{\infty}$ with sufficiently wide windows the MLEs considered in Sections 5.2 and 5.3 can be improved on asymptotically.

In practical situations, the observed data is often heavily censored. In many life testing situations only the number of failures within specified intervals are recorded. Hence, it is important to consider estimators based on grids more general than that provided by order statistics.

For general, but nonrandom, grids $\{t_{n,j}\}$ we define

$$F_n^*(x) = F_n(x_{n,j}) \quad \text{for} \quad t_{n,j} \leq x < t_{n,j+1}, \qquad (5.32)$$

and

$$K_{F_{n,\xi}^*}(x) = \int_{\xi}^{x} gG^{-1}F_n^*(u)\,du, \qquad (5.33)$$

$$t_{n,1} \leq x \leq t_{n,\infty}, \qquad F^{-1}(0) \vee t_{n,1} < \xi < t_{n,\infty} \wedge F^{-1}(1),$$

where F_n is the usual empirical distribution. Strong consistency of $\hat{r}_n$ will follow from the strong uniform consistency of $K_{F_{n,\xi}^*}$. In order to prove uniform consistency of $K_{F_{n,\xi}^*}$ it is necessary to make some additional assumptions concerning the grid. We shall say that $\{t_{n,j}\}$ *becomes dense* in an interval I if, for any pair $x_1 < x_2$ in I, there exists an integer N such that for $n \geq N$ a gridpoint $t_{n,j}$ exists with $x_1 < t_{n,j} < x_2$. It will be said that $\{t_{n,j}\}$ *becomes subexponential in the right tail* of F, if numbers N, $z > 0$ and $c > 1$ exist such that for $n \geq N$, $z \leq t_{n,j} < t_{n,j+1} < F^{-1}(1)$ implies $t_{n,j+1} \leq ct_{n,j}$. Note that if the support of F is bounded on the right, then any grid trivially becomes subexponential in the right tail of F and z may be chosen arbitrarily large. In the following lemma, a property of grids will be proved that will be needed in the sequel. Let U_n and V_n denote the infimum and the supremum of the gridpoints $t_{n,j}$ in the interval $[X_{1:n}, X_{n:n}]$.

Lemma

Let $\{t_{n,j}\}$ become dense in the support of F and subexponential in the right tail of F. Then numbers N, $z_0 \geq 0$ and $c > 1$ exist such that for $n \geq N$ and $z_0 \leq x < V_n$, $F_n^*(x) \geq F_n(x/c)$.

Proof. If $F^{-1}(1) < \infty$, the lemma is trivially true for $z_0 \geq F^{-1}(1)$ ($\geq V_n$). Assume $F^{-1}(1) = \infty$. By definition, N^1, $z > 0$ and $c > 1$ exist such that $t_{n,j+1} > t_{n,j} \geq z$ and $n \geq N^1$ imply $t_{n,j+1} \leq ct_{n,j}$. Choose $z_0 > z$ with $z_0 > F^{-1}(0)$. Since $\{t_{n,j}\}$ becomes dense in the support of F, an integer $N \geq N^1$ exists such that for $n \geq N$ a gridpoint between z and z_0 exists. Together with the definition of V_n, this implies that for $n \geq N$ and for every $x \in [z_0, V_n)$ gridpoints $t_{n,j}$ and $t_{n,j+1}$ exist with $0 < z \leq t_{n,j} \leq x < t_{n,j+1}$, and hence

$$F_n^*(x) = F_n(x_j) \geq F_n(t_{n,j+1}/c) \geq F_n(x/c). \blacksquare$$

We will make use of the transform

$$K_{F,\xi}(x) = \int_\xi^x gG^{-1}F(u)\,du.$$

Theorem 5.4

Let $F, G \in \mathscr{F}$, let

$$EX^+ = \int_0^\infty x\,dF(x) < \infty$$

and let gG^{-1} be uniformly continuous on $(0, 1)$. Assume that either $F^{-1}(1) < \infty$ or $gG^{-1}(y)/(1 - y)$ is bounded on $(0, 1)$. Then, for any fixed $\xi \in (F^{-1}(0), F^{-1}(1))$, $K_{F,\xi}(F^{-1}(1)) < \infty$. If moreover, the grid $\{t_{j,n}\}$ becomes dense in the support of F and subexponential in the right tail of F, then for $n \to \infty$

$$\sup_{\xi \leq x \leq V_n} |K_{F_n^*,\xi}(x) - K_{F,\xi}(x)| \to 0$$

almost surely.

Proof. Throughout the proofs in this section, the subscript ξ is omitted. If $F^{-1}(1) < \infty$, then $K_{F,\xi}(F^{-1}(1))$ is an integral of a bounded function over a finite interval and, therefore, finite. If $gG^{-1}(y) \leq a(1-y)$ on $(0,1)$, then

$$K_{F,\xi}(F^{-1}(1)) = \int_{\xi}^{F^{-1}(1)} gG^{-1}F(x)\,dx \leq a \int_{\xi}^{F^{-1}(1)} (1 - F(x))\,dx$$

$$= aE(X - \xi)^+ < \infty.$$

Note that U_n and V_n tend a.s. to $F^{-1}(0)$ and $F^{-1}(1)$ since we assumed that the grid becomes dense in the support of F. Hence, $K_{F_n^*,\xi}(x)$ is a.s. defined for any fixed $x \in [\xi, F^{-1}(1))$ for sufficiently large n. Since both $K_{F,\xi}$ and $K_{F_n^*,\xi}$ are nondecreasing, $K_{F,\xi}(\xi) = 0$ and $K_{F,\xi}(F^{-1}(1)) < \infty$, it is obviously sufficient to show that $K_{F_n^*,\xi}$ converges pointwise to $K_{F,\xi}$ on $[\xi, F^{-1}(1))$ and that, for any $\varepsilon > 0$, $x_0 \in (\xi, F^{-1}(1))$ can be chosen in such a way that

$$\limsup_n \int_{x_0}^{V_n} gG^{-1}F_n^*(x)\,dx < \varepsilon \tag{5.34}$$

almost surely. (The method of proof of the Glivenko–Cantelli theorem can be used to complete the argument.) The assumption that the grid becomes dense guarantees pointwise a.s. convergence of F_n^* to F on the interval where $0 < F < 1$, even though F_n^* may be defective for all n (e.g., if no gridpoint $t_{j,n} > X_n$ exists for any n). As F is a continuous probability distribution, F_n^* is nondecreasing and $0 \leq F_n^* \leq 1$, a standard argument shows that the pointwise a.s. convergence implies that

$$\sup_x |F_n^*(x) - F(x)| \xrightarrow{\text{a.s.}} 0.$$

Hence,

$$\int_{\xi}^{x} [gG^{-1}F_n^*(u) - gG^{-1}F(u)]\,du \to 0$$

almost surely by the uniform continuity of gG^{-1} on $(0,1)$.

It remains to prove (5.34), viz., for any $\varepsilon > 0$, $x_0 \in (\xi, F^{-1}(1))$ exists such that

$$\limsup_n \int_{x_0}^{V_n} gG^{-1}F_n^*(x)\,dx < \varepsilon \tag{5.35}$$

almost surely. It is only necessary to consider the case when $F^{-1}(1) = \infty$ and $gG^{-1}(y) \leq a(1-y)$ on $(0,1)$. By the above lemma, N, $z_0 > 0$ and $c > 1$ exist such that $F_n^*(x) \geq F_n(x/c)$ for $n \geq N$ and $x \in [z_0, V_n)$. Choose $x_0 > z_0$ in such a way that

$$\int_{x_0/c}^{\infty} [1 - F(x)]\,dx < \frac{\varepsilon}{ac}.$$

then

$$\limsup_n \int_{x_0}^{V_n} gG^{-1}F_n^*(x)\,dx \leq \limsup_n a \int_{x_0}^{V_n} [1 - F_n^*(x)]\,dx$$

$$\leq a \limsup_n \int_{x_0}^{V_n} \left[1 - F_n\left(\frac{x}{c}\right)\right] dx$$

$$\leq ac \limsup_n \int_{x_0/c}^{\infty} [1 - F_n(x)]\,dx$$

$$= ac \int_{x_0/c}^{\infty} [1 - F(x)]\,dx < \varepsilon,$$

by the strong law. ∎

The result for the left tail that we will need in our strong consistency Theorem 5.6 is very much weaker. Consider the special case $F = G$. Note that for $x < \xi$, $K_{G,\xi}$, $K_{G_n,\xi}$ and $K_{G_n^*,\xi}$ are negative and that $K_{G,\xi}(x) = -[G(\xi) - G(x)]$.

Theorem 5.5

Let $G \in \mathscr{F}$ and let gG^{-1} be uniformly continuous on $(0, 1)$. If, moreover, the grid $\{t_{n,j}\}$ becomes dense in the support of G, then for any fixed $\xi \in (G^{-1}(0), G^{-1}(1))$

$$\liminf_n \left[\inf_{U_n \leq x \leq \xi} \left(-K_{G_n^*,\xi}^*(x) + K_{G,\xi}(x)\right)\right] \geq 0$$

almost surely.

Proof. Let $x_0 < \xi$ with $0 < G(x_0) < \varepsilon$. Then for all $U_n \leq x \leq \xi$

$$-K_{G_n^*}(x) + K_G(x) = \int_x^\xi [gG^{-1}G_n^*(u) - gG^{-1}G(u)]\,du$$

$$\geq -\int_{x_0}^\xi |gG^{-1}G_n^*(u) - gG^{-1}G(u)|\,du - G(x_0)$$

$$\xrightarrow{\text{a.s.}} -G(x_0) \geq -\varepsilon$$

since

$$\sup_x |G_n^*(x) - G(x)| \xrightarrow{\text{a.s.}} 0,$$

and by the uniform continuity of gG^{-1} on $(0, 1)$ and the fact that $[x, \xi]$ is a.s. contained in the interval $[U_n, V_n]$, for sufficiently large n. Since $\varepsilon > 0$ may be taken arbitrarily small, the result for $K_{G_n^*,\xi}^*$ follows. ∎

From the definition of F_n^* (5.32) and from (5.27) we see that $\hat{r}_n(x)$ can be rewritten as

$$\hat{r}_n(x) = \min_{\substack{s \geq i+1 \\ t_{n,s} \leq X_{n:n}}} \max_{\substack{r \leq i \\ t_{n,r} \geq X_{1:n}}} \frac{F_n(t_{n,s}) - F_n(t_{n,r})}{K_{F_n^*}(t_{n,s}) - K_{F_n^*}(t_{n,r})} \qquad (5.36)$$

for $X_{1:n} \leq t_{n,i} \leq x < t_{n,i+1} < X_{n:n}$. We can now prove the general strong consistency for isotonic estimators of the generalized failure rate function.

Theorem 5.6

For $F, G \in \mathcal{F}$, let $F \underset{c}{<} G$, let

$$\int_0^\infty x \, dG(x) < \infty$$

and let gG^{-1} be uniformly continuous on $(0, 1)$. Assume that either $G^{-1}(1) < \infty$ or $gG^{-1}(y)/(1-y)$ is bounded on $(0, 1)$. Let $\{t_{n,j}\}$ be a nonrandom grid that becomes dense in $(-\infty, \infty)$ and subexponential in the right tail of G. Then for $\hat{r}_n$ defined by (5.36), for every fixed x_0 with $0 < F(x_0) < 1$,

$$r(x_0^-) \leq \varliminf \hat{r}_n(x_0) \leq \varlimsup \hat{r}_n(x_0) \leq r(x_0^+)$$

with probability one.

Proof. Let ξ be an arbitrary point satisfying $F^{-1}(0) < \xi < x_0$ and let $t_{n,a_n} \geq \xi$ be a sequence of gridpoints converging to ξ for $n \to \infty$. With probability 1, there exists an integer N such that for $n \geq N$, $\xi \leq t_{n,a_n} \leq x_0 < V_n$, and hence

$$\hat{r}_n(x_0) = \inf_{x_0 \leq t_{n,s} \leq X_{n:n}} \sup_{X_{1:n} \leq t_{n,r} \leq x_0} \frac{F_n(t_{n,s}) - F_n(t_{n,r})}{\sum_{j=r}^{s-1} gG^{-1}F_n(x_j)(t_{n,j+1} - t_{n,j})}$$

$$\geq \inf_{x_0 \leq x \leq V_n} \frac{F_n(x) - F_n(t_{n,a_n})}{K_{F_n,\xi}^*(x) - K_{F_n,\xi}^*(t_{n,a_n})}.$$

Since $F \underset{c}{<} G$, it is easily verified that

$$\int_0^\infty x \, dG(x) < \infty$$

implies that

$$\int_0^\infty x \, dF(x) < \infty.$$

Also $F^{-1}(1) < \infty$ if $G^{-1}(1) < \infty$, and it follows that if the grid becomes subexponential in the right tail of G, then it will also become subexponential in the right tail of F. Hence, the conditions on Theorem 5.4 are satisfied and as a result $K_{F_n^*,\xi}$ converges a.s. uniformly on $[\xi, V_n]$ to $K_{F,\xi}$. Since also F_n converges a.s. uniformly to F by the Glivenko–Cantelli theorem, $t_{n,a_n} \to \xi$ and $x_0 - \xi > 0$,

$$\varliminf \hat{r}_n(x_0) \geq \inf_{x_0 \leq x < F^{-1}(1)} \frac{F(x) - F(\xi)}{K_{F,\xi}(x)} \geq r(\xi)$$

almost surely, where the second inequality follows from

$$K_{F,\xi}(x) = \int_\xi^x gG^{-1}F(u)\,du = \int_\xi^x \frac{f(u)}{r(u)}\,du \leq \frac{1}{r(\xi)}[F(x) - F(\xi)]$$

for $x > \xi$ since r is nondecreasing. Since $\xi \in (F^{-1}(0), x_0)$ was arbitrary, this proves the left hand inequality of the theorem.

To prove the right hand inequality, let ξ be an arbitrary point satisfying $x_0 < \xi < F^{-1}(1)$. Let $t_{n,b_n} \leq \xi$ be a sequence of gridpoints converging to ξ for $n \to \infty$. With probability one, there exists an integer N such that for $n \geq N$, $V_n < x_0 < t_{n,b_n} \leq \xi$, and hence

$$\hat{r}_n(x_0) \leq \sup_{U_n \leq t_{n,r} \leq x_0} \frac{F_n(t_{n,b_n}) - F_n(t_{n,r})}{\sum_{j=r}^{b_n-1} gG^{-1}F_n(x_j)(t_{n,j+1} - t_{n,j})}.$$

Consider the random variables $Y_{1:n} < Y_{2:n} < \ldots$ defined by $Y_{i:n} = G^{-1}F(X_{i:n})$. These random variables are independent with distribution function G. The empirical distribution corresponding to $Y_{1:n} < Y_{2:n} < \ldots < Y_{n:n}$ is $G_n = F_n F^{-1} G$. Define $\tilde{x}_0 = G^{-1}F(x_0)$, $\tilde{\xi} = G^{-1}F(\xi)$, $\tilde{t}_{n,j} = G^{-1}F(t_{n,j})$, $\tilde{x}_j = G^{-1}F(x_j)$ and $\tilde{U}_n = G^{-1}F(U_n)$. Note that $0 < G(\tilde{x}_0) < G(\tilde{\xi}) < 1$ and that $\tilde{U}_n$ is the infimum of the transformed gridpoints $\tilde{t}_{n,j}$ in the interval $[Y_{1:n}, Y_{n:n}]$. Also $F_n(t_{n,j}) = G_n(\tilde{t}_{n,j})$, $F_n(x_j) = G_n(\tilde{x}_j)$, and for $t_{n,j+1} \leq \xi$ we have

$$\tilde{t}_{n,j+1} - \tilde{t}_{n,j} = \int_{t_{n,j}}^{t_{n,j+1}} r(x)\,dx \leq r(\xi)(t_{n,j+1} - t_{n,j}).$$

$$\hat{r}_n(x_0) \leq r(\xi) \sup_{\tilde{U}_n \leq \tilde{t}_{n,t} \leq \tilde{x}_0} \frac{G_n(\tilde{t}_{n,b_n}) - G_n(\tilde{t}_{n,r})}{\sum_{j=r}^{b_n-1} gG^{-1}G_n(\tilde{x}_j)(\tilde{t}_{n,j+1} - \tilde{t}_{n,j})}$$

$$\leq r(\xi) \sup_{\tilde{U}_n \leq x \leq \tilde{x}_0} \frac{G_n(\tilde{t}_{n,b_n}) - G_n(x)}{K_{G_n^*,\xi}(\tilde{t}_{n,b_n}) - K_{G_n^*,\xi}(x)},$$

where $G_n^*(x) = G_n(\tilde{x}_j)$ for $\tilde{t}_{n,j} \leq x < \tilde{t}_{n,j+1}$. Thus, the problem has been transformed to the case when $F = G$. If $\{t_{n,j}\}$ becomes dense in $(-\infty, \infty)$, then $\{\tilde{t}_{n,j}\}$ becomes dense in the support of G. If $t_{n,j} = X_{j:n}$, then $\tilde{t}_{n,j} = Y_{j:n}$. Hence, Theorem 5.5, the Glivenko–Cantelli theorem, the convergence of $\tilde{t}_{n,b_n}$ to ξ and the fact that $\xi - x_0 > 0$, yield

$$\limsup_n \hat{r}_n(x_0) \leq r(\xi) \sup_{G^{-1}(0) < x \leq \tilde{x}_0} \frac{G(\xi) - G(x)}{K_{G,\xi}(x)} = r(\xi)$$

almost surely, since $K_{G,\xi}(x) = G(\xi) - G(x)$. Since $\xi \in (x_0, F^{-1}(1))$ was arbitrary, the proof of the theorem is completed. ∎

Asymptotic distributions

In order to make asymptotic comparisons, grids with spacings of the type $t_{n,i+1} - t_{n,i} = cn^{-\alpha}$ for $c > 0$ and $0 < \alpha < 1$ are considered. Although the isotonic estimator is defined for all $x \in [X_{1:n}, X_{n:n}]$, it will be mathematically convenient to consider the asymptotic distribution of $\hat{r}_n(x)$ for fixed x only. Also, for mathematical convenience, it will be assumed that $x = (t_{n,i+1} + t_{n,i})/2$; i.e., that x is the (fixed) midpoint of a grid spacing.

It can be shown that the "naive" estimator, $\hat{\rho}_n(x)$, given by (5.26) is asymptotically normal with mean

$$r(x) + \frac{r(x)f''(x)}{24f(x)} c^2 n^{-2\alpha} + O(n^{-3\alpha}) \tag{5.37}$$

and variance

$$\frac{r^2(x)}{f(x)} n^{\alpha - 1} + O(n^{-1}). \tag{5.38}$$

Hence, the mean square error (MSE) of $\hat{\rho}_n(x)$ is approximately

$$\text{MSE}[\hat{\rho}_n(x)] \approx \frac{r^2(x) n^{\alpha - 1}}{cf(x)} + \left[\frac{c^2 r(x) f''(x)}{24 f(x)}\right]^2 n^{-4\alpha}. \tag{5.39}$$

For $\frac{1}{5} < \alpha < 1$

$$\frac{[cf(x)]^{\frac{1}{2}}}{r(x)} n^{(1-\alpha)/2} [r_n(x) - r(x)] \tag{5.40}$$

has asymptotically a $N(0, 1)$ distribution, while for $\frac{1}{7} < \alpha < \frac{1}{5}$

$$\frac{[cf(x)]^{\frac{1}{2}}}{r(x)} n^{(1-\alpha)/2} \left[r_n(x) - r(x) - \frac{c^2 n^{-2\alpha} r(x)}{24} f''(x)\right] \tag{5.41}$$

is asymptotically $N(0, 1)$. These results follow from the asymptotic normality of related multinomial random variables (cf. Barlow and van Zwet (1969)).

It will be shown that in the "wide window" case (i.e., grid spacings like $cn^{-\alpha}$ where $0 < \alpha < \frac{1}{3}$) $\hat{\rho}_n(x)$ and $\hat{r}_n(x)$ are asymptotically equivalent. Hence $\hat{r}_n(x)$ properly normalized will have, asymptotically, a normal distribution. The asymptotic distribution of $\hat{r}_n(x)$ in the narrow window case (i.e., $\frac{1}{3} \leq \alpha \leq 1$) is vastly more complicated. It will be argued later, in fact, that we should choose $\alpha = \frac{1}{5}$ in order to minimize mean square error and also to ensure obtaining an asymptotically normal distribution.

Theorem 5.7

If $F, G \in \mathscr{F}$ and

(i) $r(x) = \dfrac{f(x)}{g[G^{-1}F(x)]}$ is nondecreasing in $x \geq 0$;

(ii) r is continuously differentiable and f'' exists in a neighbourhood of x;

(iii) $r'(x) > 0$;

(iv) $t_{n,i+1} - t_{n,i} = cn^{-\alpha}$ and $0 < \alpha < \frac{1}{3}$;

then

$$\lim_{n \to \infty} \Pr\left[\hat{r}_n(x) = r_n(x)\right] = 1.$$

Only the heuristic basis for the proof will be given; the reader is referred to Barlow and van Zwet (1969) for a rigorous proof.

The mean square error in estimating $r(x)$, using the naive estimator $\hat{\rho}_n(x)$, is

$$\text{MSE}\left[\hat{\rho}_n(x)\right] = \text{Var}\left[\hat{\rho}_n(x)\right] + b^2[\hat{\rho}_n(x)] = O(n^{\alpha-1}) + O(n^{-4\alpha}),$$

where $b[\hat{\rho}_n(x)]$ is the bias term. If $(\text{MSE }[\hat{\rho}_n(x)])^{\frac{1}{2}}$ is asymptotically smaller than the window size (which is $O(n^{-\alpha})$), then it is intuitively clear that asymptotically $\hat{\rho}_n(t_{n,i}) < \hat{\rho}_n(t_{n,i+1})$ since $|r(t_{n,i+1}) - r(t_{n,i})| = O(n^{-\alpha})$ (recall that r has a positive first derivative in a neighbourhood of x). Hence, asymptotically, we do not isotonize and $\hat{\rho}_n(x)$ and $\hat{r}_n(x)$ will have the same asymptotic distribution in this case.

Clearly, $(\text{MSE }[\hat{\rho}_n(x)])^{\frac{1}{2}} = O(n^{-\alpha})$ if

$$\sigma[\hat{\rho}_n(x)] = (\text{Var }[\hat{\rho}_n(x)])^{\frac{1}{2}} = O(n^{(\alpha-1)/2}) = O(n^{-\alpha});$$

i.e., if $(\alpha - 1)/2 < -\alpha$ or $\alpha < \frac{1}{3}$. The rigorous proof is, of course, more complicated but this argument is at the heart of the proof. ∎

Theorem 5.8

Assume the conditions of Theorem 5.7. If $\frac{1}{5} < \alpha < \frac{1}{3}$, then

$$\frac{[cf(x)]^{\frac{1}{2}}}{r(x)} n^{(1-\alpha)/2} \{\hat{r}_n(x) - r(x)\}$$

is asymptotically $N(0, 1)$.
For $\frac{1}{7} < \alpha < \frac{1}{5}$,

$$\frac{[cf(x)]^{\frac{1}{2}}}{r(x)} n^{(1-\alpha)/2} \left\{ \hat{r}_n(x) - r(x) - \frac{c^2 n^{-2\alpha}}{24} \frac{f''(x) r(x)}{f(x)} \right\}$$

is asymptotically $N(0, 1)$.

The proof follows immediately from (5.37), (5.38) and Theorem 5.7.

The situation in the narrow window case (i.e., $\frac{1}{3} \leq \alpha \leq 1$) is considerably more complicated. The asymptotic distribution of the MLE in the case of monotone densities and monotone failure rate functions was obtained by Prakasa Rao (1970). His result also holds in the generalized failure rate case for isotonic window estimators when the grid spacings satisfy $t_{n,i+1} - t_{n,i} = cn^{-\alpha}$ and $\frac{1}{3} < \alpha \leq 1$.

Theorem 5.9

If $F, G \in \mathcal{F}$ and

(i) $r(x) = \dfrac{f(x)}{g[G^{-1}F(x)]}$ *is nondecreasing in $x \geq 0$;*

(ii) *r is continuously differentiable and f'' exists in a neighbourhood of x;*

(iii) $r'(x) > 0$;

(iv) $t_{n,i+1} - t_{n,i} = cn^{-\alpha}$, $\frac{1}{3} < \alpha \leq 1$;

(v) $\hat{r}_n(x) = \min\limits_{s \geq i+1} \max\limits_{r \leq i} \dfrac{F_n(t_{n,s}) - F_n(t_{n,r})}{\sum_{j=r}^{s} gG^{-1}F_n(x_j)(t_{n,j+1} - t_{n,j})}$

where $t_{n,i} \leq x < t_{n,i+1}$, then the asymptotic distribution o

$$\left[\frac{2nf(x)}{r'(x) r^2(x)} \right]^{\frac{1}{3}} [\hat{r}_n(x) - r(x)]$$

has density $\frac{1}{2}\psi(u/2)$ where ψ is the density of the minimum value of $W(t) + t^2$

and $W(t)$ is a two-sided Wiener–Levy process with mean 0 and variance 1 per unit t and $W(0) = 0$.

A proof may be found in Barlow and van Zwet (1969).

Recommendations for the window size

If one is willing to assume sufficient smoothness in the generalized failure rate function, say the existence of a second derivative, then recommendations on window size can be made in terms of the mean square error. For $\alpha < \frac{1}{3}$,

$$\text{MSE}\,[\hat{r}_n(x)] \approx \frac{r^2(x)n^{\alpha-1}}{cf(x)} + \left[\frac{r(x)f''(x)}{24}\right]^2 c^4 n^{-4\alpha}$$

$$= \frac{A}{nh} + Bh^4$$

where $h = n^{-\alpha}$. The minimum as a function of h is achieved for $h = n^{-1/5}$. Hence, $\alpha = \frac{1}{5}$ should be chosen. However, this is not helpful since the optimum choice of c in the window size still depends on the true generalized failure rate function which is, of course, unknown. The MLE takes care of this problem by choosing $t_{i,n} = X_{i:n}$, the ith order statistic from a sample of size n. The MLE might be modified by choosing $t_{i,n} = X_{[in^\beta]:n}$ where [ ] denotes the greatest integer in the quantity within the brackets and $\beta = 1 - \alpha$. Hence,

$$X_{[(i+1)n]:n} - X_{[in]:n} = O_p\left(\frac{n^\beta}{n}\right) = O_p(n^{-\alpha})$$

where O_p means "big O" in probability, and by choosing $\alpha = \frac{1}{5}$ the recommended requirement is realized.

5.5 ISOTONIC ESTIMATORS FOR STAR-ORDERED FAMILIES OF DISTRIBUTIONS

In the previous section we discussed estimators for distributions convex ordered with respect to a specified distribution, G. Weakening the convexity condition $g[\lambda x_0 + (1 - \lambda)x_1] \leq \lambda g(x_0) + (1 - \lambda)g(x_1)$, for $0 \leq \lambda \leq 1$, a function g defined on the interval $I \subset [0, \infty)$ is called *starshaped* on I if $g(\lambda x) \leq \lambda g(x)$ whenever $x \in I$, $\lambda x \in I$ and $0 \leq \lambda \leq 1$. If $I = (0, \infty)$, then the graph of g initially lies on or below any straight line through the origin, and then lies on or above it. On the class of distributions $\mathscr{F}$ with $F(0) = 0$, the following partial ordering is defined.

Definition 5.5

$F \underset{*}{\leq} G$ (F star-precedes G) if $G^{-1}F(x)$ is starshaped on $\{x|0 < F(x) < 1\}$, where $G^{-1}(u) = \inf\{x|G(x) \geq u\}$.

Note that if $G^{-1}F$ is starshaped on $\{x|0 < F(x) < 1\}$, then $G^{-1}F(x)/x$ is nondecreasing on $\{x|0 < F(x) < 1\}$. If $G(x) = 1 - e^{-x}$ for $x \geq 0$, then $F \underset{*}{\leq} G$ implies that F is IFRA (increasing failure rate average); i.e., if F has density f and failure rate r, then

$$\frac{-\log[1 - F(x)]}{x} = \int_0^x r(u)\, du/x$$

is nondecreasing in $x \geq 0$. This family of probability distributions is important in reliability theory because of its connection with coherent structures. Such structures include parallel and series structures and all combinations thereof. The IFRA class is the smallest class of failure distributions containing the exponential distributions which is closed under the formation of coherent structures and limits in distribution.

In Sections 5.2 and 5.3, maximum likelihood estimators were obtained which turned out to be isotonic regressions of certain basic estimators. In Section 5.4, more general isotonic estimators motivated by the maximum likelihood estimators were considered. Maximum likelihood estimators can also be computed for IFRA distributions and distributions, F, such that $F \underset{*}{\leq} G$ where $G(x) = x$ for $0 \leq x \leq 1$. The maximum likelihood estimators in these cases are *not* isotonic estimators. This is perhaps not particularly surprising. The surprising thing is that the maximum likelihood estimators are *not* consistent. They converge but to the wrong distributions! Isotonic regression, however, comes to the rescue with consistent estimators which also have other desirable properties.

Example 5.1 **The MLE *for starshaped distributions***

Let $0 \leq X_{1:n} \leq X_{2:n} \leq \ldots \leq X_{n:n}$ be an ordered sample from a distribution F which is starshaped on an interval $[0, b)$; i.e., $F(0) = 0$ and $F(x)/x$ is nondecreasing on $[0, b)$. For example, distributions with increasing densities on $[0, b)$ would have this property. Since such distributions can have a countable number of jumps, the maximum likelihood estimator jumps at each of the order statistics. This can be seen using the generalized definition of maximum likelihood due to Wald and Wolfowitz (cf. Barlow (1968a)).

However, the starshaped property precludes the MLE from being flat between order statistics. The likelihood can be written as

$$L(X_{1:n}, X_{2:n}, \ldots, X_{n:n}|F) = [F(X_{1:n}) - F(X_{1:n}^-)][F(X_{2:n}) - F(X_{2:n}^-)]$$

$$\ldots [F(X_{n:n}) - F(X_{n:n}^-)]$$

$$= F(X_{1:n})\left[F(X_{2:n}) - \frac{X_{2:n}}{X_{1:n}} F(X_{1:n})\right]$$

$$\ldots \left[F(X_{n:n}) - \frac{X_{n:n}}{X_{n-1:n}} F(X_{n-1:n})\right]. \quad (5.42)$$

The second expression for the likelihood follows from the fact that F must be starshaped and also that the likelihood is maximized by placing the maximum amount of probability at the order statistics.

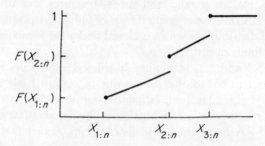

Figure 5.3   MLE for a starshaped distribution, F, given F at the order statistics

Figure 5.3 illustrates the MLE when $n = 3$. The maximization problem reduces to estimating F at the order statistics subject to the starshaped restriction.

Let

$$\lambda_i \stackrel{\text{def}}{=} \lambda(X_{i:n}) \stackrel{\text{def}}{=} F(X_{i:n})/X_{i:n} \quad \text{and} \quad \Delta_i = \lambda_i - \lambda_{i-1}$$

where $\lambda_0 \equiv 0$ so that

$$L(X_{1:n}, X_{2:n}, \ldots, X_{n:n}|F) = X_{1:n}\Delta_1 X_{2:n}\Delta_2 X_{3:n}\Delta_3 \ldots X_{n:n}\Delta_n$$

$$= \left[\prod_{i=1}^n X_{i:n} \prod_{i=1}^n \Delta_i\right].$$

Clearly, λ_i nondecreasing in i implies $\Delta_i \geq 0$. Also $\Delta_1 + \Delta_2 + \ldots + \Delta_n = F(X_{n:n})/X_{n:n} \leq 1/X_{n:n}$. To maximize the likelihood we clearly want

$\hat{F}_n(X_{n:n}) = 1$ where $\hat{F}_n$ is the MLE. Hence the maximum likelihood estimation problem reduces to:

$$\text{Maximize} \quad \prod_{i=1}^{n} \Delta_i$$
$$\text{subject to} \quad \Delta_i \geq 0 \quad i = 1, 2, \ldots, n$$

and

$$\Delta_1 + \Delta_2 + \ldots + \Delta_n = \frac{1}{X_{n:n}}.$$

Since the geometric mean is no greater than the arithmetic mean, with equality if and only if all terms are equal, it follows that

$$\Delta_1^* = \Delta_2^* = \ldots = \Delta_n^* = 1/(nX_{n:n})$$

solves the problem. Since

$$\Delta_i^* = \frac{\hat{F}_n(X_{i:n})}{X_{i:n}} - \frac{\hat{F}_n(X_{i-1:n})}{X_{i-1:n}}$$

we see that

$$\hat{F}_n(x) = \begin{cases} 0 & \text{for} \quad x \leq X_{1:n} \\ \frac{ix}{nX_{n:n}} & \text{for} \quad X_{i:n} \leq x < X_{i+1:n} \quad (1 \leq i \leq n-1) \\ 1 & \text{for} \quad x \geq X_{n:n} \end{cases} \quad (5.43)$$

is the maximum likelihood estimator for F assuming F starshaped on $\{x | 0 < F(x) < 1\}$.

Inconsistency

Suppose we have n ordered observations from the uniform distribution on $[0, 1]$. Clearly $F(x)/x = x/x = 1$ is nondecreasing on $[0, 1]$. We wish to show that in this case, $\hat{F}_n(x) \to x^2$ for $0 \leq x \leq 1$ with probability one as $n \to \infty$. Fix x and let r be defined by

$$X_{r:n} \leq x < X_{r+1:n}.$$

Note that

$$\frac{rX_{r:n}}{nX_{n:n}} = \hat{F}_n(X_{r:n}) \leq \hat{F}_n(x) \leq \hat{F}_n(X_{r+1:n}) = \frac{(r+1)X_{r+1:n}}{nX_{n:n}}.$$

Let $Y_1, Y_2, \ldots, Y_n, Y_{n+1}, \ldots$ be independent, identically distributed exponential random variables. Since $X_{r:n}$ is the rth order statistic from a uniform distribution on $[0, 1]$, it follows that

$$X_{r:n} \stackrel{\text{st}}{=} \frac{Y_1 + Y_2 + \ldots + Y_r}{Y_1 + Y_2 + \ldots + Y_{n+1}} \qquad 1 \leq r \leq n$$

where $\stackrel{\text{st}}{=}$ means stochastically equal to (see also Section 4.3). Let $n \to \infty$ and $r/n \to x$ so that $X_{r:n} \to x$ with probability one by the strong law of large numbers. Since $X_{n:n} \to 1$ with probability one it follows that $\hat{F}_n(x) \to x^2$ with probability one as claimed.

The inconsistency of the MLE in this case is rather remarkable since the empirical distribution, which is the MLE in the absence of any a priori information, is of course consistent. Also, as we saw in Sections 5.2, 5.3 and 5.4, the MLE is consistent if the stronger convex ordering assumption is made. Marshall and Proschan (1965) have shown in a similar fashion that the MLE for IFRA distributions is inconsistent. These results dramatize the need for alternative estimators such as the isotonic estimator to be considered next.

Isotonic estimators for star-ordered families of distributions

Let G be a specified distribution on $[0, \infty)$. Let $G^{-1}F$ be starshaped on $\{x | 0 < F(x) < 1\}$; i.e., $F \underset{*}{<} G$. Let $0 \equiv X_{0:n} < X_{1:n} < \ldots < X_{n:n}$ be an ordered sample from F and let F_n be the empirical distribution. Define

$$\lambda_n(X_{i:n}) = G^{-1}F_n(X_{i:n})/X_{i:n}.$$

If $\lambda_n(X_{i:n})$ is nondecreasing in i, then

$$\hat{F}_n(t) = \begin{cases} 0 & \text{for} \quad t < X_{1:n} \\ G[\lambda_n(X_{i:n})t] & \text{for} \quad X_{i:n} \leq t < X_{i+1:n} \quad 1 \leq i \leq n-1 \\ 1 & \text{for} \quad t \geq X_{n:n} \end{cases}$$

satisfies the restriction that $\hat{F}_n \underset{*}{<} G$ and $\hat{F}_n$ estimates F.

If $\lambda_n(X_{i:n})$ is not nondecreasing in i, then we form the isotonic regression with respect to weights $w_i > 0$ which we otherwise leave unspecified. Let

$$\lambda_n^*(X_{i:n}) = \min_{t \geq i} \max_{s \leq i} \frac{\sum_{j=s}^{t} \lambda_n(X_{j:n}) w_j}{\sum_{j=s}^{t} w_j} \tag{5.44}$$

and

$$\lambda_n^*(t) = \begin{cases} 0 & \text{for} & t < X_{1:n} \\ \lambda_n^*(X_{i:n}) & \text{for} & X_{i:n} \leq t < X_{i+1:n} \\ 1 & \text{for} & t \geq X_{n:n}. \end{cases}$$

Consistency of the isotonic estimator

Strong consistency of the isotonic estimator follows from the strong consistency of the empirical distribution. To see this, fix x and define r by $X_{r:n} \leq x < X_{r+1:n}$; then from the monotonicity of G^{-1},

$$\lambda_n(X_{r:n}) \leq \lambda_n(x) \leq \lambda_n(X_{r+1:n}) = \frac{G^{-1}F_n(X_{r+1:n})}{X_{r+1:n}}.$$

It follows that $\lambda_n(x) \to \lambda(x) = G^{-1}F(x)/x$ with probability one from the strong consistency of F_n and of the order statistics. From Theorem 1.6 of Chapter 1 it follows that

$$|\lambda_n(X_{i:n}) - \lambda(X_{i:n})| \leq \varepsilon \quad \text{for} \quad i = 1, 2, \ldots, n$$

implies

$$|\lambda_n^*(X_{i:n}) - \lambda(X_{i:n})| \leq \varepsilon \quad \text{for} \quad i = 1, 2, \ldots, n.$$

Hence $\lambda_n^*(x) \to \lambda(x)$ with probability one from the strong consistency of $\lambda_n(x)$.

In practice it would seem reasonable to choose

$$w_j = F_n(X_{j:n}) - F_n(X_{j:n}^-) = 1/n.$$

Asymptotically, of course, it does not seem to matter. Besides the virtue of consistency, our estimator has a least squares property, namely

$$\sum_{i=1}^n [\lambda(X_{i:n}) - \lambda_n^*(X_{i:n})]^2 w_i \leq \sum_{i=1}^n [\lambda(X_{i:n}) - \lambda_n(X_{i:n})]^2 w_i$$

where $\lambda(x) = G^{-1}F(x)/x$. However, the estimator is really only an *ad hoc* estimator. As will be seen below, there is another isotonic estimator which, in some respects, may be superior.

Percentile estimators in the star-ordering case

As before, assume that $G^{-1}F$ is starshaped on $\{x|0 < F(x) < 1\}$ where G is specified. Let $x(p) = \inf\{x|F(x) \geq p\}$, i.e., $x(p) = F^{-1}(p)$, so that $F(x(p)) = P$

when F is continuous. Let $X_{1:n} \leq X_{2:n} \leq \ldots \leq X_{n:n}$ be an ordered sample from F. The usual nonparametric estimator for $x(p)$ is the $[np]$th order statistic, $X_{[np]}$, where [ ] denotes the greatest integer contained within the quantity in brackets. For the sake of generality, let $\xi_n(p)$ be any estimator of $x(p)$. Given positive weights, $w(i/n) > 0$ consider the following problem:

$$\text{Minimize} \quad \sum_{i=1}^{n} [\xi(i/n) - \xi_n(i/n)]^2 w(i/n)$$

$$\text{subject to} \quad \frac{\xi(1/n)}{G^{-1}(1/n)} \geq \frac{\xi(2/n)}{G^{-1}(2/n)} \geq \ldots \geq \frac{\xi(1)}{G^{-1}(1)} \right\} \quad (5.45)$$

and

$$\xi(1/n) \leq \xi(2/n) \leq \ldots \leq \xi(1).$$

Theorem 5.10

If $F(0) = 0$ and $G^{-1}F$ is starshaped on $\{x | 0 < F(x) < 1\}$, then

$$\hat{\xi}_n(i/n) = \left[\min_{s \leq i} \max_{t \geq i} \frac{\sum_{j=s}^{t} \xi_n(j/n)[G^{-1}(j/n)]^2 w(j/n)}{\sum_{j=s}^{t} [G^{-1}(j/n)]^2 w(j/n)} \right] G^{-1}(i/n)$$

solves (5.45).

Proof. This is an immediate consequence of Theorem 1.2 of Chapter 1 if we let

$$X = \left\{ \frac{1}{n}, \frac{2}{n}, \ldots, \frac{n}{n} \right\},$$

$$g(i/n) = \xi_n(i/n),$$

$$c(i/n) = G^{-1}(i/n)$$

and

$$w(i/n) > 0$$

are given positive weights. ∎

If we choose $\xi_n(i/n) = X_{i:n}$; i.e., the ith order statistic, the strong consistency of $\hat{\xi}_n(p)$ will follow from the strong consistency of the order statistic estimator if $F^{-1}(p)$ is a continuity point of F. A convenient choice would be to let $w(i/n) = 1$ for $i = 1, 2, \ldots, n$. Of course $\hat{\xi}_n(i/n)$ can also be used to provide an estimate of F.

5.6 COMPLEMENTS

The derivation of the MLE for unimodal densities (Section 5.2) with known mode follows that of Prakasa Rao (1969). The idea of using Lemma B preceding Theorem 5.1 to prove consistency is due to Albert W. Marshall.

The maximum likelihood estimator for unimodal densities is a conditional expectation with respect to a special σ-lattice. (See Chapter 7 for definitions and further background.) The following development is due to Robertson (1967) who attributes the original idea to Pyke. Suppose the mode, M, is known. Consider the set $\mathscr{L}$ consisting of all intervals containing M, together with the empty set. It is easy to verify that this is a σ-lattice since it is closed under arbitrary unions and intersections. Let $X_{1:n} \leq X_{2:n} \leq \ldots \leq X_{n:n}$ be an ordered sample from a decreasing density. Let $L(X_{i:n})$ be the smallest member of $\mathscr{L}$ containing $X_{i:n}$. Form the sets

$$B_0 = \bigcap_{i=1}^{n} L(X_{i:n}),$$

$$B_1 = \bigcap_{i \neq n} L(X_{i:n}),$$

$$B_2 = \bigcap_{i \neq n-1} L(X_{i:n}), \ldots, B_n = \bigcap_{i \neq 1} L(X_{i:n}),$$

$$B_{n+1} = \bigcap_{i \neq n-1, n} L(X_{i:n}), \ldots, B_H = L(X_{1:n}).$$

Let $A_0 = B_0$ and

$$A_i = B_i - \bigcup_{j<i} B_j = B_i - \sum_{j<i} A_j$$

for $i = 1, 2, \ldots, H$. Suppose that k of the sets $A_0, A_1, \ldots, A_H$ are nonempty and label them $A_1, A_2, \ldots, A_k$. Let $g_n(i) = n_i[n\lambda(A_i)]^{-1}$ ($i = 1, 2, \ldots, k$) where n_i is the number of observations in A_i and $\lambda(A_i)$ is the Lebesgue measure of A_i. Let $\Omega^* = \{1, 2, \ldots, k\}$ and form the σ-lattice $\mathscr{L}^*$ of subsets of Ω^* as follows: $\phi \in \mathscr{L}^*$ and $T \in \mathscr{L}^*$ if and only if

$$\bigcup_{i \in T} A_i \in \mathscr{L}.$$

Let $\lambda^*(i) = \lambda(A_i)$. Then the MLE can be represented as the conditional expectation of g_n given $\mathscr{L}^*$; i.e.,

$$\hat{f}_n = E_{\lambda^*}[g_n | \mathscr{L}^*].$$

The MLE for IFR distributions (Section 5.3) was first obtained by Grenander (1956). This was extended and consistency was proved by Marshall and Proschan (1965). The derivation of the MLE given here follows Marshall and

Proschan. The consistency proof here is a combination of ideas due to Robertson (1967), Barlow and van Zwet (1970) and Albert W. Marshall.

The notion of a convex ordering (Section 5.4) on the space of distribution functions is due to van Zwet (1964). Isotonic window estimators for the generalized failure rate function were first considered in Barlow and van Zwet (1969, 1970).

Optimal choice of window size for nonparametric estimators of the density and failure rate function have been investigated by Parzen (1962), Watson and Leadbetter (1964), and Weiss and Wolfowitz (1967).

Maximum likelihood estimators for star-ordered families of distributions (Section 5.5) were first investigated by Marshall and Proschan (unpublished) who also noted their inconsistency.

Isotonic estimators for star-ordered families of distributions were proposed and investigated by Barlow and Scheuer (1971). Percentile estimators in the star-ordering case grew out of conversations with W. R. van Zwet.

Birnbaum, Esary and Marshall (1966) characterized the class of IFRA distributions in terms of closure under the formation of coherent systems. This initiated interest in the subject of classes of distributions based on star-ordering.

CHAPTER 6

Isotonic Tests for Goodness of Fit

6.1 INTRODUCTION

A considerable part of the literature in applied statistics and operations research is based on the exponential distribution. The question of validating an exponential model has inspired a large number of papers on omnibus tests for exponentiality. Since alternatives are often not specified, the question of optimality usually cannot be answered. By restricting alternatives in a rather natural way, it is possible to prove asymptotic minimax optimality (Section 6.3) for a statistic which is called the *cumulative total time on test statistic*.

This statistic has a very old history (see complements) and has a very convenient distribution under the exponential hypothesis (Table 6.1). The test based on this statistic also has the desirable properties of unbiasedness and isotonic power with respect to star-ordering (Section 6.2). We illustrate the application of this statistic to data on air conditioning equipment failures (Section 6.2) and to the problem of debugging large systems (Section 6.4). In Section 6.4, the analogue of this statistic for tests of trend in a series of events is discussed. Also, in this section, the concept of total time on test is used to extend the unbiasedness results to tests for monotone failure rate based on incomplete data.

Again in this chapter we use increasing for nondecreasing and decreasing for nonincreasing.

6.2 TESTS FOR EXPONENTIALITY AGAINST MONOTONE FAILURE RATE ALTERNATIVES

In Section 5.3 of Chapter 5, we considered the problem of estimating the failure rate function,

$$r(t) = \frac{f(t)}{1 - F(t)},$$

assumed increasing. In this section some of the statistics and theory developed in Section 5.3 are used to test the hypothesis of constant versus increasing failure rate. In Barlow and Doksum (1972) this is generalized to the problem of testing $F = G$ versus $F < G$ where $<$ stands for convex ordering and G is specified. However, throughout this section and the next it is assumed that $G(x) = 1 - e^{-x}$ for $x \geq 0$.

Suppose that the first k ordered observations out of a sample of n test items from a distribution F are obtained:

$$X_{1:n} \leq X_{2:n} \leq \ldots \leq X_{k:n} \qquad (1 \leq k \leq n).$$

Let $D_{i:n} = (n - i + 1)(X_{i:n} - X_{i-1:n})$ (where $X_{0:n} \equiv 0$) be the normalized sample spacings. It is well known that if $F(x) = 1 - e^{-\lambda x}$ for $x \geq 0$, then $D_{i:n}$ ($i = 1, 2, \ldots, n$) are independent and exponentially distributed random variables with mean $1/\lambda$. If F is IFR, then we claim that

$$D_{1:n} \underset{st}{\geq} D_{2:n} \underset{st}{\geq} \ldots \underset{st}{\geq} D_{n:n}$$

where $\underset{st}{\geq}$ indicates stochastic (or probabilistic) ordering. This is the basis for many tests for exponentiality versus increasing (or decreasing) failure rate. (Note that the isotonic regression, $D^*_{i:n}$, on $D_{i:n}$ with respect to unit weights and the ordering $1 \succ 2 \succ 3 \succ \ldots \succ n$ gives the maximum likelihood estimate for $1/r(t)$ when $r(t)$ is assumed increasing, Section 5.3, Chapter 5.)

It will be convenient to use the notation $\bar{F}(x)$ for $1 - F(x)$. The following useful lemma will be needed.

Lemma

Let μ be a signed measure of bounded variation on $[a, b]$, $-\infty \leq a < b \leq \infty$ such that

$$\int_a^b d\mu(u) < \infty.$$

Then

$$\int_a^b \phi(u)\, d\mu(u) \geq 0$$

for all bounded increasing ϕ if and only if

$$\int_x^b d\mu(u) \geq 0$$

for all $x \in [a, b]$ and

$$\int_a^b d\mu(u) = 0.$$

Proof. Suppose first that
$$\int_a^b \phi(x)\, d\mu(x) \geq 0$$
for all ϕ increasing. Then with
$$\phi(x) = H_z(x) = \begin{cases} 0 & x \leq z \\ 1 & x > z \end{cases}$$
it follows that
$$\int_z^b d\mu(u) \geq 0.$$
Also $\phi(x) \equiv 1$ and $\phi(x) \equiv -1$ imply
$$\int_a^b d\mu(u) = 0.$$

Next suppose
$$\int_x^b d\mu(u) \geq 0$$
for all $x \in [a, b]$, and suppose in addition $\phi \geq 0$. Approximate ϕ by an increasing sequence $\{\phi_k\}$ of increasing step functions. Since
$$\phi_k(z) = \sum_{i=1}^n a_i H_{x_i}(z)$$
and $a_i > 0$ it is concluded that
$$\int_a^b \phi_k(u)\, d\mu(u) \geq 0.$$
The lemma follows from the Lebesgue monotone convergence theorem. If $\phi \leq 0$,
$$\int_a^b -\phi(u)\, d\mu(u) \leq 0$$
is obtained by an identical argument. Finally note that any increasing ϕ can be decomposed into $\phi_1 + \phi_2$ where $\phi_1 \geq 0$, $\phi_2 \leq 0$ and ϕ_i is increasing $(i = 1, 2)$. ∎

Theorem 6.1

If F is IFR (DFR), then $D_{i:n} = (n - i + 1)(X_{i:n} - X_{i-1:n})$ is stochastically increasing (decreasing) in $n \geq i$ for fixed i.

Proof. Assume F is IFR. Let $F_{i:n}(x) = \Pr[X_{i:n} \leq x]$ and $F_u(x) = [F(x + u) - F(u)]/\bar{F}(u)$. Then

$$\Pr[(n - i)(X_{i+1:n} - X_{i:n}) > x]$$
$$= \int_0^\infty \{\bar{F}_u[x/(n - i)]\}^{n-i} \, dF_{i:n}(u)$$
$$\leq \int_0^\infty \{\bar{F}_u[x/(n + 1 - i)]\}^{n+1-i} \, dF_{i:n}(u), \quad (6.1)$$

since $[\bar{F}(t)]^{1/t}$ is decreasing in t for F IFR. (F IFR implies $\log \bar{F}(t)$ concave or $[\log \bar{F}(t)]/t$ decreasing since $\log \bar{F}(0) = 0$.) Also since $\bar{F}_u(x)$ is decreasing in u for F IFR and $F_{i:n}(x) \leq F_{i:n+1}(x)$ for all F,

$$\int_0^\infty \{\bar{F}_u[x/(n + 1 - i)]\}^{n+1-i} \, dF_{i:n}(u)$$
$$\leq \int_0^\infty \{\bar{F}_u[x/(n + 1 - i)]\}^{n+1-i} \, dF_{i:n+1}(u)$$
$$= \Pr[(n + 1 - i)(X_{i+1:n+1} - X_{i:n+1}) > x] \quad (6.2)$$

by the above lemma. (Let $\mu(u) = F_{i:n}(u) - F_{i:n+1}(u)$ and $-\phi(u) = \{\bar{F}_u[x/(n + 1 - i)]\}^{n+1-i}$ and note that

$$\int_x^\infty d\mu(u) = F_{i:n}(x) - F_{i:n+1}(x) \geq 0.)$$

Inequalities (6.1) and (6.2) are reversed when F is DFR. ∎

Corollary

If F is IFR (DFR), then $D_{i:n} = (n - i + 1)(X_{i:n} - X_{i-1:n})$ is stochastically decreasing (increasing) in $i = 1, 2, \ldots, n$ for fixed n.

Proof. Assume F is IFR. First it will be shown that

$$(n - 1)(X_{2:n} - X_{1:n}) \underset{\text{st}}{\leq} nX_{1:n}.$$

Given $X_{1:n}$, $X_{2:n} - X_{1:n}$ is the minimum of $n - 1$ random variables each stochastically less than $X_{1:n}$. Hence,

$$X_{2:n} - X_{1:n} \underset{\text{st}}{\leq} X_{1:n-1}.$$

By Theorem 6.1,

$$(n - 1)X_{1:n-1} \underset{\text{st}}{\leq} nX_{1:n},$$

so that
$$(n-1)(X_{2:n} - X_{1:n}) \underset{\text{st}}{\leq} nX_{1:n}.$$

The result follows by repeated conditioning.

An analogous argument applies in the DFR case. ∎

Since under the exponential assumption
$$\Pr[D_{i:n} > D_{j:n}] = \tfrac{1}{2}, \quad i \neq j$$
while
$$D_{1:n} \underset{\text{st}}{\geq} D_{2:n} \underset{\text{st}}{\geq} \ldots \underset{\text{st}}{\geq} D_{n:n}$$

under the IFR assumption, many tests have been suggested based on reversals in the assumed ordering. (See Bickel and Doksum (1969) and Proschan and Pyke (1967).) The fact that under the IFR assumption, $D_{i:n}$ tends to be larger than $D_{j:n}$ for $i < j$ should be reflected by a positive slope in the linear regression of the $D_{n-i+1:n}$ on the values $i - 1$. Under exponentiality, of course, this regression should have zero slope. When the slope is nonzero, it will be sensitive to changes in scale. In order to make the statistic scale invariant, the slope is divided by the average of the $D_{i:n}$. The resulting statistic is linearly equivalent to

$$V_n \stackrel{\text{def}}{=} n^{-1} \sum_{i=1}^{n} (i-1) D_{n-i+1:n} \Big/ n^{-1} \sum_{i=1}^{n} D_{i:n} \tag{6.3}$$

which is the same as

$$V_n = n^{-1} \sum_{i=1}^{n-1} \left[\sum_{j=1}^{i} D_{j:n} \right] \Big/ n^{-1} \sum_{i=1}^{n} D_{i:n} \tag{6.4}$$

if the numerator of the right hand side of (6.3) is summed by parts. Thus

$$T_n(X_{i:n}) = \sum_{j=1}^{i} D_{j:n}$$

is recognized as the total time on test statistic and for this reason the following definition is made.

Definition 6.1

Given the first k ($1 \leq k \leq n$) ordered observations out of a sample of n test items,

$$V_k = k^{-1} \sum_{i=1}^{k-1} \left[\sum_{j=1}^{i} D_{j:n} \right] \Big/ k^{-1} \sum_{i=1}^{k} D_{i:n} \tag{6.5}$$

is called the *cumulative total time on test statistic*.

If $k = n$, (6.5) agrees with (6.4). Under exponentiality the distribution of V_k will depend only on k, the observed number of failures, and *not* on n, the sample size.

We use the statistic V_k for the problem of testing

$$H_0: F \text{ exponential}$$

versus

$$H_1: F \text{ IFR and not exponential.}$$

Definition 6.2

Let ψ be the test corresponding to V_k where

$$\psi(X_{1:n}, X_{2:n}, \ldots, X_{k:n}) = \begin{cases} 1 & \text{if } V_k \geq c_{k,1-\alpha} \\ 0 & \text{otherwise} \end{cases}$$

where $\Pr_G[V_k \geq c_{k,1-\alpha}] = \alpha$ and $G(x) = 1 - \exp(-x)$ for $x \geq 0$. The following well-known result is stated without proof.

Theorem 6.2

If $F(x) = 1 - \exp(-\lambda x)$ for $x \geq 0$, then

$$V_k \underset{\text{st}}{=} U_1 + U_2 + \ldots + U_{k-1} \tag{6.6}$$

where $\underset{\text{st}}{=}$ indicates stochastic equivalence and U_i ($i = 1, 2, \ldots, k-1$) are independent uniform random variables on $[0, 1]$.

It follows that

$$\mathscr{L}[(12(k-1))^{\frac{1}{2}}\{(k-1)^{-1}V_k - \tfrac{1}{2}\}] \to N(0, 1)$$

when $F(x) = 1 - \exp(-\lambda x)$ for $x \geq 0$.

Table 6.1 provides critical numbers for performing the test, ψ. If the observations are from an IFR distribution then V_k will tend to be large. The table provides percentage points such that if the true distribution is exponential then

$$\Pr[V_k \leq c_{k,1-\alpha}] = 1 - \alpha$$

where $c_{k,1-\alpha}$ is the number tabulated. For example, if $1 - \alpha = 0.90$ and $k = 5$ we look in row $k - 1 = 4$ and find $c_{4,0.90} = 2.753$. If the computed statistic V_k exceeds the number 2.753 it is concluded that the true

Table 6.1 Percentiles $c_{k,1-\alpha}$ of the Cumulative Total Time on Test Statistic, V_k, under H_0

$k-1$	$1-\alpha$				
	0.900	0.950	0.975	0.990	0.995
2	1.553	1.684	1.776	1.859	1.900
3	2.157	2.331	2.469	2.609	2.689
4	2.753	2.953	3.120	3.300	3.411
5	3.339	3.565	3.754	3.963	4.097
6	3.917	4.166	4.367	4.610	4.762
7	4.489	4.759	4.988	5.244	5.413
8	5.056	5.346	5.592	5.869	6.053
9	5.619	5.927	6.189	6.487	6.683
10	6.178	6.504	6.781	7.097	7.307
11	6.735	7.077	7.369	7.702	7.924
12	7.289	7.647	7.953	8.302	8.535

$k =$ number of failures observed in incomplete sample.

distribution is IFR. There is, of course, a 10% chance that an error has been made; i.e.,

$$\Pr[V_k \geq 2.753] = 0.10$$

when the true distribution is exponential and $k = 5$.

Example 6.1 Air conditioning failures

In Table 6.2 we list the times between air conditioner failures on selected aircraft. After roughly 2000 hours of service the planes received major overhaul; the failure interval containing major overhaul is omitted from the listing since the length of that failure interval may have been affected by major overhaul.

We should like to determine if the intervals between failures have an exponential distribution or if there is a wearout trend as the equipment ages. In the event that there is a wearout trend maintenance should be scheduled according to equipment *age* rather than the present policy.

The computations associated with the above data are given in Table 6.3. Computations for data from a given plane were made as if they were independent observations on different aircraft of the same type.

Since the sample size for plane 7908 exceeds the range of Table 6.1, we use the fact that

$$Z = [12(k-1)]^{\frac{1}{2}}[(k-1)^{-1}V_k - \tfrac{1}{2}]$$

Table 6.2 *Intervals Between Failures of Air Conditioning Equipment on Jet Aircraft*

7907	7908	7915	7916	8044
194	413	359	50	487
15	14	9	254	18
41	58	12	5	100
29	37	270	283	7
33	100	603	35	98
181	65	3	12	5
	9	104		85
	169	2		91
	447	438		43
	184			230
	36			3
	201			130
	118			
	*			
	34			
	31			
	18			
	18			
	67			
	57			
	62			
	7			
	22			
	34			

* Major overhaul.

Table 6.3

Plane	Sample size k	Statistic V_k	Conclusion
7907	6	$V_6 = 2.243$	Exponential, i.e., V_6 *not* significant at the 10% level
7908	23	$V_{23} = 8.829$ $Z = -1.1309$	Exponential, i.e., V_{23} *not* significant at the 10% level
7915	9	$V_9 = 2.80$	Decreasing failure rate, i.e., V_9 significant at the 10% level
7916	6	$V_6 = 1.67$	Exponential, i.e., V_6 *not* significant at the 10% level
8044	12	$V_{12} = 4.22$	Exponential, i.e., V_{12} *not* significant at the 10% level

is approximately normally distributed with mean 0 and variance 1 when k is large and the observations are from an exponential distribution.

We conclude that plane 7915 has a decreasing failure rate distribution (DFR) since V_9 is small. We use the fact that

$$V_k \underset{\text{st}}{=} U_1 + U_2 + \ldots + U_{k-1}$$

is symmetric about $(k-1)/2$; i.e.,

$$\Pr[V_k - (k-1)/2 \geq x] = \Pr[V_k - (k-1)/2 \leq -x].$$

From Table 6.1 it is seen that $x = 1.056$ if $k = 9$ will give a probability of 0.10. Hence a lower critical number for V_k is $(k-1)/2 - x = 2.944$. Since V_k is less than this number in this case it is concluded that the data is DFR.

The total time on test transformation

In Section 5.3 of Chapter 5, the total time on test transformation

$$H_F^{-1}(t) = \int_0^{F^{-1}(t)} [1 - F(u)] \, du \quad \text{for} \quad 0 \leq t \leq 1$$

was introduced and we noted that its inverse, H_F, is a distribution function on $[F^{-1}(0), \mu]$ where

$$\mu = \int_0^\infty x \, dF(x) = H^{-1}(1) < \infty$$

(if F is IFR). If $G(x) = 1 - \exp(-x)$ for $x \geq 0$, then $H_G(t) = t$ for $0 \leq t \leq 1$. Since H_F is convex if F is IFR (Section 5.3) the problem of testing

$$H_0 : F \text{ exponential}$$

versus

$$H_1 : F \text{ IFR and not exponential}$$

is equivalent to testing

$$H_0 : H_F \text{ linear on } 0 < H_F < 1$$

versus

$$H_1 : H_F \text{ convex and not linear on } 0 < H_F < 1.$$

This latter problem is somewhat more tractable. However estimates of H_F are needed. The natural estimator of H_F^{-1} is

$$H_{F_n}^{-1}(t) \stackrel{\text{def}}{=} H_n^{-1}(t) = \int_0^{F_n^{-1}} (t)[1 - F_n(u)] \, du.$$

As was pointed out in Chapter 5

$$H_n^{-1}\left(\frac{i}{n}\right) = n^{-1}\sum_{j=1}^{i} D_{j:n} = n^{-1}\left[\sum_{j=1}^{i}(n-j+1)(X_{j:n} - X_{j-1:n})\right]$$

so that the cumulative total time on test statistic can be written as

$$V_k = \sum_{i=1}^{k} H_n^{-1}\left(\frac{i}{n}\right) \Big/ H_n^{-1}\left(\frac{k}{n}\right).$$

Let $U_{i:n}$ be the ith order statistic from the uniform distribution on $[0, 1]$. Then

$$Z_{i:n} \stackrel{\text{def}}{=} H_F^{-1}(U_{i:n}) \stackrel{\text{def}}{=} \int_0^{F^{-1}(U_{i:n})} [1 - F(u)]\, du \stackrel{\ }{=}_{\text{st}} \int_0^{X_{i:n}} [1 - F(u)]\, du$$

will be distributed as the ith order statistic in a sample of size n from the distribution H_F. From Lemma A in Section 5.3 we see that

$$\sup_{1 \le i \le n} \left| H_n^{-1}\left(\frac{i}{n}\right) - Z_{i:n} \right| \to 0 \qquad (6.7)$$

almost surely. Hence, in this rather strong way, $H_n^{-1}(i/n)$ for $i = 1, 2, \ldots, n$, behave asymptotically like order statistics from H. However their joint distribution does not agree asymptotically with the joint distribution of $Z_{i:n}$ ($i = 1, 2, \ldots, n$) except in the exponential case.

It follows from (6.7) that

$$n^{-1}\sum_{i=1}^{n} H_n^{-1}\left(\frac{i}{n}\right) \to \int_0^1 H_F^{-1}(u)\, du$$

almost surely as $n \to \infty$. If $G(x) = 1 - \exp(-x)$ for $x \ge 0$, then

$$\int_0^1 H_G^{-1}(u)\, du = \tfrac{1}{2}.$$

Since

$$\frac{H_F^{-1}(0)}{H_F^{-1}(1)} = \frac{F^{-1}(0)}{\mu} \ge H_G^{-1}(0) = 0 \qquad (6.8)$$

and

$$\frac{H_F^{-1}(1)}{H_F^{-1}(1)} = H_G^{-1}(1) = 1 \qquad (6.9)$$

it follows that F IFR and not exponential implies

$$\frac{\int_0^1 H_F^{-1}(u)\, du}{H_F^{-1}(1)} > \tfrac{1}{2}.$$

Hence in this case

$$n^{-1}V_n \to \int_0^1 H^{-1}(u)\,du/H^{-1}(1) > \tfrac{1}{2}$$

almost surely as $n \to \infty$.

Isotonic tests with respect to star ordering

Since the problem is clearly scale invariant, we consider tests based on the studentized statistics

$$W_{F_n}\left(\frac{i}{n}\right) \stackrel{\text{def}}{=} W_{i:n} \stackrel{\text{def}}{=} H_n^{-1}\left(\frac{i}{n}\right)\Big/H_n^{-1}(1).$$

In Section 5.5 of Chapter 5, starshaped functions and star ordering ($\underset{*}{<}$) on the class of distribution functions on $[0, \infty)$ were introduced. (Convex ordering implies star ordering.) If $G(x) = 1 - \exp(-x)$ for $x \geq 0$ and $F \underset{*}{<} G$, then the failure rate function, r, of F is increasing on the average; i.e.,

$$\int_0^t r(u)\,du/t$$

is increasing in $t \geq 0$. F is called IFRA for short. Because of the wide application in reliability theory for such distributions we show that the statistics, W_{F_n}, are isotonic with respect to star ordering in the sense that $F \underset{*}{<} K$ implies

$$W_{F_n} \underset{\text{st}}{\geq} W_{K_n}$$

where F_n (K_n) are the empirical distributions corresponding to $F(K)$. To do this the following lemma is needed.

Lemma

If β_i/α_i is increasing in i and $\alpha_i > 0$ is increasing in i $(1 \leq i \leq n)$, then

$$\sum_{i=1}^r (n - i + 1)(\beta_i - \beta_{i-1}) \Big/ \sum_{i=1}^r (n - i + 1)(\alpha_i - \alpha_{i-1})$$

is increasing in r $(1 \leq r \leq n)$ where $\alpha_0 = \beta_0 = 0$.

Proof. Define $\psi(0) = 0$, $\psi(\alpha_1 + \ldots + \alpha_i) = \beta_1 + \ldots + \beta_i$ for $i = 1, 2, \ldots, n$. Define ψ elsewhere on $[0, \alpha_1 + \alpha_2 + \ldots + \alpha_n]$ by linear interpolation. Now β_i/α_i increasing implies ψ is convex on $[0, \alpha_1 + \ldots + \alpha_n]$. Let $\psi_r = \psi$ on $[0, \alpha_1 + \ldots + \alpha_r]$ and elsewhere agree with the left hand tangent to ψ at

$$A_r \stackrel{\text{def}}{=} \alpha_1 + \ldots + \alpha_r.$$

Let $\psi_{r+1} = \psi$ on $[0, \alpha_1 + \ldots + \alpha_{r+1}]$ and elsewhere agree with the left hand tangent to ψ at

$$A_{r+1} \stackrel{\text{def}}{=} \alpha_1 + \ldots + \alpha_{r+1}.$$

Since ψ is convex, $\psi_r(x) \leq \psi_{r+1}(x)$ for $x \geq 0$. Since α_i is increasing in i, $A_r + (n-r)\alpha_r \leq A_{r+1} + (n-r-1)\alpha_{r+1}$. Hence, since ψ_{r+1} is starshaped

$$\frac{\psi_r(A_r + (n-r)\alpha_r)}{A_r + (n-r)\alpha_r} \leq \frac{\psi_{r+1}(A_r + (n-r)\alpha_r)}{A_r + (n-r)\alpha_r} \leq \frac{\psi_{r+1}(A_{r+1} + (n-r-1)\alpha_{r+1})}{A_{r+1} + (n-r-1)\alpha_{r+1}}.$$

(6.10)

Note that

$$\sum_{i=1}^{r}(n-i+1)(\beta_i - \beta_{i-1}) = \beta_1 + \ldots + \beta_r + \underbrace{\beta_r + \ldots + \beta_r}_{n-r}$$

and likewise

$$\sum_{i=1}^{r}(n-i+1)(\alpha_i - \alpha_{i-1}) = \alpha_1 + \ldots + \alpha_r + \underbrace{\alpha_r + \ldots + \alpha_r}_{n-r}.$$

Hence

$$\frac{\psi_r(A_r + (n-r)\alpha_r)}{A_r + (n-r)\alpha_r} = \frac{\sum_{i=1}^{r}(n-i+1)(\beta_i - \beta_{i-1})}{\sum_{i=1}^{r}(n-i+1)(\alpha_i - \alpha_{i-1})}$$

and

$$\frac{\psi_{r+1}(A_{r+1} + (n-r-1)\alpha_{r+1})}{A_{r+1} + (n-r-1)\alpha_{r+1}} = \frac{\sum_{i=1}^{r+1}(n-i+1)(\beta_i - \beta_{i-1})}{\sum_{i+1}^{r+1}(n-i+1)(\alpha_i - \alpha_{i-1})}.$$

The lemma follows from (6.10). ∎

Theorem 6.3

If $F \underset{*}{<} G$ and $X_{1:n} \leq \ldots \leq X_{n:n}$ ($Y_{1:n} \leq \ldots \leq Y_{n:n}$) is an ordered sample from F (G), then

$$\frac{\sum_{j=1}^{i}(n-j+1)(X_{j:n} - X_{j-1:n})}{\sum_{j=1}^{k}(n-j+1)(X_{j:n} - X_{j-1:n})} \underset{st}{\geq} \frac{\sum_{j=1}^{i}(n-j+1)(Y_{j:n} - Y_{j-1:n})}{\sum_{j=1}^{k}(n-j+1)(Y_{j:n} - Y_{j-1:n})} \quad (6.11)$$

or $0 \leq i \leq k \leq n$ and $X_{0:n} = Y_{0:n} = 0$.

Proof. Let $Y'_{i:n} = G^{-1}F(X_{i:n})$. Note $X_{i:n}$ is increasing in i. Since $F \underset{*}{<} G$, $Y'_{i:n}/X_{i:n}$ is increasing in $i = 1, 2, \ldots, n$. Let $\beta_i = Y'_{i:n}$ and $\alpha_i = X_{i:n}$ in the above lemma. It follows that

$$\frac{\sum_{j=1}^{i}(n-j+1)(X_{j:n} - X_{j-1:n})}{\sum_{j=1}^{k}(n-j+1)(X_{j:n} - X_{j-1:n})} \geq \frac{\sum_{j=1}^{i}(n-j+1)(Y'_{j:n} - Y'_{j-1:n})}{\sum_{j=1}^{k}(n-j+1)(Y'_{j:n} - Y'_{j-1:n})}$$

for $0 \leq i \leq k$ and $X_{0:n} = Y_{0:n} = 0$. The theorem follows from the observation that

$$(Y_{1:n}, \ldots, Y_{n:n}) \underset{st}{=} (Y'_{1:n}, \ldots, Y'_{n:n}). \blacksquare$$

Corollary

If $F \underset{*}{<} K$, then

(i) $W_{F_n} \underset{st}{\geq} W_{K_n}$;

(ii) $\Pr[V_k \geq x|F] \geq \Pr[V_k \geq x|K]$ for all $x \geq 0$.

Proof. Both (i) and (ii) follow directly from Theorem 6.3. $\blacksquare$

Definition 6.3

A test ϕ based on $X_1, X_2, \ldots, X_n$ is *monotonic* if

$$\phi(X_1, \ldots, X_n) = \begin{cases} 1 & \text{if } T(X_1, \ldots, X_n) > c_{n,\alpha} \\ 0 & \text{otherwise} \end{cases}$$

where T is coordinatewise increasing.

Definition 6.4

A test ϕ is *isotonic* with respect to star ordering if $F_1 <_* F_2$ and $\mathbf{X} = (X_1, \ldots, X_n)$ ($\mathbf{Y} = (Y_1, \ldots, Y_n)$) is a random sample from F_1 (F_2) implies

$$\phi(\mathbf{X}) \underset{\text{st}}{\geq} \phi(\mathbf{Y}).$$

Theorem 6.4

Let $G(x) = 1 - \exp(-x)$ for $x \geq 0$. Then monotonic tests based on $W_{1:n}, W_{2:n}, \ldots, W_{n:n}$ are isotonic tests with respect to star ordering. Isotonic tests with respect to star ordering have isotonic power with respect to star ordering; i.e., $F_1 <_* F_2 <_* G$ implies

$$\beta_\phi(F_1) \geq \beta_\phi(F_2) \geq \beta_\phi(G)$$

where $\beta_\phi(F)$ is the power of ϕ when the true distribution is F.

Proof. This follows immediately from the above corollary and the definitions. ∎

It is clear that the test, ψ, based on the statistic, V_k, would provide an isotonic test for the problem of testing

$$H_0 : F = G$$

versus

$$H_1 : F <_* G \quad \text{and} \quad F \neq G$$

where $G(x) = 1 - \exp(-x)$. It can be shown that $F <_* G$ implies H star-shaped. Hence an equivalent problem is to test

$$H_0 : H \text{ linear on } 0 < H < 1$$

versus

$$H_1 : H \text{ starshaped and not linear on } 0 < H < 1.$$

Asymptotic distribution of V_n

It has previously been observed that

$$n^{-1} V_n = n^{-1} \sum_{i=1}^{n} H_n^{-1}\left(\frac{i}{n}\right) \bigg/ H_n^{-1}(1) \to \int_0^1 H^{-1}(u) \, du \big/ H^{-1}(1)$$

almost surely as $n \to \infty$. For convenience let

$$\mu(H) \stackrel{\text{def}}{=} \int_0^1 H^{-1}(u) \, du / H^{-1}(1). \tag{6.12}$$

We seek the asymptotic distribution of

$$T_n = n^{\frac{1}{2}}[n^{-1} V_n - \mu(H)]$$

under the general alternative distribution, F. First we prove the following lemma.

Lemma

If

$$\int_0^\infty x \, dF(x) < \infty,$$

then

$$\int_0^1 H^{-1}(u) \, du = 2 \int_0^\infty x[1 - F(x)] \, dF(x).$$

Proof. Integrate by parts to obtain

$$\int_0^1 H^{-1}(u) \, du = \int_0^1 F^{-1}(u) \, du - \int_0^1 t(1-t) \, dF^{-1}(t).$$

Integrate by parts again to obtain

$$-\int_0^1 t(1-t) \, dF^{-1}(t) = \int_0^1 F^{-1}(u)(1-2u) \, du$$

so that

$$\int_0^1 H^{-1}(t) \, dt = 2 \int_0^1 (1-u) F^{-1}(u) \, du = 2 \int_0^\infty x[1-F(x)] \, dF(x)$$

by a change of variable. ∎

To obtain the asymptotic distribution of T_n the following result is used [cf. Moore (1968)].

Let $X_{1:n} \leq X_{2:n} \leq \cdots \leq X_{n:n}$ be the order statistics from F and

$$S_n = n^{-1} \sum_{i=1}^n L(i/n) X_{i:n},$$

$$\sigma^2 = \sigma^2(F) = 2 \iint_{s<t} L[F(s)] L[F(t)] F(s)[1 - F(t)] \, ds \, dt$$

Theorem 6.5 [*D. S. Moore (1968)*]

If $\sigma^2 < \infty$ and

(a) $E|X| = \int_0^1 |F^{-1}(u)|\, du < \infty$;

(b) L is continuous on $[0, 1]$ except for jump discontinuities at $a_1, \ldots, a_M$, and L' is continuous and of bounded variation on $[0, 1] - \{a_1, \ldots, a_M\}$,

then

$$\mathscr{L}\left\{n^{\frac{1}{2}}\left[S_n - \int_{-\infty}^{\infty} xL[F(x)]\, dF(x)\right]\right\} \to N(0, \sigma^2).$$

To use this result note that

$$H_n^{-1}(i/n) = \int_0^{F_n^{-1}(i/n)} [1 - F_n(u)]\, du$$

$$= \int_0^{X_{i:n}} [1 - F_n(u)]\, du$$

$$= X_{1:n} + (1 - 1/n)(X_{2:n} - X_{1:n}) + \ldots$$

$$+ \left(1 - \frac{(i-1)}{n}\right)(X_{i:n} - X_{i-1:n}). \tag{6.13}$$

Let

$$S_n = n^{-1} \sum_{i=1}^{n-1} H_n^{-1}(i/n) - \mu(H) H_n^{-1}(1)$$

or

$$S_n = n^{-1} \sum_{i=1}^{n} [2(1 - i/n) - \mu(H)] X_{i:n}$$

using (6.13). Let $L(i/n) = 2(1 - i/n) - \mu(H)$ so that

$$S_n = n^{-1} \sum_{i=1}^{n} L(i/n) X_{i:n}.$$

By the previous lemma,

$$\int_0^{\infty} xL[F(x)]\, dF(x) = \int_0^{\infty} x\{2[1 - F(x)] - \mu(H)\}\, dF(x) = 0.$$

By Theorem 6.5

$$\mathscr{L}\{n^{\frac{1}{2}} S_n\} \to N(0, \sigma^2(F))$$

where

$$\sigma^2(F) = 2 \iint\limits_{s<t} \{2[1 - F(s)] - \mu(H)\}$$
$$\times \{2[1 - F(t)] - \mu(H)\} F(s)[1 - F(t)] \, ds \, dt. \quad (6.14)$$

Since

$$T_n = S_n / H_n^{-1}(1)$$

an application of Slutsky's theorem gives the following result.

Theorem 6.6

If

$$\int_0^\infty x \, dF(x) < \infty, \qquad \sigma^2(F) < \infty$$

then

$$\mathscr{L}\left\{n^{\frac{1}{2}}\left[n^{-1}\sum_{i=1}^{n-1} H_n^{-1}(i/n)/H_n^{-1}(1) - \mu(H)\right]\right\} \to N(0, \sigma^2(F)/[H^{-1}(1)]^2)$$

where $\sigma^2(F)$ is given by (6.14).

This result was first obtained by Nadler and Eilbott (1967) by a different and more tedious argument.

6.3 ASYMPTOTIC MINIMAX PROPERTY OF THE CUMULATIVE TOTAL TIME ON TEST STATISTIC

In this section it is assumed, for mathematical convenience, that

$$H^{-1}(1) = \int_0^\infty x \, dF(x) = 1.$$

Consider the problem of testing

$$H_0 : F \underset{c}{=} G$$

versus

$$H_1 : F \underset{c}{<} G \quad \text{and} \quad F \underset{c}{\neq} G$$

(i.e., F IFR), where $G(x) = 1 - e^{-x}$ for $x \geq 0$. If we use the total time on test transformation,

$$H_F^{-1}(t) = \int_0^{F^{-1}(t)} [1 - F(u)]\, du,$$

the problem becomes that of testing

$$H_0: H(t) = t \qquad 0 \leq t \leq 1$$

versus

$$H_1: H(t) \text{ convex and not linear on } [0, 1].$$

Note that $H(t) \leq t$ under H_1.

If C is a class of level α tests for this problem and Ω is a class of alternatives H with H convex, then $\psi \in C$ is said to be *minimax* over Ω and C if and only if it maximizes the minimum power, i.e., if and only if

$$\inf_{H \in \Omega} \beta_\psi(H) = \sup_{\phi \in C} [\inf_{H \in \Omega} \beta_\phi(H)]. \tag{6.15}$$

It is clear that Ω cannot be taken to be all H with H convex since, for this class, the infima in (6.15) would be α and all tests in C would be minimax. Thus, the alternatives in Ω must be "separated". Birnbaum (1953), Chapman (1958) and Doksum (1966) among others have considered alternatives separated by the Kolmogorov distance, i.e., alternatives H, with H convex in this case, and

$$\sup_{t \in [0,1]} [t - H(t)] \geq \Delta.$$

Here, $\Omega(\Delta)$ will denote the class of H with H convex and

$$\sup_{t \in [0,1]} [t - H(t)] \geq \Delta.$$

Extremal classes

The following distributions have Kolmogorov distance Δ, i.e.

$$\sup_{t \in [0,1]} [t - H(t)] = \Delta,$$

and are convex on $[0, 1]$ with $H^{-1}(1) = 1$:

$$H_{u,\Delta}(t) = \begin{cases} a_1 t & 0 \leq t \leq u \\ 1 - a_2(1-t) & u \leq t \leq 1 \end{cases} \quad (\Delta \leq u \leq 1) \tag{6.16}$$

where

$$a_1 = \frac{u - \Delta}{u}, \qquad a_2 = 1 + \frac{\Delta}{1 - u}.$$

(See Figure 6.1.) We shall also need to use

$$H_{u,\Delta}^{-1}(y) = \begin{cases} y/a_1 & 0 \leq y \leq u - \Delta \\ 1 - (1 - y)/a_2 & u - \Delta \leq y \leq 1 \end{cases} \quad (\Delta \leq u \leq 1).$$

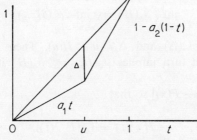

Figure 6.1 Graph of $H_{u,\Delta}$ (Originally published by the University of California Press; reprinted by permission of The Regents of the University of California)

Let

$$h_{u,\Delta}(t) = \begin{cases} a_1 & 0 \leq t \leq u \\ a_2 & u < t \leq 1 \end{cases}$$

denote the density of $H_{u,\Delta}(t)$. The distribution (F IFR) corresponding to $H_{u,\Delta}$ has the form

$$G_{u,\Delta}(x) = \begin{cases} G(a_1 x) & 0 \leq x \leq x_0 \\ G[a_1 x_0 + a_2(x - x_0)] & x_0 \leq x < \infty \end{cases} \quad (6.17)$$

where $x_0 = G^{-1}(u - \Delta)/a_1$. To verify this, compute

$$H_{G_{u,\Delta}}^{-1}(y) = \int_0^{G_{u,\Delta}^{-1}(y)} g G^{-1} G_{u,\Delta}(x)\, dx = \int_0^{G_{n,\Delta}^{-1}(y)} g(a_1 x)\, dx$$

$$= \frac{G_{u,\Delta} G_{u,\Delta}^{-1}(y)}{a_1} = y/a_1$$

for $0 \leq y \leq u - \Delta$. A similar calculation verifies the assertion for $u - \Delta \leq y \leq 1$.

The following lemma is a consequence of the fact that

$$\inf_{H \in \Omega(\Delta)} H(t) = \inf_{\Delta \leq u \leq 1} H_{u,\Delta}(t).$$

Lemma

The distributions $\{H_{u,\Delta}\}$, $0 \leq \Delta \leq u \leq 1$ are least favourable in $\Omega(\Delta)$ for the class of monotone tests in the sense that if ϕ is a monotone test, based on observations $Z_1, Z_2, \ldots, Z_n$ from H, then

$$\inf_{H \in \Omega(\Delta)} \beta_\phi(H) = \inf_{\Delta \leq u \leq 1} \beta_\phi(H_{u,\Delta}).$$

Proof. Suppose $H \in \Omega(\Delta)$ and $\Delta = u - H(u)$. Then $H_{u,\Delta}(x) \geq H(x)$ for $0 \leq x \leq 1$, which in turn implies $\beta_\phi(H_{u,\Delta}) \leq \beta_\phi(H)$ if ϕ is a monotone test. ∎

Let $r(x) = f(x)/[1 - F(x)]$ so that

$$\frac{d}{dt} H^{-1}(t) = 1/r(F^{-1}(t)).$$

We claim that the Kolmogorov distance applied to transforms of distributions provides a reasonable way of separating distributions having different failure rate variation. Suppose that F has transform $H_F \in \Omega(\Delta)$ and $\Delta = u - H(u)$. Then since $H^{-1}(t)$ is concave

$$\sup_{0 \leq t \leq 1} r(F^{-1}(t)) - \inf_{0 \leq t \leq 1} r(F^{-1}(t))$$

$$= \sup_{0 \leq t \leq 1} \left[\frac{d}{dt} H^{-1}(t)\right]^{-1} - \inf_{0 \leq t \leq 1} \left[\frac{d}{dt} H^{-1}(t)\right]^{-1}$$

$$\geq \sup_{0 \leq t \leq 1} \left[\frac{d}{dt} H_{u,\Delta}^{-1}(t)\right]^{-1} - \inf_{0 \leq t \leq 1} \left[\frac{d}{dt} H_{u,\Delta}^{-1}(t)\right]^{-1}$$

$$= a_2 - a_1 = \Delta/u(1 - u) \geq 4\Delta.$$

Hence, large values of Δ correspond to large failure rate variation.

Contiguity

The concept of contiguous alternatives plays a crucial role in this section. (See LeCam (1966), Hájek and Sidák (1967).) Let $\{H_\nu, K_\nu\}$, $\nu \geq 1$, be a sequence of similar testing problems. In this sequence the νth testing problem concerns n_ν observations $X_1, \ldots, X_{n_\nu}$ with $n_\nu \to \infty$. In our set-up H_ν depends on ν through n_ν only, whereas K_ν depends on the parameters Δ_{n_ν}, in addition. The problem is to determine

$$\lim_{\nu \to \infty} \beta(\alpha, H_\nu, K_\nu) = \beta(\alpha) \qquad 0 \leq \alpha \leq 1 \qquad (6.18)$$

where the sequence $\{\Delta_n\}$ will be chosen so that

$$\alpha < \beta(\alpha) < 1.$$

The concept of contiguous alternatives will be useful in computing (6.18).

Definition 6.5

A sequence $\{g_{u,\Delta_n}\}$ is said to be *contiguous* to $g_{u,0}$ (in the sense of LeCam–Hájek) if for any sequence of random variables $R_n(X_1, \ldots, X_n)$, $R_n \to 0$ in P_0 probability implies $R_n \to 0$ in P_{Δ_n} probability where P_θ denotes the probability distribution of $X_1, \ldots, X_n$ if $g_{u,\theta}$ is true.

The following conditions implying contiguity for sequences when

$$\lim_{n \to \infty} n^{\frac{1}{2}} \Delta_n = c, \text{ for some } c, 0 \leq c < \infty,$$

can be found in Bickel and Doksum (1969):

$$\left.\begin{array}{l}(a) \ \partial g_{u,\Delta}(x)/\partial \Delta \neq 0 \quad \text{whenever} \quad g_{u,\Delta}(x) > 0 \\ (b) \ \int_0^\infty \sup\{[\partial g_{u,\Delta}(x)/\partial \Delta]^2 [g_{u,\Delta}(x)]^{-1} : 0 \leq \Delta \leq \delta\} \, dx < \infty\end{array}\right\} \qquad (6.19)$$

for some $\delta > 0$.

It is easy to verify that (b) holds for

$$g_{u,\Delta}(x) = \frac{\partial G_{u,\Delta}(x)}{\partial x}$$

(where $G_{u,\Delta}$ is defined by (6.17) and $G(x) = 1 - e^{-x}$) for some $\delta > 0$ such that $0 < \delta < u < 1$. Hence, by condition (6.19) $\{g_{u,\Delta_n}\}$ are contiguous alternatives to $g_{u,0}$ if $0 < u < 1$. This fact will be needed later.

General scores statistics

If we consider the problem in which the alternative to the uniform distribution is specified then the power can be maximized by using the Neyman–Pearson Lemma. Let

$$h(t) = \frac{d}{dt} H(t).$$

If $Z_{1:n} \leq Z_{2:n} \leq \ldots \leq Z_{n:n}$ are the order statistics from H, then the most powerful (MP) level α test would reject when

$$\sum_{i=1}^{n} \log h(Z_{i:n}) > k_{n,\alpha}.$$

Since $W_{i:n}$ $(1 \leq i \leq n)$ "behave" asymptotically like order statistics from H we are led to consider statistics of the form

$$T_n(J) = n^{-1} \sum_{i=1}^{n} J[W_{i:n}],$$

where J is an increasing function on $[0, 1]$. (Note that since H is convex, h is increasing and so is $J(x) = \log h(x)$.) The corresponding test would reject H_0 for large values of the statistic. Tests based on such statistics are isotonic and hence have isotonic power by Theorem 6.4.

Definition 6.6

The test ψ corresponding to $J(x) = x$, for which

$$\psi[W_{1:n}, \ldots, W_{n:n}] = 1 \quad \text{if } n^{-1} \sum_{i=1}^{n} W_{i:n} > k_{n,\alpha} \quad (6.20)$$

$$= 0 \quad \text{otherwise}$$

is called the *uniform scores* test and

$$n^{-1} \sum_{i=1}^{n} W_{i:n} \quad \left(\text{or } n^{-1} \sum_{i=1}^{n-1} W_{i:n} \text{ since } W_{n:n} \equiv 1\right)$$

is called the *cumulative total time on test statistic*.

Other general scores tests are Fisher's test for the problem of combining tests with

$$J(x) = \log x,$$

the Pearson or exponential scores test with

$$J(x) = -\log(1 - x),$$

and the normal scores test with

$$J(x) = \Phi^{-1}(x)$$

where Φ is the $N(0, 1)$ distribution.

It will be shown that the uniform scores test is asymptotically minimax over a certain natural class of alternatives determined by the Kolmogorov distance and with respect to a class of tests including all of the above examples. Tests based on general scores statistics where J is increasing on $[0, 1]$ are clearly unbiased since they have isotonic power as noted previously.

Theorem 6.7

If $F(0) = 0$,

$$\mu = \int_0^\infty x \, dF(x) < \infty,$$

J is uniformly continuous on $[0, 1]$ and

$$\int_0^1 J\left[\frac{H^{-1}(u)}{H^{-1}(1)}\right] du < \infty$$

then

$$n^{-1} \sum_{i=1}^n J[W_{i:n}] \to \int_0^1 J\left[\frac{H^{-1}(u)}{H^{-1}(1)}\right] du$$

almost surely as $n \to \infty$.

Proof. Without loss of generality it may be assumed that $H^{-1}(1) = 1$. Let $Z_{1:n} \leq \ldots \leq Z_{n:n}$ be order statistics from H. By the strong law of large numbers

$$n^{-1} \sum_{i=1}^n J[Z_{i:n}] \to \int_0^1 J\left[\frac{H^{-1}(u)}{H^{-1}(1)}\right] du$$

almost surely as $n \to \infty$. Since J is uniformly continuous and $|W_{i:n} - Z_{i:n}| \to 0$ uniformly in i/n and almost surely as $n \to \infty$,

$$n^{-1} \sum_{i=1}^n \{J[W_{i:n}] - J[Z_{i:n}]\} \to 0$$

almost surely as $n \to \infty$. ∎

To show consistency of general scores tests it will first be shown that the total time on test transformation converts convex ordering to stochastic ordering.

Theorem 6.8

If $F_1 \underset{c}{<} F_2 \underset{c}{<} G$ and F_1, F_2 have densities f_1 and f_2 respectively, then

$$\frac{H_{F_1}^{-1}(t)}{H_{F_1}^{-1}(1)} \geq \frac{H_{F_2}^{-1}(t)}{H_{F_2}^{-1}(1)} \geq t \qquad (0 \leq t \leq 1). \tag{6.21}$$

Proof. Note that $F_1 \underset{c}{<} F_2$ implies

$$\frac{f_1(x)}{f_2[F_2^{-1}F_1(x)]}$$

is increasing in x or

$$\frac{f_1[F_1^{-1}(u)]}{f_2[F_2^{-1}(u)]}$$

is increasing in u. Hence,

$$\frac{H_{F_1}^{-1}(t)}{H_{F_1}^{-1}(1)} - \frac{H_{F_2}^{-1}(t)}{H_{F_2}^{-1}(1)} = \int_0^t \left[\frac{1}{H_{F_1}^{-1}(1)} \frac{(1-u)}{f_1[F_1^{-1}(u)]} - \frac{1}{H_{F_2}^{-1}(1)} \frac{(1-u)}{f_2[F_2^{-1}(u)]} \right] du$$

$$= \int_0^t \left[\frac{1}{H_{F_1}^{-1}(1)} \frac{f_2 F_2^{-1}(u)}{f_1 F_1^{-1}(u)} - \frac{1}{H_{F_2}^{-1}(1)} \frac{(1-u)}{f_2 F_2^{-1}(u)} \right] du$$

$$\overset{\text{def}}{=} \int_0^t h(u) \frac{(1-u)}{f_2 F_2^{-1}(u)} du.$$

Since

$$\int_0^1 h(u) \frac{(1-u)}{f_2 F_2^{-1}(u)} du = 0$$

and $h(u)$ changes sign at most once and from positive to negative values if at all, it follows that

$$\int_0^t h(u) \frac{(1-u)}{f_2 F_2^{-1}(u)} du \geq 0.$$

The second inequality follows from $H_G^{-1}(t) = t$. ∎

Consistency of general scores tests follows from Theorem 6.7 and the observation that $F \underset{c}{<} G$ and $F \neq G$ implies

$$\int_0^1 J\left[\frac{H^{-1}(u)}{H^{-1}(1)}\right] du > \int_0^1 J(u)\, du$$

by Theorem 6.8 (if J is strictly increasing).

Recall that

$$\mu(H) = \int_0^1 \frac{H^{-1}(u)}{H^{-1}(1)}\, du = \tfrac{1}{2}$$

when $F \underset{c}{=} G$.

Asymptotic normality and efficiency of statistics based on total time on test statistics

Bickel and Doksum (1969) and Bickel (1969) considered four classes of statistics for testing $H_0: F(x) = G_\lambda(x) = 1 - \exp(-\lambda x)$ against IFR alternatives. These four types of statistics were shown to be asymptotically equivalent and it was shown that each of the classes contain asymptotically most powerful statistics for parametric alternatives. It is now shown that the statistics

$$T_n(J) = n^{-1} \sum_{i=1}^{n} J(W_{i:n}) \tag{6.22}$$

based on the total time on test statistics are asymptotically equivalent to these four classes of statistics. Consequently, for a given parametric family $\{F_\theta\}$ of IFR distributions, it is possible to find a $J = J_{F_\theta}$ such that the test that rejects H_0 for large values of $T_n(J)$ is asymptotically most powerful.

Recall that under H_0, $W_{1:n}, \ldots, W_{n-1:n}$ are distributed as the order statistics of a sample of size $n - 1$ from the uniform distribution on $[0, 1]$. Using the following results of Moore (1968) it will be shown that $T_n(J)$ is asymptotically equivalent to

$$S_n(J) = n^{-1} \sum_{i=1}^{n} J'\left(\frac{i}{n}\right) W_{i:n} \tag{6.23}$$

under H_0.

In the course of proving Theorem 6.5, Moore (1968) also proved the following very useful results. As before, let $X_1, X_2, \ldots, X_n$ be a random sample from F and $X_{1:n} \leq X_{2:n} \leq \ldots \leq X_{n:n}$ the corresponding order statistics.

Theorem 6.9 [D. S. Moore (1968)]

Let F and L satisfy the conditions of Theorem 6.5. Then

$$n^{\frac{1}{2}}\left\{\left[n^{-1}\sum_{i=1}^{n} L\left(\frac{i}{n}\right)X_{i:n} - \int_{-\infty}^{\infty} xL[F(x)]\,dF(x)\right] - \left[n^{-1}\sum_{i=1}^{n} B_F(X_i)\right]\right\} \to 0$$

in probability where

$$B_F(x) = -\int_{x}^{\infty} L[F(t)]\,dt + \int_{-\infty}^{\infty} F(t)L[F(t)]\,dt.$$

The reader is referred to Moore's paper for the proof of this theorem. The following corollary, which is an immediate consequence of Theorem 6.9, will be needed. Let $U_1, U_2, \ldots, U_n$ be a random sample from the uniform distribution on $[0, 1]$ and $U_{1:n} \leq U_{2:n} \leq \ldots \leq U_{n:n}$ the corresponding order statistics.

Corollary

Let $F(x) = x$ for $0 \leq x \leq 1$ and $L = J'$ satisfy the conditions of Theorem 6.5. Then

$$n^{\frac{1}{2}}\left\{\left[n^{-1}\sum_{i=1}^{n} J'\left(\frac{i}{n}\right) U_{i:n} - (J(1) - \mu_J)\right] - \left[n^{-1}\sum_{i=1}^{n} J(U_i) - \mu_J\right]\right\} \to 0$$

in probability where

$$\mu_J = \int_0^1 J(x)\, dx.$$

Proof. Substituting $F(x) = x$ in Theorem 6.9 it is seen that

$$\int_{-\infty}^{\infty} xJ'[F(x)]\, dF(x) = \int_0^1 xJ'(x)\, dx = J(1) - \mu_J,$$

while

$$B_F(x) = -\int_x^1 J'(x)\, dx + \int_0^1 xJ'(x)\, dx$$
$$= J(x) - \mu_J. \blacksquare$$

If we define

$$\mu_J = \int_0^1 J(x)\, dx \quad \text{and} \quad \sigma_J^2 = \int_0^1 J^2(x)\, dx - \mu_J^2, \quad (6.24)$$

then the following lemma is obtained.

Lemma

Suppose that $0 < \sigma_J^2 < \infty$, and that J' satisfies condition (b) of Theorem 6.5, then

$$\sqrt{n}\{[S_n(J) - (J(1) - \mu_J)] - [T_n(J) - \mu_J]\} \quad (6.25)$$

converges to zero in probability under H_0, where $S_n(J)$ is defined by (6.23).

The proof is immediate from the above corollary.

Next note that since

$$n\bar{X}W_{i:n} = \sum_{j=1}^{i}(n - j + 1)(X_{j:n} - X_{j-1:n}),$$

if we set $D_{j:n} = (n - j + 1)(X_{j:n} - X_{j-1:n})$ and

$$V_n(J) = -(n\bar{X})^{-1} \sum_{i=1}^{n} J\left(\frac{i}{n}\right) D_{i:n}, \qquad (6.26)$$

then

$$\sqrt{n}\{S_n(J) - [V_n(J) + J(1)]\}$$

tends to zero in probability under H_0. For a given parametric family $\{F_\theta\}$ of distributions, it is shown in Bickel (1969) that there exists a function $a(u) = a_{F_\theta}(u)$ [see Bickel (1969), Equation (2.9)] such that the test that rejects H_0 for large values of $V_n(a)$ is asymptotically most powerful for $\{F_\theta\}$. Using this, the above lemma, and the definition of contiguity, the following theorem is obtained.

Theorem 6.10

If $J = a$ satisfies the conditions of the above lemma and $\{F_\theta\}$ satisfies the conditions of Corollary 2.1 of Bickel (1969), then the test that rejects H_0 for large values of $T_n(a)$ is asymptotically most powerful among all similar tests.

Note that we can write

$$-\bar{X} V_n(J) = n^{-1} \sum_{i=1}^{n} \left\{ (1 - (i-1)/n) \right.$$
$$\left. \times \left[n\left(J\left(\frac{i}{n}\right) - J\left(\frac{i+1}{n}\right)\right) \right] + J\left(\frac{i+1}{n}\right) \right\} X_{i:n}. \quad (6.27)$$

It follows that if we define

$$L(u) = L_J(u) = (1 - u)J'(u) - J(u) \qquad (6.28)$$

and

$$W_n(J) = n^{-1} \sum_{i=1}^{n} L(i/n) X_{i:n}, \qquad (6.29)$$

then

$$\sqrt{n}(W_n(J) - \bar{X} V_n(J))$$

and

$$\sqrt{n}([W_n(J)/\bar{X}] - V_n(J))$$

tend to zero in probability under H_0 where

$$\bar{X} = n^{-1} \sum_{i=1}^{n} X_i.$$

Let

$$\xi_W \stackrel{\text{def}}{=} \int_0^\infty xL[F(x)]\,dF(x) \qquad (6.30)$$

and

$$\tau_W^2 \stackrel{\text{def}}{=} 2 \iint_{0<s<t<\infty} L[F(s)]L[F(t)]F(s)[1-F(t)]\,ds\,dt.$$

If Theorem 6.9 is applied to $W_n(J)$ the following lemma is obtained.

Lemma A

If $\tau_W^2 < \infty$, if $\mu_{F^{-1}} < \infty$, and if L satisfies condition (b) of Theorem 6.5, then $\sqrt{n}\{[W_n(J) - \xi_W] - Q_n(J)\}$ tends to zero in probability, where

$$Q_n(J) = n^{-1} \sum_{i=1}^n B_F(X_i) \qquad (6.31)$$

and

$$B_F(x) = -\int_x^\infty L[F(t)]\,dt + \int_0^\infty F(t)L[F(t)]\,dt.$$

This result establishes the asymptotic equivalence under contiguous alternatives of all the statistics in this section with sums of independent, identically distributed random variables. For the computations of asymptotic power some lemmas are needed.

Lemma B

If the conditions of Lemma A hold, then $E[B_F(X)|F] = 0$.

Proof. If we define $1_x(t) = 1\,(0)$ if $x \leq t\,(x > t)$, then

$$B_F(x) = \int_0^\infty [F(t) - 1_x(t)]L[F(t)]\,dt.$$

The result follows since $E[1_x(t)|F] = F(t)$. ∎

Lemma C

If $F(x) = 1 - \exp(-x)$, then

$$B_F(x) = J[F(x)] - \int_0^x J[F(t)]\,dt.$$

Proof. Note that $d[(1 - F(x))J[F(x)]]/dx = L[F(x)]f(x)$ and that $f(x) = 1 - F(x)$. Thus, integrating by parts

$$\int_0^x F(t)L[F(t)]\,dt = \int_0^x F(t)[1 - F(t)]^{-1}\,d[(1 - F(t))J[F(t)]]$$

$$= F(x)J[F(x)] - \int_0^x [1 - F(t)]J[F(t)]\,d[F(t)[1 - F(t)]^{-1}]$$

$$= F(x)J[F(x)] - \int_0^x J[F(x)]\,dx$$

where the last equality follows from $F(t)[1 - F(t)]^{-1} = e^t - 1$.

Similarly,

$$\int_x^\infty [1 - F(t)]L[f(t)]\,dt$$

$$= \int_x^\infty d[(1 - F(t))J[F(t)]] = -(1 - F(x))J[F(x)]. \blacksquare$$

Lemma D

(i) If $F(x) = G_\lambda(x) = 1 - \exp(-\lambda x)$, then $B_{G_\lambda}(x) = B_{G_1}(\lambda x)/\lambda$.

(ii) If $J(u) = u$, then $B_{G_1}(x) = 2G_1(x) - x$.

Proof.

(i) Follows by setting $x = \lambda t$ in the definition of $B_{G_\lambda}(x)$.

(ii) Is immediate. $\blacksquare$

Lemma E

If $t[1 - F(t)]J[F(t)] \to 0$ as $t \to \infty$, then

(i) $\mu_W = -\int_0^\infty [1 - F(t)]J[F(t)]\,dt$.

(ii) If $F(x) = G_1(x)$, then $\mu_W = -\mu_J$.

Proof.

$$\xi_W = \int_0^\infty t[1 - F(t)]J'[F(t)]\,dF(t) - \int_0^\infty tJ[F(t)]\,dF(t).$$

But

$$\int_0^\infty t[1-F(t)]J'[F(t)]\,dF(t) = \int_0^\infty t[1-F(t)]\,dJ[F(t)]$$
$$= -\int_0^\infty [1-F(t)]J[F(t)]\,dt + \int_0^\infty tJ[F(t)]\,dF(t)$$

by integration by parts; (i) follows.

To show (ii), note that if $u = F(t) = G_1(t)$, then

$$du/dt = \exp(-t) = 1-u;$$

thus

$$\int_0^\infty [1-F(t)]J[F(t)]\,dt = \int_0^1 J(u)\,du. \quad \blacksquare$$

If the results of this section are now put together, we have that

$$\sqrt{n}[T_n(J) - \mu_J] - \sqrt{n}[\bar{X}^{-1}(Q_n(J) - \mu_J) + \mu_J] \qquad (6.32)$$

converges to zero under G_1 and under contiguous alternatives, where

$$Q_n(J) = n^{-1}\sum_{i=1}^n B_{G_1}(X_i). \qquad (6.33)$$

Let

$$\mu(F, J) = \lim_{n\to\infty} \{\bar{X}^{-1}[Q_n(J) - \mu_J] + \mu_J\}$$
$$= [\mu(F^{-1})]^{-1}\left[\int_0^\infty J[G_1(t)]\,dF(t)\right.$$
$$\left. - \int_0^\infty [1-F(t)]J[G_1(t)]\,dt - \mu_J\right] + \mu_J. \qquad (6.34)$$

The following theorem has been proved.

Theorem 6.11

If the conditions of the lemma preceding Theorem 6.10 and Lemma A above are satisfied, and if $\{F_n\}$ is a sequence of alternative distributions contiguous to G_1, then

$$\frac{\sqrt{n}[T_n(J) - \mu_J - \mu(F_n, J)]}{\sigma_J}$$

converges in law to a standard normal variable.

Let
$$\phi = \begin{cases} 1 & \text{if } \sqrt{n}[T_n(J) - \mu(J)]/\sigma_J > k_{n,\alpha} \\ 0 & \text{otherwise.} \end{cases}$$

where $k_{n,\alpha}$ is defined by $\beta_\phi(U) = \alpha$ and U is the uniform distribution on $[0, 1]$. Let

$$\beta_\phi(G_{u,\Delta}) = \Pr_{G_{u,\Delta}} \left\{ \frac{\sqrt{n}\,[T_n(J) - \mu_J - \mu(G_{u,\Delta}, J)]}{\sigma_J} > k_{n,\alpha} - \frac{\mu(G_{u,\Delta}, J)}{\sigma_J} \right\}, \quad (6.35)$$

where $G_{u,\Delta}$ is defined by (6.17) and $G(x) = 1 - e^{-x}$. If

$$\lim_{n \to \infty} n^{\frac{1}{2}} \Delta_n = c > 0,$$

then using Theorem 6.11 it is easy to verify that

$$\lim_{n \to \infty} \beta_\phi(G_{u,\Delta_n})$$

$$= \Phi \left(-k_\alpha + c \frac{\left[-\frac{1}{u} \int_0^u J(x)\,dx + \frac{1}{1-u} \int_u^1 J(x)\,dx \right]}{\sigma_J} \right). \quad (6.36)$$

Asymptotic minimax property

Let F have transform H and

$$H_1^{-1}(t) = H^{-1}(t)/H^{-1}(1).$$

Let $\Omega_1(\Delta)$ be the class of distributions for which

$$\sup_{\Delta < u < 1} [u - H_1(u)] \geq \Delta$$

and H is convex on $[0, 1]$. Note that if $F \in \Omega_1(\Delta)$ there does not necessarily exist $u \in [\Delta, 1]$ such that $F \underset{c}{<} G_{u,\Delta}$. If the discussion is restricted to totally ordered classes of distributions in $\Omega_1(\Delta)$ containing some $G_{u,\Delta}$, then $F \in \Omega_1(\Delta)$ implies $F \underset{c}{<} G_{u,\Delta}$. The union of these classes is contained in the class $\Gamma(\Delta)$ of F in $\Omega_1(\Delta)$ for which there exists a $u \in [\Delta, 1]$ such that $F \underset{c}{<} G_{u,\Delta}$. Suppose that $F \in \Gamma(\Delta)$. Since ϕ is an isotonic test it is known by Theorem 6.4 that

$$\beta_\phi(F) = \Pr\,[n^{\frac{1}{2}}[T_n(J) - \mu_J]/\sigma_J > k_{n,\alpha}|F]$$
$$\geq \Pr\,[n^{\frac{1}{2}}[T_n(J) - \mu_J]/\sigma_J > k_{n,\alpha}|G_{u,\Delta}] = \beta_\phi(G_{u,\Delta})$$

for some $u \in [\Delta, 1]$. Hence

$$\inf [\beta_\phi(F): F \in \Gamma(\Delta)] = \inf [\beta_\phi(G_{u,\Delta}): u \in [\Delta, 1]]. \tag{6.37}$$

Let

$$T_n^{(1)} = n^{-1} \sum_{i=1}^{n-1} W_{i:n}$$

and

$$\psi = \begin{cases} 1 & \text{if } (12n)^{\frac{1}{2}}(T_n^{(1)} - \tfrac{1}{2}) \geq k_{n,\alpha} \\ 0 & \text{otherwise;} \end{cases}$$

i.e., the test based on the cumulative total time on test statistic. Let

$$\mu(H_1) = [H^{-1}(1)]^{-1} \int_0^1 H^{-1}(u)\, du \quad \text{and} \quad \sigma_1^2(H) = \sigma^2(F)[H^{-1}(1)]^{-2}$$

where $\sigma^2(F)$ is given by (6.14).

Lemma A

If

$$\lim_{n \to \infty} n^{\frac{1}{2}} \Delta_n = c,$$

then

$$\lim_{n \to \infty} [\inf \beta_\psi(F): F \in \Gamma(\Delta_n)] = \Phi(-k_\alpha + \sqrt{3}c). \tag{6.38}$$

Proof. Assume $H^{-1}(1) = 1$. By definition

$$\begin{aligned} \beta_\psi(G_{u,\Delta}) &= \Pr_{u,\Delta}[(12n)^{\frac{1}{2}}(T_n^{(1)} - \tfrac{1}{2}) \geq k_{n,\alpha}] \\ &= \Pr_{u,\Delta}[(12n)^{\frac{1}{2}}[(T_n^{(1)} - \tfrac{1}{2}) - (\mu(H_{u,\Delta}) - \tfrac{1}{2})] \\ &\geq k_{n,\alpha} - (12n)^{\frac{1}{2}}(\mu(H_{u,\Delta}) - \tfrac{1}{2})]. \end{aligned}$$

For the distribution G_{u,Δ_n}, $\Delta_n = c/\sqrt{n}$, and the statistic

$$S_n^{(1)} = n^{-1} \sum_{i=1}^n J_0\left(\frac{i}{n}\right) X_{i:n}, \qquad J_0(t) = 2(1-t)$$

it is found that the error term I_{2n} of Moore (1968) is zero, while the second one, I_{3n}, tends to zero in probability uniformly in $u \in [\Delta_n, 1]$, i.e.,

$$\sup_u \Pr_{u,\Delta_n}(|I_{3n}| > \varepsilon) \to 0 \quad \text{as} \quad n \to \infty$$

for each $\varepsilon > 0$. Thus, $n^{-1}[S_n^{(1)} - \mu(H_{u,\Delta})]$ can be expressed as a sum of independent, identically distributed random variables with third moments plus a term that tends to zero uniformly in u. Using the representation

$$\sqrt{n}[T_n^{(1)} - \mu(H_{u,\Delta})] = \frac{\sqrt{n}[S_n^{(1)} - \bar{X}\mu(H_{u,\Delta})]}{\bar{X}}$$

it is found that the same thing is true for

$$\sqrt{n}[T_n^{(1)} - \mu(H_{u,\Delta})].$$

If a uniform version of Slutsky's theorem and the Berry–Essen Theorem are now applied,

$$\sup_{u \in [\Delta_n, 1]} |\beta_\psi(G_{u,\Delta_n}) - \{1 - \Phi[k_{n,\alpha} - (12n)^{\frac{1}{2}}(\mu(H_{u,\Delta_n}) - \tfrac{1}{2})]\}| \to 0 \quad \text{as} \quad n \to \infty$$

is obtained. The result now follows by straightforward computations. ∎

To show that ψ is asymptotically minimax it only remains to show that

$$\lim_{n \to \infty} [\inf \beta_\psi(F) : F \in \Gamma(\Delta_n)] \geq \overline{\lim_{n \to \infty}} [\inf \beta_\phi(F) : F \in \Gamma(\Delta_n)].$$

By (6.38) and (6.36) it will be sufficient to show that

$$\inf_{0 < u < 1} \left[\frac{-\dfrac{1}{u} \int_0^u J(x)\,dx + \dfrac{1}{1-u} \int_u^1 J(x)\,dx}{\sigma_J} \right] \leq \sqrt{3}.$$

Lemma B

$$A_J = \inf_{0 < u < 1} \left[\frac{-u^{-1} \int_0^u J(x)\,dx + (1-u)^{-1} \int_u^1 J(x)\,dx}{\sigma_J} \right]$$

is maximized among all square integrable J on $(0, 1)$ by $J(x) = x$ where

$$\sigma_J^2 = \int_0^1 J^2(x)\,dx - \left[\int_0^1 J(x)\,dx \right]^2.$$

Proof. Since the value of A_J remains unchanged if J is replaced by $aJ + b$, $a > 0$, it may be assumed that

$$\int_0^1 J(x)\,dx = \int_0^1 x\,dx = \tfrac{1}{2}$$

and

$$\int_0^1 J^2(x)\,dx = \int_0^1 x^2\,dx = \tfrac{1}{3}.$$

Let
$$J(x) = x + K(x).$$

Then
$$\int_0^1 K(x)\,dx = 0 \tag{6.39}$$

and
$$\int_0^1 [K^2(x) + 2xK(x)]\,dx = 0, \tag{6.40}$$

$$A_J = A_I + \inf_{0<u<1} \frac{-u^{-1}\int_0^u K(x)\,dx + (1-u)^{-1}\int_u^1 K(x)\,dx}{(1/12)^{\frac{1}{2}}}$$

where $I(x) = x$ and $A_I = \sqrt{3}$ (for $J = I$ the infimum is assumed at every u). Suppose the proposition were false; then there would be, for some K satisfying (6.39) and (6.40),

$$\inf_{0<u<1}\left[-u^{-1}\int_0^u K(x)\,dx + (1-u)^{-1}\int_u^1 K(x)\,dx\right]$$
$$= \inf_{0<u<1} \frac{1}{u(1-u)}\int_u^1 K(x)\,dx > 0,$$

and hence
$$\int_u^1 K(x)\,dx > 0 \quad \text{for all} \quad 0 < u < 1. \tag{6.41}$$

However,
$$\int_0^1 xK(x)\,dx = \int_0^1 dx \int_x^1 K(y)\,dy$$

and hence (6.41) would imply that
$$\int_0^1 xK(x)\,dx > 0$$

which contradicts (6.40). ∎

The basic minimax theorem has now been proved.

Theorem 6.12

If the conditions of the lemma preceding Theorem 6.10 and Lemma A following Theorem 6.10 are satisfied, then

$$\lim_{n \to \infty} [\inf \beta_\psi(F) : F \in \Gamma(\Delta_n)] \geq \overline{\lim_{n \to \infty}} [\inf \beta_\phi(F) : F \in \Gamma(\Delta_n)]$$

for each sequence $\{\Delta_n\}$ *satisfying*

$$\lim_{n \to \infty} n^{\frac{1}{2}} \Delta_n = c,$$

for some $c \in [0, \infty)$.

Thus, it has been shown that the cumulative total time on test statistic is asymptotically minimax over $\Gamma(\Delta)$ in each of the classes of statistics of Bickel and Doksum (1969) and of this section.

Since it was not necessary to assume J increasing in Theorem 6.12, it follows that ψ is also asymptotically minimax for the problem of testing for exponentiality versus decreasing failure rate. (In this case ψ would reject for small values of the statistic.)

6.4 GENERALIZATIONS

Example 6.1 System debugging

In many situations data is observed sequentially in time. A good example of this type of data occurs in the "debugging" of a large system. It is common practice after installing a new complex system such as that involving a missile, aircraft, computer, etc. to "debug" it during the initial portion of its total life. During this debugging period, failures and errors are corrected as they occur, with resulting improvement in subsequent system performance. One mathematical idealization of this process leads to the assumption that the "failure intensity function", say $\lambda(t)$, is decreasing with time. In practice, the debugging phase is considered completed when $\lambda(t)$ reaches an equilibrium or constant value.

Suppose failures occur according to a nonhomogeneous Poisson process with intensity $\lambda(t)$; i.e., if $N(t)$ is the number of failures in $[0, t]$, then

$$\Pr[N(t) = k] = \frac{[\Lambda(t)]^k}{k!} e^{-\Lambda(t)}$$

where

$$\Lambda(t) = \int_0^t \lambda(x)\, dx.$$

The time required for repairs is not counted in describing this process. Given observed failure times $T_1, T_2, \ldots, T_n$ and assuming $\lambda(t)$ is decreasing, we may wish to estimate $\lambda(t)$.

The common sense procedure often used in this situation to estimate $\lambda(t)$ is to draw a graph in which the cumulative number of failures is plotted against elapsed operational time as in Figure 6.2. This data, describing

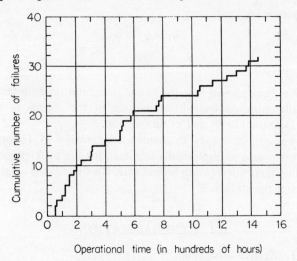

Figure 6.2 Cumulative failures vs. time

failures of an IBM computer, was analysed by Rosner (1961). For example, the failure rate for the interval [600, 800] would be 3/200. Thus, slopes of successive secants represent failure rates over successive time periods.

Let $0 \equiv t_0 < t_1 < t_2 < \ldots < t_k$ be successive interval endpoints. Let n be the number of failures in $[t_{i-1}, t_i)$. Then $n_i/(t_i - t_{i-1})$ estimates the computer failure rate in the ith interval. However, as can be seen from Figure 6.2, these estimates will in general exhibit statistical fluctuations and will not be everywhere decreasing as is believed to be the case. We may wish to smooth our estimates in a least squares sense; i.e., we may seek to

$$\text{minimize} \quad \sum_{i=1}^{k} \left[\frac{n_i}{t_i - t_{i-1}} - x_i \right]^2 (t_i - t_{i-1})$$

subject to $\quad x_1 \geq x_2 \geq \ldots \geq x_k$.

The weights, $t_i - t_{i-1}$, are introduced to reflect possible differences in interval lengths. The solution is of course the isotonic regression discussed in Section 1.2 of Chapter 1. At some stage, we may also wish to test for decreasing trend.

Tests for trend in a series of events

Let $0 \leq T_1 \leq T_2 \leq \ldots \leq T_n$ be observations on a nonhomogeneous Poisson process $\{N(t); t \geq 0\}$ where

$$\Pr[N(t) = k] = \frac{[\Lambda(t)]^k}{k!} e^{-\Lambda(t)} \qquad k = 0, 1, 2, \ldots$$

$$\Lambda(t) = \int_0^t \lambda(x)\, dx \quad \text{and} \quad \lambda(x) > 0$$

is increasing. Note that $\Lambda(t)$ is the mean number of events in $[0, t]$. It is desired to test

$$H_0 : \lambda(t) \text{ constant}$$

versus

$$H_1 : \lambda(t) \text{ increasing}.$$

Let Λ_1 and Λ_2 be the mean functions for two such processes. As in Section 5.5 of Chapter 5, a star-ordering can be introduced on the class of mean functions.

Definition 6.7

$\Lambda_1 \underset{*}{<} \Lambda_2$ if and only if $\Lambda_2^{-1}\Lambda_1(t)/t$ is increasing in $t \geq 0$.

Let $\Lambda_0(t) = \lambda t$ for some $\lambda > 0$. If $\lambda(t) > 0$ is increasing, then clearly

$$\Lambda(t) = \int_0^t \lambda(x)\, dx$$

is convex increasing and $\Lambda \underset{*}{<} \Lambda_0$ since $\Lambda(t)/t$ is increasing in $t \geq 0$.

Theorem 6.13

Let $S_1 \leq S_2 \leq \ldots \leq S_n$ $(T_1 \leq T_2 \leq \ldots \leq T_n)$ be observations from processes with mean functions Λ_1 (Λ_2). If $\Lambda_1 \underset{*}{<} \Lambda_2$, then

$$\left(\frac{S_1}{S_n}, \frac{S_2}{S_n}, \ldots, \frac{S_{n-1}}{S_n}\right) \underset{\text{st}}{\geq} \left(\frac{T_1}{T_n}, \frac{T_2}{T_n}, \ldots, \frac{T_{n-1}}{T_n}\right). \qquad (6.42)$$

Proof. Let $\Lambda_2^{-1}\Lambda_1(S_i) = T_i'$. Clearly, $\Lambda_2^{-1}\Lambda_1(S_i)/S_i$ increasing in i implies

$$\frac{T_i'}{S_i} \underset{\text{st}}{\leq} \frac{T_j'}{S_j}$$

for $i \leq j$ so that in particular

$$\frac{T'_i}{T'_n} \underset{\text{st}}{\leq} \frac{S_i}{S_n} \quad \text{for} \quad i = 1, 2, \ldots, n. \tag{6.43}$$

Also

$$\Pr[T'_i > t] = \Pr[\Lambda_2^{-1}\Lambda_1(S_i) > t]$$
$$= \Pr[S_i > \Lambda_2\Lambda_1^{-1}(t)]$$
$$= \sum_{j=0}^{i-1} \frac{[\Lambda_2(t)]^j}{j!} e^{-\Lambda_2(t)}$$

and similarly the joint distribution of $(T'_1, T'_2, \ldots, T'_n)$ is the same as that of $(T_1, T_2, \ldots, T_n)$. The result follows from (6.43). ∎

The *cumulative total time on test statistic* (normalized) for this model is

$$V_n = \sum_{i=1}^{n-1} T_i/T_n. \tag{6.44}$$

Its distribution under H_0 is given in Table 6.1 for $n \leq 13$.

Definition 6.8

$$\psi(T_1, T_2, \ldots, T_n) = \begin{cases} 1 & \text{if } \sum_{i=1}^{n-1} T_i/T_n > c_{n,\alpha} \\ 0 & \text{otherwise,} \end{cases}$$

where $\Pr_{\Lambda_0}[V_n \geq c_{n,\alpha}] = \alpha$ and $\Lambda_0(t) = t$ for $t \geq 0$.

It follows from Theorem 6.13 that ψ is an isotonic test for the problem with respect to star-ordering on mean functions; i.e., $\Lambda_1 \underset{*}{<} \Lambda_2$ implies

$$\beta_\psi(\Lambda_1) \geq \beta_\psi(\Lambda_2).$$

Cox (1955) and Bartholomew (1956) considered the problem of testing

$$H_0: \beta = 0$$

versus

$$H_1: \beta > 0$$

for the model $\lambda(t) = \alpha e^{\beta t}$. Cox shows that ψ is the best test for this problem. Bartholomew compares an analogue of the Kolmogorov D_n^+ test with ψ. The asymptotic minimax result of Section 6.3 suggests that a similar result may be true for the ψ of Definition 6.8.

Tests for monotone failure rate based on incomplete data

There is a common principle underlying the estimators of Chapter 5 and the tests of this chapter. It is the *total time on test* idea. This concept which is basic to exponential life test theory is also useful in nonexponential situations as was seen in Chapter 5. It is even more useful in analysing incomplete data.

Suppose n units are put on life test at time $t = 0$. Unit i has age a_i when testing commences. It is withdrawn at time $L_i > a_i$ if it does not fail in the interval (a_i, L_i). It is further assumed that $L_i = \infty$ for $i = 1, 2, \ldots, k$ ($1 \leq k \leq n$), so that we are assured of observing at least k failures. (k is fixed in this discussion.) Testing stops at the kth failure.

Let $n(u)$ be the number of units under observation at time u. Since $n(u)$ depends on the failure history to time u, $\{n(u); u \geq 0\}$ is a *stochastic process*. Note, that with historical data, *time* on a given unit is *measured with respect to the birth date of that particular unit*. Let $0 \leq Z_{(1)} \leq Z_{(2)} \leq \ldots \leq Z_{(k)}$ be the observed failure times.

Theorem 6.14

Suppose that n units are sampled. Then for any distribution F ($F(0) = 0$) with failure rate $r(t)$, and fixed k ($1 \leq k \leq n$)

$$Y_i = \int_{Z_{(i-1)}}^{Z_{(i)}} r(u)n(u)\, du, \tag{6.45}$$

$i = 1, 2, \ldots, k$, *are independently distributed with density* $\exp(-y)$.

Proof. Let

$$Y_1 = \int_0^{Z_{(1)}} r(u)n(u)\, du \quad \text{and} \quad S_0(t) = \int_0^t r(u)n(u)\, du.$$

Note that $S_0(t)$ is well defined up to the time of the first observed failure, $Z_{(1)}$ since $n(u)$ depends only on the specified ages at which observation begins and the specified truncation times which are less than $Z_{(1)}$. Then

$$\begin{aligned}
\Pr[Y_1 > y_1] &= \Pr[S_0(Z_{(1)}) > y_1] \\
&= \Pr[Z_{(1)} > S_0^{-1}(y_1)] \\
&= \exp[-S_0(S_0^{-1}(y_1))] \\
&= \exp(-y_1).
\end{aligned}$$

Thus, Y_1 has density $\exp(-y_1)$.

Now let

$$Y_2 = \int_{Z_{(1)}}^{Z_{(2)}} r(u)n(u)\, du \quad \text{and} \quad S_{x_1}(t) = \int_{x_1}^{t} r(u)n(u)\, du.$$

Note that conditionally on $Z_{(1)} = x_1$, S_{x_1} is well defined for $x_1 \leq t < Z_{(2)}$. Hence,

$$\begin{aligned}
\Pr[Y_2 > y_2 | Z_{(1)} = x_1] &= \Pr[S_{x_1}(Z_{(2)}) > y_2 | Z_{(1)} = x_1] \\
&= \Pr[Z_{(2)} > S_{x_1}^{-1}(y_2) | Z_{(1)} = x_1] \\
&= \exp[-S_{x_1}(S_{x_1}^{-1}(y_2))] \\
&= \exp(-y_2).
\end{aligned}$$

Thus, Y_2 is independent of Y_1 and also exponentially distributed with mean 1. ∎

Conditioning in the same manner, the following corollary is established.

Corollary

Suppose that n units are sampled. If $F(x) = 1 - \exp(-\lambda x)$, then

$$Y_i = \int_{Z_{(i-1)}}^{Z_{(i)}} n(u)\, du, \quad i = 1, 2, \ldots, k$$

are independent, exponentially distributed with density $\lambda \exp(-\lambda y)$.

Proof. Let $r(u) = \lambda$ in Theorem 6.14. ∎

If we wish to test

$$H_0 : r \text{ constant}$$

versus

$$H_1 : r \text{ increasing}$$

as in Section 6.2, then, as before, we can use the cumulative total time on test statistic

$$V_k = \sum_{i=1}^{k-1} \int_0^{Z_{(i)}} n(u)\, du \Big/ \int_0^{Z_{(k)}} n(u)\, du. \tag{6.46}$$

Under H_0, V_k will be distributed as the sum of $k - 1$ independent uniform random variables by the above corollary. Again the test, ψ, rejects H_0 for large values of V_k.

Unbiasedness under IFRA alternatives

We cannot show that ψ has isotonic power with respect to star-ordering as was done in Theorem 6.4 because of the possibility of incomplete observations. However, we can show that ψ is unbiased.

Define

$$R(t) = \int_0^t r(u)\,du \quad \text{and} \quad T(t) = \int_0^t n(u)\,du.$$

Lemma

If $R(t)/t$ is nondecreasing in $t \geq 0$, $n(t) \geq 0$ and $T(t)/t$ is nonincreasing in $t \geq 0$, then

(i) $r(t) \geq \int_0^t r(u)\,du/t \geq \int_0^t r(u)\,dT(u)/T(t)$

(ii) $\int_0^t r(u)\,dT(u)/T(t)$ *is nondecreasing in $t \geq 0$,*

when the indicated integrals exist.

Proof. To show (i): the first inequality follows from differentiating $R(t)/t$. Since $R(t)/t \geq 0$ is nondecreasing in $t \geq 0$, $R(t)$ can be approximated arbitrarily closely from below by a positive linear combination of functions of the form

$$R(t) = 0, \quad 0 \leq t < x,$$
$$= t, \quad t \geq x.$$

If the second inequality in (i) can be established for functions $R(t)$ of this type, then the second inequality in (i) will hold in general, by the Lebesgue monotone convergence theorem. For $t < x$, both sides of the second inequality of (i) are zero. For $t \geq x$,

$$\int_0^t n(u)\,dR(u)/T(t) = \left[n(x)x + \int_x^t n(u)\,du\right]\bigg/T(t)$$
$$= 1 + [xn(x) - T(x)]/T(t).$$

Thus, the left side of the second inequality of (i) equals one while the right side is less than one since $xn(x) - T(x) \leq 0$, a consequence of $T(x)/x$ being nonincreasing in $x \geq 0$.

To show (ii): clearly,

$$\frac{d}{dt}\left[\int_0^t r(u)n(u)\,du \Big/ \int_0^t n(u)\,du\right] \geq 0$$

if and only if

$$r(t)n(t)\int_0^t n(u)\,du \geq n(t)\int_0^t r(u)n(u)\,du,$$

which follows from (i). ∎

Note that if $r(t)$ is nondecreasing in $t \geq 0$, then (ii) follows for all $n(t) \geq 0$; i.e., the assumption that $T(t)/t$ is nondecreasing may be dropped.

Theorem 6.15

If F is IFRA with failure rate $r(t)$ and $Z_{(1)} \leq Z_{(2)} \leq \ldots \leq Z_{(k)}$ are the observed failure times, $n(t) \geq 0$ for $t \geq 0$ and $T(t)/t \geq 0$ is nonincreasing in $t \geq 0$, then (conditional on k),

$$\sum_{i=1}^k \Delta_i \int_0^{Z_{(i)}} n(u)\,du \Big/ \int_0^{Z_{(k)}} n(u)\,du \underset{\text{st}}{\geq} \sum_{i=1}^k \Delta_i \left[\sum_{j=1}^i Y_j\right] \Big/ \sum_{j=1}^k Y_j \quad (6.47)$$

when $\Delta_i \geq 0$ ($i = 1, 2, \ldots, k$), and $Y_1, Y_2, \ldots, Y_k$ are independent exponential random variables with unit mean.

Proof. By the above lemma

$$\frac{\int_0^{Z_{(1)}} r(u)n(u)\,du}{\int_0^{Z_{(1)}} n(u)\,du} \leq \frac{\int_0^{Z_{(k)}} r(u)n(u)\,du}{\int_0^{Z_{(k)}} n(u)\,du}.$$

Hence,

$$\frac{\int_0^{Z_{(1)}} n(u)\,du}{\int_0^{Z_{(k)}} n(u)\,du} \geq \frac{\int_0^{Z_{(1)}} r(u)n(u)\,du}{\int_0^{Z_{(k)}} r(u)n(u)\,du} \underset{\text{st}}{=} \frac{Y_1 + \ldots + Y_i}{Y_1 + \ldots + Y_k}$$

by Theorem 6.14. (6.47) is now clear. ∎

Statistics of the form

$$\sum_{i=1}^k \Delta_i \int_0^{Z_{(i)}} n(u)\,du \Big/ \int_0^{Z_{(k)}} n(u)\,du$$

are equivalent to the linear spacings tests considered by Bickel and Doksum (1969) in the complete data case. The cumulative total time on test statistic corresponds to $\Delta_i = 1$ $(i = 1, 2, \ldots, n)$. Clearly, $F \underset{*}{<} G$ implies

$$\beta_\psi(F) \geq \beta_\psi(G)$$

when $G(x) = 1 - \exp(-x)$ for $x \geq 0$.

6.5 COMPLEMENTS

The problem of testing the hypothesis that F is a negative exponential distribution with unknown scale parameter against the alternative that F has monotone increasing nonconstant failure rate (Section 6.2) has been studied by a number of authors; namely, Proschan and Pyke (1967), Nadler and Eilbott (1967), Barlow (1968a), Bickel and Doksum (1969) and Bickel (1969). Bickel and Doksum show that the test proposed by Proschan and Pyke is asymptotically inadmissible. They then take an essentially parametric approach to the problem. In particular, they obtain the studentized asymptotically most powerful linear spacings tests for selected parametric families of distributions which are IFR when the parameter $\theta > 0$ and exponential when $\theta = 0$. Bickel (1969) proves that these tests are actually asymptotically equivalent to the level α tests which are most powerful among all tests which are similar and level α (for the associated parametric problems).

The analogue of the cumulative total time on test statistic (Section 6.4) for tests for trend in a series of events has been described as the oldest known statistical test having been put forward by Laplace in 1773 to test whether comets originate in the solar system. Cox (1955) and Cox and Lewis (1966) have discussed this statistic and its optimality with respect to an interesting parametric alternative.

Epstein (1960a) proposed the test (Definition 6.2) among others for life testing problems. Properties of normalized spacings from IFR distributions (Section 6.2) can be found in Barlow and Proschan (1966). The interpretation of V_n as the slope of the linear regression of the normalized spacings is due to Nadler and Eilbott (1967). The data in Example 6.1 was taken from Proschan (1963). The proof of the asymptotic normality of V_n given in Section 6.2 can be found in Barlow and Doksum (1972). This result was first obtained by Nadler and Eilbott (1967) by a different and more tedious argument. The total time on test transformation first appeared in Marshall and Proschan (1965). However, they made no use of it! The proof of asymptotic minimax optimality (Section 6.3) essentially follows Barlow and Doksum (1972). The proof of Lemma B preceding Theorem 6.12 is due to W. R. van Zwet.

The debugging problem (Section 6.4) was discussed in Barlow, Proschan and Scheuer (1971). The conditional likelihood ratio test for trend was studied by M. T. Boswell (1966). He obtained the asymptotic distribution of the likelihood ratio under the null hypothesis. Unfortunately, it is very complicated. It is related to the likelihood ratio test for testing exponentiality versus IFR alternatives. Barlow (1968a) showed through Monte Carlo studies that the cumulative total time on test statistic is much better than the likelihood ratio statistic. However, no asymptotic comparisons have been made. The incomplete data results were obtained by Barlow and Proschan (1969).

CHAPTER 7

Conditional Expectation Given a σ-Lattice

7.1 INTRODUCTION

If, for each x in a certain set X, there is a random variable $\tilde{y}$ with finite variance, then the conditional expectation of $\tilde{y}$ given $\tilde{x}$ is also the regression function of $\tilde{y}$ on $\tilde{x}$. Placed in an appropriate measure-theoretic context, the conditional expectation of $\tilde{y}$ given $\tilde{x}$ is the conditional expectation of $\tilde{y}$ given the σ-field generated by $\tilde{x}$. Similarly, the isotonic regression of $\tilde{y}$ on $\tilde{x}$ is the conditional expectation of $\tilde{y}$ given a certain σ-lattice of sets. This concept is defined and its properties are studied in this chapter. We shall see in Sections 7.2 and 7.3 that it does give the isotonic regression.

Conditional expectation given a σ-lattice may be defined in L_2, the class of square-integrable functions on a measure space, as projection. This development is given in Section 7.3. It may be defined in L_1, the class of integrable functions, by extension from L_2 and also via an extended Radon–Nikodym theory. The latter will be discussed in Section 7.4. The two definitions coincide in $L_1 \cap L_2$. Each seems to be useful for an appropriate class of extremum problems, though either would suffice for most of the extremum problems which have occurred in applications.

7.2 FORMULATION OF THE ISOTONIC REGRESSION PROBLEM IN TERMS OF PROJECTION IN L_2

In Chapter 1 we supposed given a finite set X, a function g on X, and a weight function w, and set the problem of finding the function f in a certain class of functions on X which minimizes

$$\sum_x [g(x) - f(x)]^2 w(x).$$

Since the set X was finite, real valued functions on X could be interpreted as points in a finite-dimensional space, and the weighted sum of squares as squared distance, giving the regression problem a geometric interpretation. Our purpose in this section is to formulate the regression problem more generally, and to interpret it as the problem of finding the closest point of a given subset of a particular Hilbert space, L_2, to a given point of the space. There are several possible approaches to the regression problem. Three will be sketched, and it will be shown that all may be encompassed within the following formalism. Let $(\Omega, \mathscr{S}, \lambda)$ be a measure space. Even though λ need not be a probability measure, it is convenient to speak of real valued functions on Ω as random variables. Let $L_2 = L_2(\Omega, \mathscr{S}, \lambda)$ denote the class of square-integrable random variables. We suppose that a certain subset A of L_2 and a random variable $\tilde{y} \in L_2$ are given. The regression problem will be interpreted as a projection problem in L_2: that of finding the point in A closest to $\tilde{y}$.

A space $(\Omega, \mathscr{S}, \lambda)$ may enter a regression problem in any of several ways. First, it may be given initially. That is, in the initial formulation of the problem, $\tilde{y}$ may be given as a random variable on a measure space $(\Omega, \mathscr{S}, \lambda)$, and $\tilde{x}$ as a mapping from Ω into a space X; and it may be required to find that real valued function of $\tilde{x}$ in a certain class A of functions of $\tilde{x}$ which is closest to $\tilde{y}$ in the L_2 norm: to minimize

$$\int \{y(\omega) - f[x(\omega)]\}^2 \lambda(d\omega).$$

Another way in which a regression problem may be presented is closer to the situation discussed in the early pages of Section 1.2. A measure space $(X, \mathscr{A}, w)$ is given, and a probability distribution (of a random variable $\tilde{y}$) is associated with each $\tilde{x} \in X$. In Chapter 1, X was finite, $\mathscr{A}$ was the class of all subsets of X, w was a given weight function; the measure w could have been defined by

$$w(A) = \sum_{x \in A} w(x)$$

for $A \in \mathscr{A}$. In such a situation there are two natural ways of formulating the regression problem as a problem of projection in a space $L_2(\Omega, \mathscr{S}, \lambda)$, and both are sketched briefly.

In the approach of Chapter 1 the measure space $(X, \mathscr{A}, w)$ itself plays the role of $(\Omega, \mathscr{S}, \lambda)$; $\tilde{x}$ may be taken as the identity mapping, and the role of $\tilde{y}$ is played by a given function on X, after a preliminary reduction. This reduction is accomplished in the general case as follows. A regression problem is to choose f in a given class of functions on X so as to minimize

$$\int_X E([\tilde{y} - f(x)]^2 | \tilde{x} = x) w(dx).$$

Set $\mu(x) = E(\tilde{y} | \tilde{x} = x)$ for $x \in X$. Then for each $x \in X$,

$$E\{[\tilde{y} - f(x)]^2 | \tilde{x} = x\} = E\{[\tilde{y} - \mu(x)]^2 | \tilde{x} = x\} + [\mu(x) - f(x)]^2.$$

Since the first term on the right does not involve f at all, an equivalent formulation of the regression problem is to minimize

$$\int_X [\mu(x) - f(x)]^2 w(dx)$$

in the given class of functions on X. This is an instance of the general problem formulated above, in which the measure space $(\Omega, \mathcal{S}, \lambda)$ is $(X, \mathcal{A}, w)$, the random variable $\tilde{x}$ is the identity mapping on X, and the role of the given random variable $\tilde{y}$ is played by the function μ on X, which corresponds to the function g in Chapter 1.

An alternative formulation involves directly, rather than through its mean μ, a random variable $\tilde{y}$ having the prescribed probability distribution given x. Again, we suppose the measure space $(X, \mathcal{A}, w)$ is given, as well as a conditional distribution $\nu(\cdot|x)$ for each $x \in X$. We suppose that this conditional distribution, which need not actually be a probability (normed) distribution, is a *transition measure*, i.e.,

for each $x \in X$, $\nu(\cdot|x)$ is a positive measure on the class $\mathcal{B}$ of Borel sets of real numbers;

for each $B \in \mathcal{B}$, $\nu(B|\cdot)$ is an $\mathcal{A}$-measurable function on X.

Let Ω be the Cartesian product space $\Omega = X \times R$, where R is the set of real numbers: Ω is the set of ordered pairs (x, y), $x \in X$, $y \in R$. Let $\mathcal{S} = \mathcal{A} \otimes \mathcal{B}$, the product σ-field: $\mathcal{S}$ is the σ-field generated by the rectangles in Ω, $A \times B$, $A \in \mathcal{A}$, $B \in \mathcal{B}$. It can be shown (for the case of probability measures, cf. Neveu (1965, p. 74)) that there is a positive measure λ on $(\Omega, \mathcal{S})$ such that

$$\lambda(A \times B) = \int_A \nu(B|x) w(dx), \qquad A \in \mathcal{A}, \qquad B \in \mathcal{B}.$$

The coordinate mappings on $\Omega = X \times R$ are denoted by $\tilde{x}$ and $\tilde{y}$: the values associated with (x, y) in Ω by $\tilde{x}$ and $\tilde{y}$ are $x \in X$ and $y \in R$ respectively. If ν is a transition *probability* then it furnishes the conditional probability distribution of $\tilde{y}$ for given x: $\nu(B|x) = \Pr[\tilde{y} \in B | \tilde{x} = x]$. Then

$$\int_\Omega [y - f(x)]^2 \lambda(d(x, y)) = \int_X \int_R [y - f(x)]^2 \nu(dy|x) w(dx)$$
$$= \int_X E([\tilde{y} - f(x)]^2 | \tilde{x} = x) w(dx),$$

so that the regression problem in $(\Omega, \mathcal{S}, \lambda)$ coincides with the given regression problem of minimizing

$$\int_X E([\tilde{y} - f(x)]^2 | \tilde{x} = x) w(dx).$$

Thus in either case a given regression problem can be interpreted as the problem of choosing a point $f(\tilde{x})$ in a subset A of $L_2(\Omega, \mathscr{S}, \lambda)$ consisting of functions of $\tilde{x}$, so as to minimize the squared distance from a given point $\tilde{y}$,

$$\int_\Omega \{y(\omega) - f[x(\omega)]\}^2 \lambda(d\omega) = ||\tilde{y} - f(\tilde{x})||^2,$$

where $||\cdot||$ denotes the norm in $L_2(\Omega, \mathscr{S}, \lambda)$.

Finally, then, an *isotonic* regression problem may be interpreted as follows. There are given a measure space $(\Omega, \mathscr{S}, \lambda)$, a random variable $\tilde{y} \in L_2(\Omega, \mathscr{S}, \lambda)$, and an $\mathscr{S}$-measurable mapping $\tilde{x} = x(\cdot)$ from $(\Omega, \mathscr{S})$ into $(X, \mathscr{A})$. The space X is provided with a quasi-order "$\lesssim$". For every measurable function $f(\cdot)$ from $(X, \mathscr{A})$ into $(R, \mathscr{B})$, the composite function $f[x(\cdot)]$ is a measurable mapping from $(\Omega, \mathscr{S})$ into $(R, \mathscr{B})$, and thus a random variable $f(\tilde{x})$. *An isotonic regression problem is to determine f so as to minimize*

$$\int \{y(\omega) - f[x(\omega)]\}^2 \lambda(d\omega)$$

in the class of isotonic functions f on X such that $f(\tilde{x}) \in L_2(\Omega, \mathscr{S}, \lambda)$.

The remainder of this section is devoted to interpreting this regression problem as that of finding the point of a certain subset $L_2(\mathscr{U})$ of L_2 which is closest to $\tilde{y}$.

Let $\mathscr{U}_X$ denote the class of sets in $\mathscr{A}$ whose indicator functions are isotonic: $U \in \mathscr{U}_X$ if and only if $U \in \mathscr{A}$ and 1_U is isotonic, i.e.,

$$x_1 \lesssim x_2 \Rightarrow 1_U(x_1) \leq 1_U(x_2).$$

Equivalently $U \in \mathscr{U}_X$ if and only if $U \in \mathscr{A}$ and

$$x_1 \lesssim x_2, \, x_1 \in U \Rightarrow x_2 \in U.$$

Sets in $\mathscr{U}_X$ are called *upper sets*. Note that $\phi \in \mathscr{U}_X$ and $X \in \mathscr{U}_X$.

If the requirement $U \in \mathscr{A}$ were omitted, the class defined would be a complete lattice: closed under arbitrary unions and intersections. But since $\mathscr{A}$ is presumed only to be a σ-field, it can be concluded only that $\mathscr{U}_X$ is a σ-*lattice*: closed under *countable* union and intersection. Of course in special cases, such as that of Chapter 1 in which $\mathscr{A}$ is closed under arbitrary union and intersection, so also will be $\mathscr{U}_X$.

Definition 7.1

Let $\mathscr{U}_0$ be a σ-lattice of subsets of a space Ω_0. Then an extended-real-valued function u on Ω_0 is $\mathscr{U}_0$-*measurable* if

$$[u > a] \stackrel{\text{def}}{=} \{\omega : \omega \in \Omega_0, u(\omega) > a\} \in \mathscr{U}_0$$

for all real a; or equivalently if

$$[u \geq a] \in \mathscr{U}_0$$

for all real a. The class of $\mathscr{U}_0$-measurable functions u, $-\infty \leq u(\omega) \leq \infty$ for $\omega \in \Omega_0$, is denoted by $R(\mathscr{U}_0)$.

If $u_i \in R(\mathscr{U}_0)$, $i = 1, 2$, then

$$[u_1 + u_2 > a] = \bigcup_r \{[u_1 > r] \cap [u_2 > a - r]\} \in \mathscr{U}_0$$

for all real a, the union being taken over rationals r. Thus $u_1 + u_2 \in R(\mathscr{U}_0)$. Also, if $u \in R(\mathscr{U}_0)$ and $k \geq 0$, then $ku \in R(\mathscr{U}_0)$. That is, $R(\mathscr{U}_0)$ is a *convex cone*. Further, if $u_i \in R(\mathscr{U}_0)$, $i = 1, 2$, then

$$[u_1 \vee u_2 > a] = [u_1 > a] \cup [u_2 > a] \in \mathscr{U}_0$$

and

$$[u_1 \wedge u_2 > a] = [u_1 > a] \cap [u_2 > a] \in \mathscr{U}_0$$

for all real a, so that $u_1 \vee u_2 \in R(\mathscr{U}_0)$ and $u_1 \wedge u_2 \in R(\mathscr{U}_0)$. That is, $R(\mathscr{U}_0)$ is a *lattice*. Suppose now $u_1 \leq u_2 \leq \ldots \leq u_n \in R(\mathscr{U}_0)$, $n = 1, 2, \ldots$, and

$$\lim_n u_n(\omega) = u(\omega), \qquad \omega \in \Omega_0.$$

Then

$$[u > a] = \bigcup_n [u_n > a] \in \mathscr{U}_0$$

for all real a, hence $u \in R(\mathscr{U}_0)$. Suppose again that $u_1 \geq u_2 \geq \ldots \geq u_n \in R(\mathscr{U}_0)$, $n = 1, 2, \ldots$, and

$$\lim_n u_n(\omega) = u(\omega), \qquad \omega \in \Omega_0.$$

Then

$$[u \geq a] = \bigcap_n [u_n \geq a] \in \mathscr{U}_0$$

for all real a, hence again $u \in R(\mathscr{U}_0)$. That is, $R(\mathscr{U}_0)$ is a σ-complete lattice.

We note that $f \in R(\mathscr{U}_x)$ if and only if f is isotonic and $\mathscr{A}$-measurable. Thus the class of functions f in which we seek to minimize

$$\int \{y(\omega) - f[x(\omega)]\}^2 \lambda(d\omega)$$

is the class of functions f in $R(\mathscr{U}_x)$ such that $f(\tilde{x}) \in L_2 = L_2(\Omega, \mathscr{S}, \lambda)$. Set

$$A = \{f(\tilde{x}) : f \in R(\mathscr{U}_x) \cap L_2\}.$$

The problem may be separated formally into two parts: first, find the point of A closest to $\tilde{y}$ in the L_2-norm; this point is $f(\tilde{x})$ for some function $f \in R(\mathscr{U}_X)$, which is then the desired solution of the problem. The set A will now be described in terms of a sub-σ-lattice of $\mathscr{S}$ induced by $\mathscr{U}_X$.

Set

$$\mathscr{U}_\Omega = x^{-1}(\mathscr{U}_X) = \{x^{-1}(B): B \in \mathscr{U}_X\}$$

and

$$L_2(\mathscr{U}_\Omega) = R(\mathscr{U}_\Omega) \cap L_2.$$

Note that $\mathscr{U}_\Omega \subset \mathscr{S}$ since $\mathscr{U}_X \subset \mathscr{A}$, and $\tilde{x}$ is $\mathscr{S}$-measurable.

Theorem 7.1

$L_2(\mathscr{U}_\Omega)$ *is a closed convex cone and σ-complete lattice in L_2; and $A = L_2(\mathscr{U}_\Omega)$.*

With the proof of this theorem we shall have completed the description of the isotonic regression problem as that of finding the projection (closest point) of a given element $\tilde{y}$ of the Hilbert space $L_2 = L_2(\Omega, \mathscr{S}, \lambda)$ on the closed convex cone and σ-complete lattice $L_2(\mathscr{U}_\Omega)$.

Proof. Since both $R(\mathscr{U}_\Omega)$ and L_2 are convex cones and lattices, so is their intersection $L_2(\mathscr{U}_\Omega)$. Suppose now that $\{u_n\}$ is a sequence of elements of $L_2(\mathscr{U}_\Omega)$ converging in L_2 to $u \in L_2$:

$$\int [u_n(\omega) - u(\omega)]^2 \lambda(d\omega) \to 0 \quad \text{as} \quad n \to \infty.$$

Then there is a subsequence, say $\{u_{n_k}\}$, converging a.e. (λ) to u. Then

$$[u \geq a] = \bigcap_{s=1}^\infty \bigcup_{r=1}^\infty \bigcap_{k=r}^\infty [u_{n_k} > a - (1/s)] \in \mathscr{U}_\Omega$$

for each real a, so that $u \in R(\mathscr{U}_\Omega)$ and hence $u \in L_2(\mathscr{U}_\Omega)$. Thus $L_2(\mathscr{U}_\Omega)$ is *closed* in the L_2 metric. Since $R(\mathscr{U}_\Omega)$ is a σ-complete lattice, so also is $L_2(\mathscr{U}_\Omega)$; that is, if $\{u_n\}$ is a monotonic sequence of elements of L_2 converging in L_2 to $u \in L_2$, then $u \in L_2(\mathscr{U}_\Omega)$. It remains to show that $L_2(\mathscr{U}_\Omega) = A$, i.e., that

$$R(\mathscr{U}_\Omega) \cap L_2 = \{f(\tilde{x}): f \in R(\mathscr{U}_X)\} \cap L_2.$$

It clearly suffices to show that

$$R(\mathscr{U}_\Omega) = \{f(\tilde{x}): f \in R(\mathscr{U}_X)\}.$$

Set

$$C = \{f(\tilde{x}): f \in R(\mathscr{U}_X)\}.$$

One inclusion is immediate: if $f \in R(\mathscr{U}_X)$, then

$$[f(x) > a] = \{\omega : f[x(\omega)] > a\} = x^{-1}\{x : f(x) > a\} \in \mathscr{U}_\Omega$$

by definition, for each real a. Thus

$$C = \{f(\tilde{x}) : f \in R(\mathscr{U}_X)\} \subset R(\mathscr{U}_\Omega).$$

We complete the proof that $u \in R(\mathscr{U}_\Omega)$ implies $u \in C$ in four steps.

(1) *C contains the indicator of each set U in $\mathscr{U}_\Omega$.* By the definition of $\mathscr{U}_\Omega$, if $U \in \mathscr{U}_\Omega$, there is a set B in $\mathscr{U}_X$ such that

$$U = x^{-1}(B) = \{\omega : x(\omega) \in B\}.$$

Then

$$1_U(\omega) = 1_B[x(\omega)] \quad \text{for} \quad \omega \in \Omega,$$

i.e., $1_U = f(\tilde{x})$ where $f = 1_B \in R(\mathscr{U}_X)$. Thus $1_U \in C$.

(2) *C is a convex cone and σ-complete lattice.* That C is a convex cone and lattice follows immediately from the fact that $R(\mathscr{U}_X)$ is. Now suppose $f_n[x(\cdot)]$ is a nondecreasing sequence of extended-real-valued functions converging on Ω to an extended-real-valued function $g(\cdot)$, with $f_n \in R(\mathscr{U}_X)$, $n = 1, 2, \ldots$. Set

$$h_n(x) = \bigvee_{j \leq n} f_j(x), \quad x \in X.$$

Then $h_n \in R(\mathscr{U}_X)$, $n = 1, 2, \ldots$, and $f_n[x(\omega)] = h_n[x(\omega)]$ for $\omega \in \Omega$. Set

$$h = \bigvee_{n \geq 1} h_n = \lim_n h_n;$$

then $h \in R(\mathscr{U}_X)$, and

$$g(\omega) = \lim_n f_n[x(\omega)] = \lim_n h_n[x(\omega)] = h[x(\omega)].$$

Thus $g \in C$. The proof for a nonincreasing sequence is symmetric.

(3) *C contains each nonnegative function in $R(\mathscr{U}_\Omega)$.* Let $u \in R(\mathscr{U}_\Omega)$, $u \geq 0$. For $n = 1, 2, \ldots$, set

$$u_n = \frac{1}{2^n} \sum_{k=1}^{n2^n} 1_{[u \geq k/2^n]}.$$

Since $u \in R(\mathscr{U}_\Omega)$, for each k and n, $1_{[u \geq k/2^n]} \in C$, by (1). Then since C is a convex cone, $u_n \in C$ for each n. Finally, u_n increases to u, so that $u \in C$ by (2).

(4) *C contains $R(\mathscr{U}_\Omega)$.* It is clear from the definition that C contains every constant function. If $u \in R(\mathscr{U}_\Omega)$ and if u is bounded below by the constant function m, then $u - m \geq 0$ so that $u - m \in C$ by (3); then $u \in C$ since C is

closed under addition. Now if u is an arbitrary member of $R(\mathcal{U}_\Omega)$ and m is an integer, $u \vee m$ is a function in $R(\mathcal{U}_\Omega)$ bounded below by m, and so $u \vee m \in C$. Finally, $u \vee m$ decreases to u as m decreases to $-\infty$, so that $u \in C$ by (2). ∎

7.3 CONDITIONAL EXPECTATION GIVEN A σ-LATTICE AS PROJECTION IN L_2

Let H be a complete real inner product space: a complete metric space with distance between elements f, g given by $||f - g||$, where $||h|| = (h, h)^{\frac{1}{2}}$ for $h \in H$ and where $(\cdot, \cdot)$ denotes the inner product. An example of such a space is furnished by $L_2(\Omega, \mathscr{S}, \lambda)$, with inner product

$$(u, v) = \int u(\omega) v(\omega) \lambda(d\omega).$$

If $u \in H$, $v \in H$, then seg uv will denote the segment joining u and v:

$$\text{seg } uv \stackrel{\text{def}}{=} \{h : h \in H, \exists \alpha \in [0, 1] \text{ such that } h = (1 - \alpha)u + \alpha v\}.$$

A subset of H will be called *closed* if it is closed in the topology of the metric $||u - v||$. A subset A of H is *convex* if $u \in A$ and $v \in A$ imply seg $uv \subset A$.

The existence, stated in Theorem 7.2, of a projection of a given point of H onto a given closed convex set is well known and easily proved using the identity

$$||u + v||^2 + ||u - v||^2 = 2(||u||^2 + ||v||^2), \tag{7.1}$$

which, in turn, is easily verified.

Theorem 7.2

Let A be a closed convex subset of H, and let $y \in H$. Then there is a unique closest point $y^* = P(y|A)$ of A to y.

Proof. Let

$$\gamma = \inf_{u \in A} ||y - u||.$$

Let $\{u_n\}$ be a sequence of points in A such that

$$\lim_n ||u_n - y|| = \gamma.$$

From (7.1),

$$\tfrac{1}{4}||u_m - u_n||^2 = \tfrac{1}{2}||u_n - y||^2 + \tfrac{1}{2}||u_m - y||^2$$
$$- ||\tfrac{1}{2}(u_n + u_m) - y||^2. \tag{7.2}$$

Since A is convex, $\frac{1}{2}(u_m + u_n) \in A$, so that $||\frac{1}{2}(u_m + u_n) - y||^2 \geq \gamma$ for all positive integers m and n. But $||u_n - y||^2 \to \gamma$ as $n \to \infty$ and $||u_m - y||^2 \to \gamma$ as $m \to \infty$. It follows that $\{u_n\}$ is a Cauchy sequence in H and has a limit, which is denoted by y^* and also by $P(y|A)$. To prove uniqueness, suppose there are two closest points, y_1^* and y_2^*: $y_i^* \in A$, $||y_i^* - y|| = \gamma$, $i = 1, 2$. Applying (7.2) with u_m and u_n replaced by y_1^* and y_2^* respectively we find that $||y_1^* - y_2^*|| = 0$, i.e., $y_1^* = y_2^*$. ∎

Corollary

If $y \in L_2(\Omega, \mathscr{S}, \lambda)$ and if $\mathscr{U}$ is a sub-σ-lattice of $\mathscr{S}$, then there exists a unique (up to λ-equivalence) $\mathscr{U}$-measurable function $y^* = E(y|\mathscr{U}) \in L_2$ minimizing in the class of $\mathscr{U}$-measurable functions u on L_2 the integral $\int(y - u)^2 \, d\lambda$.

Theorem 7.3

If $y \in H$ and if A is a closed convex subset of H then an element v in A is the closest point $y^* = P(y|A)$ of A to y if and only if

$$(y - v, v - h) \geq 0 \quad \text{for all} \quad h \in A. \tag{7.3}$$

Proof. For every $h \in A$, the closest point y^* of A to y is also the closest point to y of seg y^*h. If $u \in \text{seg } y^*h$, there exists α such that $0 \leq \alpha \leq 1$ and $u = (1 - \alpha)y^* + \alpha h$. Then

$$||y - u||^2 = ||(y - y^*) - \alpha(h - y^*)||^2$$
$$= ||y - y^*||^2 + 2\alpha(y - y^*, y^* - h) + \alpha^2 ||y^* - h||^2.$$

This function of α achieves its minimum on $0 \leq \alpha \leq 1$ at $\alpha = 0$ ($u = y^*$) hence its derivative at $\alpha = 0$, $2(y - y^*, y^* - h)$, is non-negative.

Conversely, if $v \in A$ and if $(y - v, v - h) \geq 0$ for every $h \in A$, then for every $h \in A$, v is the closest point of seg vh to y. In particular, v is the closest point of seg vy^* to y. Since by Theorem 7.2 the closest point is unique, v and y^* must coincide. ∎

Theorem 7.4

If $y \in H$, if A is a closed convex subset of H, then $-A = \{u: -u \in A\}$ is a closed convex subset of H and $P(-y|-A) = -P(y|A)$.

Proof. Set $y^* = P(y|A)$. If $h \in -A$ then $-h \in A$ and

$$((-y) - (-y^*), (-y^*) - h) = (y - y^*, y^* - (-h)) \geq 0.$$

Theorem 7.3 then implies that $-y^* = P(-y|-A)$. ∎

In the following immediate corollary of Theorem 7.4, $\mathscr{U}^c$ represents the σ-lattice consisting of sets which are complements of sets in the σ-lattice $\mathscr{U}$.

Corollary

If $y \in L_2(\Omega, \mathscr{S}, \lambda)$, if $\mathscr{U}$ is a sub-σ-lattice of $\mathscr{S}$, then $E(-y|\mathscr{U}^c) = -E(y|\mathscr{U})$.

Proof. Let A denote the closed convex subset of $L_2(\Omega, \mathscr{S}, \lambda)$ consisting of $\mathscr{U}$-measurable functions. Then $-A$ is the subset of L_2 consisting of $\mathscr{U}^c$-measurable functions, and the conclusion of the corollary follows from Theorem 7.4. ∎

The term "operator" is used in this section to denote a mapping from H into itself; it need not be linear. Thus projection on a closed convex set is an operator. A *unitary* operator is by definition a *linear* operator which preserves the inner product (and hence distance):

$$(Ty, Tz) = (y, z).$$

Theorem 7.5

If $y \in H$, if A is a closed convex subset of H, and if T is a unitary operator on H, then TA is a closed convex set and $P(Ty|TA) = TP(y|A)$. If k is real then $kA = \{ku : u \in A\}$ is a closed convex set and $P(ky|kA) = kP(y|A)$. If a is real, $ak > 0$, then

$$P(ay + (k-a)P(y|A)|kA) = kP(y|A). \tag{7.4}$$

Loosely expressed, this theorem states that projection is preserved under stretchings and under distance preserving transformations.

Proof. Let $y^* = P(y|A)$. Then $(Ty - Ty^*, Ty^* - Th) = (y - y^*, y^* - h) \geq 0$ for all $h \in A$. By Theorem 7.3, $Ty^* = P(Ty|TA)$. Also $(ay + (k-a)y^* - ky^*, ky^* - kh) = ak(y - y^*, y^* - h) \geq 0$ for all $h \in A$, hence $P(ay + (k-a)y^*|kA) = ky^*$; the preceding equality in the statement of Theorem 7.5 is obtained on setting $a = k$. ∎

Corollary

If $y \in L_2(\Omega, \mathscr{S}, \lambda)$, if $\mathscr{U}$ is a sub-σ-lattice of $\mathscr{S}$, and if $k \geq 0$ then $E(ky|\mathscr{U}) = kE(y|\mathscr{U})$.

Proof. Let A denote the subset of $L_2(\Omega, \mathscr{S}, \lambda)$ consisting of $\mathscr{U}$-measurable functions. Then $kA = A$ and the above corollary is an immediate consequence of Theorem 7.5. ∎

Theorem 7.6

Projection on a closed convex set is a distance reducing operator.

Proof. Let $y_i \in H$, $i = 1, 2$, let A be a closed convex subset of H, and let $y_i^* = P(y_i|A)$, $i = 1, 2$. We assert that $||y_1^* - y_2^*|| \leq ||y_1 - y_2||$. But this is immediate from Theorem 7.3 and the equations

$$\begin{aligned}
||y_2 - y_1||^2 &= ||y_2 - y_2^* + y_2^* - y_1^* + y_1^* - y_1||^2 \\
&= ||y_2 - y_2^* + y_1^* - y_1||^2 + ||y_2^* - y_1^*||^2 \\
&\quad + 2(y_1 - y_1^*, y_1^* - y_2^*) + 2(y_2 - y_2^*, y_2^* - y_1^*) \\
&\geq ||y_2^* - y_1^*||^2. \blacksquare
\end{aligned}$$

Corollary

If $y_i \in L_2(\Omega, \mathscr{S}, \lambda)$, $i = 1, 2$, and if $\mathscr{U}$ is a sub-σ-lattice of $\mathscr{S}$, then $\int (y_2^* - y_1^*)^2 \, d\lambda \leq \int (y_2 - y_1)^2 \, d\lambda$.

Definition 7.2

An operator T on Hilbert space H is *strictly monotonic at z* if $Ty \neq Tz$ implies $(y - z, Ty - Tz) > 0$. It is *strictly monotonic* if strictly monotonic at each $z \in H$.

Theorem 7.7

If A is a closed convex subset of H then $P(\cdot|A)$ is strictly monotonic. Indeed, if $y_i \in H$, $y_i^* = P(y_i|A)$, $i = 1, 2$, then

$$(y_1 - y_2, y_1^* - y_2^*) \geq ||y_1^* - y_2^*||^2. \tag{7.5}$$

Proof. Let $y_i^* = P(y_i|A)$, $i = 1, 2$. Then

$$\begin{aligned}
(y_1 - y_2, y_1^* - y_2^*) &= (y_1 - y_1^*, y_1^* - y_2^*) + (y_1^*, y_1^* - y_2^*) \\
&\quad - (y_2 - y_2^*, y_1^* - y_2^*) - (y_2^*, y_1^* - y_2^*) \\
&= (y_1 - y_1^*, y_1^* - y_2^*) + (y_2 - y_2^*, y_2^* - y_1^*) \\
&\quad + ||y_1^* - y_2^*||^2.
\end{aligned}$$

Thus, by Theorem 7.3,

$$(y_1 - y_2, y_1^* - y_2^*) \geq ||y_1^* - y_2^*||^2 \geq 0. \blacksquare$$

We remark in passing that an alternative proof of Theorem 7.6 is furnished by (7.5) and the Schwarz–Buniakowski inequality:

$$||y_1^* - y_2^*||^2 \leq (y_1 - y_2, y_1^* - y_2^*) \leq ||y_1 - y_2|| \cdot ||y_1^* - y_2^*||.$$

Theorem 7.8

If $y \in H$ and if A is a closed convex cone in H, then $v = P(y|A)$ if and only if $v \in A$,

$$(y - v, v) = 0 \qquad (7.6)$$

and

$$(y - v, h) \leq 0 \qquad (7.7)$$

for all h in A. Also if $y^* = P(y|A)$ and if $a \geq -1$ then

$$P(y + ay^*|A) = (a + 1)y^*. \qquad (7.8)$$

Proof. Suppose first that $v = y^* = P(y|A)$. Apply Theorem 7.3 with $h = kv$ where $k > 1$. We know from (7.3) that $(y - v, v) \leq 0$. With $0 < k < 1$ we find that $(y - v, v) \geq 0$, verifying (7.6). Then (7.7) follows from (7.3) and (7.6). Conversely, suppose (7.6) and (7.7) are satisfied for some $v \in A$. Inequality (7.3) follows, so that $v = y^*$. Applying the first part of the theorem with $y + ay^*$ in place of y and $(a + 1)y^*$ in place of v yields the last statement of the theorem. ∎

Corollary A

If $y \in L_2(\Omega, \mathscr{S}, \lambda)$ and if $\mathscr{U}$ is a sub-σ-lattice of $\mathscr{S}$ then $v = y^* = E(y|\mathscr{U})$ if and only if $v \in L_2(\mathscr{U})$,

$$\int (y - v) v \, d\lambda = 0 \qquad (7.9)$$

and

$$\int (y - v) h \, d\lambda \leq 0 \qquad (7.10)$$

for all $h \in L_2(\mathscr{U})$. Also $E(y + ay^*|\mathscr{U}) = (a + 1)y^*$ for all $a \geq -1$.

The theorems preceding Theorem 7.8 have application to more general kinds of regression than isotonic regression, since A is required only to be a closed convex set and not a cone or lattice. Theorem 7.8 adds only the property of being a cone. If any closed convex cone A in $L_2(\Omega, \mathscr{S}, \lambda)$ replaces $L_2(\mathscr{U})$ in (7.9), the new version of (7.10) and (7.9) together characterize projection on A. Suppose for example that $\tilde{x}$ and $\tilde{y}$ are random variables on $(\Omega, \mathscr{S}, \lambda)$ and that A consists of functions $f(\tilde{x}) \in L_2$ such that f is convex on the range of $\tilde{x}$. Then A is a closed convex cone in L_2. Or for another example, let $\phi_1, \ldots, \phi_k$ be prescribed real functions on the range of $\tilde{x}$ such that $\phi_j(\tilde{x}) \in L_2, j = 1, 2, \ldots, k$, and let A consist of linear combinations

$$f(\tilde{x}) = \sum_{j=1}^{k} c_j \phi_j(\tilde{x}),$$

where some or all (or none) of the coefficients $c_1, \ldots, c_k$ are required to be non-negative. Again A is a closed convex cone in L_2.

If $(\Omega, \mathscr{S}, \lambda)$ is a probability space and A contains all constant functions then (7.9) and (7.10) can be expressed in terms of variance and covariance. Note first that $Ey = Ey^*$.

Corollary B

If $y \in L_2(\Omega, \mathscr{S}, \lambda)$, if A is a closed convex cone in L_2 containing the constant functions, and if $y^* = P(y|A)$ then

$$\int y \, d\lambda = \int y^* \, d\lambda. \tag{7.11}$$

Proof. This follows from (7.10), on setting h first equal to the constant function 1, and then equal to -1. ∎

Corollary C

If $(\Omega, \mathscr{S}, \lambda)$ is a probability measure space, if A is a closed convex cone in L_2 containing the constant functions, and if $y \in L_2$, then a function v in A is equal to $y^* = P(y|A)$ if and only if

$$Ev = Ey, \tag{7.12}$$

$$\mathrm{Cov}\,(v, h) \geq \mathrm{Cov}\,(y, h) \tag{7.13}$$

for all $h \in A$, and

$$\mathrm{Var}\,v = \mathrm{Cov}\,(v, y). \tag{7.14}$$

Further, for each real a, $P(y + a|A) = P(y|A) + a$.

Proof. The first conclusion is immediate from Corollaries A and B above. The second follows from the first. ∎

Corollary D

$\mathrm{Var}\,y = E(y - y^*)^2 + \mathrm{Var}\,y^*$, *hence* $\mathrm{Var}\,y^* \leq \mathrm{Var}\,y$ *with equality if and only if $y^* = y$.*

Corollary E

If $(\Omega, \mathscr{S}, \lambda)$ *is a probability measure space, if* A *is a closed convex cone in* L_2 *containing the constant functions, and if* $y \in L_2$, *then*

  (i) $y^* = P(y|A)$ *is the constant (degenerate random variable)* Ey *if and only if all nondegenerate random variables in* A *have negative correlations with* y.
  (ii) *If there exists a nondegenerate random variable in* A *having non-negative correlation with* y *then* $y^* = P(y|A)$ *is the random variable in* A *most highly correlated with* y.

Proof. We show first that $\operatorname{Var} y^* \geq \operatorname{Var} h$ for every h in A such that $\operatorname{Var} h = \operatorname{Cov}(y, h)$ and $Eh = Ey$. In order to see this, let $\mu_y = Ey = Ey^*$, and let A_1 denote the subset of A consisting of functions h in A such that $Eh = \mu_y$, $\operatorname{Cov}(y, h) = \operatorname{Var} h$. Now y^* minimizes $E(y - h)^2$ in A; since $y^* \in A_1$, y^* also minimizes, in A_1, $E(y - h)^2 = \operatorname{Var} y - 2\operatorname{Cov}(y, h) + \operatorname{Var} h = \operatorname{Var} y - \operatorname{Var} h$. That is, $\operatorname{Var} y^* \geq \operatorname{Var} h$ for $h \in A_1$. Suppose now that there exist nondegenerate random variables in A which have non-negative correlation with y. Since $\operatorname{Cov}(y, y^*) = \operatorname{Var} y^* \geq 0$, to prove (ii) it suffices to prove that the correlation between y and y^* is at least as large as that between y and h_0 for an arbitrary $h_0 \in A$ such that $\operatorname{Cov}(y, h_0) > 0$. Set $h_1 = \mu_y + k(h_0 - Eh_0)$, where $k = \operatorname{Cov}(y, h_0)/\operatorname{Var} h_0$. Then $h_1 \in A_1$, whence $\operatorname{Var} y^* \geq \operatorname{Var} h_1$. Then

$$\operatorname{Cov}(y, y^*)/\sigma_y \sigma_y^* = \operatorname{Var} y^*/\sigma_y \sigma_y^* = \sigma_y^*/\sigma_y$$

$$\geq \sigma_{h_1}/\sigma_y = \operatorname{Var} h_1/\sigma_y \sigma_{h_1} = \operatorname{Cov}(y, h_1)/\sigma_y \sigma_{h_1} = \operatorname{Cov}(y, h_0)/\sigma_y \sigma_{h_0},$$

as was to be proved. On the other hand, suppose all nondegenerate random variables in A have negative correlation with y. Since (7.14) implies $\operatorname{Cov}(y, y^*) \geq 0$, $\operatorname{Var} y^* = \operatorname{Cov}(y, y^*) = 0$, so that y^* is degenerate. By (7.12) its constant value is Ey. ∎

The space $L_2 = L_2(\Omega, \mathscr{S}, \lambda)$ is a Hilbert space possessing in addition a lattice structure: if $u, v \in L_2$, so do $u \wedge v$ and $u \vee v$. Two useful properties involving this lattice structure are:

$$(u, v) = \int uv \, d\lambda = \int (u \wedge v)(u \vee v) \, d\lambda = (u \wedge v, u \vee v)$$

and

$$u \geq 0, v \geq 0 \Rightarrow (u, v) = \int uv \, d\lambda \geq 0.$$

Let us assume such a lattice structure for the Hilbert space H. Suppose H is not only a Hilbert space but also an ordered vector lattice (Riesz space). That is, there is a partial order "$\lesssim$" on H such that every pair u, v of elements has an infimum, $u \wedge v$, and a supremum, $u \vee v$, and such that

$$u \lesssim v \Rightarrow u + h \lesssim v + h, \qquad (7.15)$$

for $u, v, h \in H$;

$$u \lesssim v \Rightarrow ku \lesssim kv \qquad (7.16)$$

for $u, v \in H$ and $k > 0$. Then it can be shown that

$$u + v = u \vee v + u \wedge v \qquad (7.17)$$

for all $u, v \in H$. It is assumed further that

$$(u, v) = (u \vee v, u \wedge v) \qquad (7.18)$$

for $u, v \in H$ and

$$u \gtrsim 0, v \gtrsim 0 \Rightarrow (u, v) \geq 0. \qquad (7.19)$$

Theorem 7.9

If $y_i \in H$, $i = 1, 2$, if A is a closed convex cone and lattice in H, if $y_1^* = P(y_i|A)$, $i = 1, 2$, and if $y_2 \gtrsim y_1$ then $y_2^* \gtrsim y_1^*$.

Proof. Set $\bar{y} = y_1^* \vee y_2^*$, $\underline{y} = y_1^* \wedge y_2^*$. We note first that by (7.3)

$$(y_1 - y_1^*, y_1^* - \underline{y}) \geq 0,$$

since $\underline{y} \in A$. Also, from (7.19)

$$(y_2 - y_1, y_1^* - \underline{y}) \geq 0.$$

Adding these inequalities yields $(y_2 - y_1^*, y_1^* - \underline{y}) \geq 0$, or, using (7.17), $(y_2, \bar{y} - y_2^*) \geq (y_1^*, y_1^* - \underline{y})$. This latter is equivalent to

$$-2(y_2, y_2^*) \geq 2||y_1^*||^2 - 2(y_2, \bar{y}) - 2(y_1^*, \underline{y}).$$

Also, (7.17) and (7.18) imply that

$$||y_1^*||^2 + ||y_2^*||^2 = ||\underline{y}||^2 + ||\bar{y}||^2,$$

so that, adding $||y_2||^2 + ||y_2^*||^2$ to each member of the inequality above,

$$||y_2 - y_2^*||^2 \geq ||y_2 - \bar{y}||^2 + ||y_1^* - \underline{y}||^2.$$

Since $\underline{y}$ and $\bar{y}$ are in A, and $u = y_2^*$ minimizes $||y_2 - u||^2$ in A, this last inequality implies $y_1^* = \underline{y}$ and hence $\bar{y} = y_2^*$. ∎

Corollary A

If $y_i \in L_2(\Omega, \mathscr{S}, \lambda)$, $i = 1, 2$, if $\mathscr{U}$ is a sub-σ-lattice of $\mathscr{S}$, if $y_i^* = E(y_i|\mathscr{U})$, and if $y_2 \geq y_1$, then $y_2^* \geq y_1^*$.

Corollary B

If $y \in L_2(\Omega, \mathscr{S}, \lambda)$, if $\mathscr{U}$ is a sub-σ-lattice of $\mathscr{S}$, if $y^* = E(y|\mathscr{U})$, if $h \in L_2(\mathscr{U})$ and if $y \leq h$ ($y \geq h$) then $y^* \leq h$ ($y^* \geq h$).

Proof. This follows from Corollary A above on noting that $h^* = h$.

We close this section with a characterization due to Dykstra (1970a) of conditional expectation given a σ-lattice as an operator on H. First a characterization of projection on a closed convex cone is given. An operator T is *positively homogeneous* if $y \in H$, $a \geq 0$ imply $T(ay) = aTy$. It is *distance reducing* if $y, z \in H$ imply $||Ty - Tz|| \leq ||y - z||$.

Proposition 7.1

Let T be a positively homogeneous, idempotent, distance reducing operator, strictly monotonic at 0, with convex range A, on a Hilbert space H. Then $T = P(\cdot|A)$.

Remark

If $T(0) = 0$ then T is strictly monotonic at 0 if and only if for each $y \in H$ there is a positive $a = a(y)$ such that $||y||^2 \geq ||y - aTy||^2$.

For if the latter holds, then $(y, Ty) \geq (a/2)||Ty||^2$, so that T is strictly monotonic at 0. On the other hand, if T is strictly monotonic at 0 and if $Ty \neq 0$ then $||y||^2 \geq ||y - aTy||^2$ where $a \leq 2(y, Ty)/||Ty||^2$. Thus strict monotonicity at 0 has an obvious geometric interpretation: if $T(0) = 0$, T is strictly monotonic at 0 if and only if for each $y \in H$ some positive multiple of Ty is at least as close to y as is 0.

Proof of Proposition 7.1. By hypothesis the range A of T,

$$A = \{f \in H : \exists\, y \in H \text{ such that } Ty = f\}$$

is convex. Since T is distance reducing, A is closed; for if $f_n \in A$, $n = 1, 2, \ldots$, $||f_n - y|| \to 0$ as $n \to \infty$, then $||f_n - Ty|| = ||Tf_n - Ty|| \leq ||f_n - y||$

$\to 0$ as $n \to \infty$, hence $y = Ty \in A$. Since T is positively homogeneous, A is a cone containing the origin. It will now be shown that $y \in H, f \in A$ imply

$$(y - Ty, f) \leq 0. \tag{7.20}$$

For if $c > 0$, $||Ty - cf||^2 = ||Ty - T(cf)||^2 \leq ||y - cf||^2$, hence

$$||Ty||^2 \leq ||y||^2 - 2c(y - Ty, f).$$

Since this holds for arbitrary positive c and since $||Ty|| < \infty$ we conclude that $(y - Ty, f) \leq 0$. Now set $T_1 = P(\cdot | A)$, projection on the closed convex cone A. We observed that

$$T_1 y = 0 \Rightarrow Ty = 0. \tag{7.21}$$

For by Theorem 7.8, $0 \geq (y - T_1 y, Ty) = (y, Ty)$ if $T_1 y = 0$. Since T is strictly monotonic at 0, it follows that $Ty = 0$.

Now from the Buniakowski–Schwarz inequality, Theorem 7.8, and (7.20) we have, for $y \in H$,

$$||T_1 y||^2 = (y, T_1 y) \leq (Ty, T_1 y) \leq ||Ty|| \, ||T_1 y||.$$

Hence $||T_1 y|| \leq ||Ty||$ with equality if and only if there is a real number a such that $T_1 y = aTy$. In this latter case $a^2 = 1$, and we can assert that in fact $Ty = T_1 y$.

For suppose $T_1 y = -Ty$; then, from (7.20),

$$0 \geq (y - Ty, T_1 y) = (y - T_1 y + 2T_1 y, T_1 y) = 2||T_1 y||^2$$

by Theorem 7.8, hence $||T_1 y|| = ||Ty|| = 0$, and $T_1 y = Ty$. Thus $||T_1 y|| \leq ||Ty||$ always, and $||T_1 y|| = ||Ty||$ if and only if $T_1 y = Ty$. Finally, from Theorem 7.8, $T_1(y - T_1 y) = 0$, so that from (7.21) also $T(y - T_1 y) = 0$. Then, since T is distance reducing,

$$||T_1 y||^2 = ||y - (y - T_1 y)||^2 \geq ||Ty - T(y - T_1 y)||^2 = ||Ty||^2.$$

But, since also $||T_1 y|| \leq ||Ty||$, $||T_1 y|| = ||Ty||$ follows and hence $T_1 y = Ty$. ∎

Definition 7.3

Let $(\Omega, \mathscr{S}, \lambda)$ be a probability space. An operator T on $L_2(\Omega, \mathscr{S}, \lambda)$ is *isotonic* if $y_i \in L_2$, $i = 1, 2$, $y_1 \leq y_2$ imply $Ty_1 \leq Ty_2$. It is *expectation invariant* if $ETy = Ey$ for all y in L_2.

Proposition 7.2

Let $(\Omega, \mathscr{S}, \lambda)$ be a probability space and let T be a positively homogeneous, isotonic, expectation invariant, idempotent, distance reducing operator on $L_2(\Omega, \mathscr{S}, \lambda)$. Let A be the range of T, and set $\mathscr{M} = \{[f > a] : f \in A, a \in R\}$. Then $\mathscr{M}$ is a sub-σ-lattice of $\mathscr{S}$, and $A = L_2(\mathscr{M})$.

Proof. 1°. A is a lattice: $y, z \in A$ imply $y \vee z, y \wedge z \in A$. To prove 1°, let y, $z \in A$. Since T is isotonic and idempotent, $T(y \vee z) \geq Ty \vee Tz = y \vee z$. Then

$$\int [T(y \vee z) - y]^2 \, d\lambda \geq \int [(y \vee z) - y]^2 \, d\lambda.$$

But since T is distance reducing, also

$$\int [(y \vee z) - y]^2 \, d\lambda \geq \int [T(y \vee z) - Ty]^2 \, d\lambda = \int [T(y \vee z) - y]^2 \, d\lambda,$$

hence

$$\int [T(y \vee z) - y]^2 \, d\lambda = \int [y \vee z - y]^2 \, d\lambda.$$

Since $T(y \vee z) \geq y \vee z \geq y$, it follows that $T(y \vee z) = y \vee z$, hence $y \vee z \in A$. Similarly it can be shown that $y \wedge z \in A$.

2°. If $f \in A$ and if c is a constant function then $c \in A$, $f + c \in A$, and $T(f + c) = Tf + c$. To prove 2°, note first that by the Buniakowski–Schwarz inequality,

$$\int [T(f + c) - Tf]^2 \, d\lambda \geq (\int [T(f + c) - Tf] \, d\lambda)^2 = (\int c \, d\lambda)^2$$

since T is expectation invariant. Thus

$$\int [T(f + c) - Tf]^2 \, d\lambda \geq \int c^2 \, d\lambda = \int [(f + c) - f]^2 \, d\lambda.$$

But the reverse inequality holds also since T is distance reducing, and this implies that there is a real number a such that $T(f + c) - Tf = a[(f + c) - f] = ac$. But again since T is expectation invariant, $T(f + c) - Tf = c$.

3°. $f \in A$ implies $1_{[f > a]} \in A$, where $1_{[f > a]}$ denotes the indicator function of the set $[f > a]$. To see this, set $f_n = [n(f - a) \vee 0] \wedge 1$ for $n = 1, 2, \ldots$; $f_n \in A$ by 1° and 2°. Then $f_n \uparrow 1_{[f > a]}$ and hence

$$\int [1_{[f > a]} - f_n]^2 \, d\lambda \to 0.$$

Since the hypothesis that T is distance reducing implies that A is closed, $1_{[f > a]} \in A$ follows.

4°. $\mathscr{M}$ is a σ-lattice. Suppose $B_n \in \mathscr{M}$, $n = 1, 2, \ldots$. Then by 3°, $1_{B_n} \in A$, $n = 1, 2, \ldots$. By 1°,

$$1_{\cup_{n=1}^{k} B_n} = \bigvee_{n=1}^{k} 1_{B_n} \in A$$

and

$$1_{\cap_{n=1}^k B_n} = \bigwedge_{n=1}^k 1_{B_n} \in A.$$

Again, since A is closed,

$$1_{\cup_{n=1}^k B_n} \uparrow 1_{\cup_{n=1}^\infty B_n}$$

implies

$$\bigcup_{n=1}^\infty B_n \in \mathcal{M},$$

and

$$1_{\cap_{n=1}^k B_n} \downarrow 1_{\cap_{n=1}^\infty B_n}$$

implies

$$\bigcap_{n=1}^\infty B_n \in \mathcal{M}.$$

5^0. $A = L_2(\mathcal{M})$. We have $A \subset L_2(\mathcal{M})$ by definition. To show $L_2(\mathcal{M}) \subset A$, suppose $f \in L_2(\mathcal{M})$. For $n = 1, 2, \ldots$, $i = 1, 2, \ldots, n2^{n+1}$, set

$$B_{ni} = [f \geq -n + i/2^n].$$

Then $B_{ni} \in \mathcal{M}$ by definition of $L_2(\mathcal{M})$. By the definition of $\mathcal{M}$, there exist $y_{ni} \in A$ and real a_{ni} such that $B_{ni} = [y_{ni} > a_{ni}]$. By 3^0, $1_{B_{ni}} \in A$. Then by 1^0 and 2^0,

$$f_n \stackrel{\text{def}}{=} \bigvee_{i=1}^{n2^{n+1}} (i/2^n) 1_{[f \geq -n + i/2^n]} - n \in A.$$

Then $f_n \to f$ and $|f_n| \leq |f| + 1$, $n = 1, 2, \ldots$, so that by the dominated convergence theorem $\int (f_n - f)^2 \, d\lambda \to 0$. Since A is closed, we have $f \in A$. ∎

Theorem 7.10

Let $(\Omega, \mathcal{S}, \lambda)$ be a probability space and let T be a positively homogeneous, isotonic, expectation invariant, idempotent, distance reducing operator on $L_2 = L_2(\Omega, \mathcal{S}, \lambda)$, strictly monotonic at 0. Let $\mathcal{M} = \{[Ty > a] : y \in L_2, a \in R\}$. Then $T = E(\cdot | \mathcal{M})$.

Proof. This is immediate from Propositions 7.1 and 7.2. ∎

The following example in which T satisfies all the hypotheses except that it is not strictly monotonic at 0, and in which the conclusion fails, is due to

Dykstra. Let $\Omega = \{1, 2\}$, $\mathscr{S} = \{\phi, \{1\}, \{2\}, \Omega\}$, $\lambda\{1\} = \lambda\{2\} = 1/2$. Each function in L_2 can be represented by a point in the plane whose coordinates are its values at 1 and at 2 respectively. If $y = (y_1, y_2)$, $y_1 < y_2$, set $Ty = (y_2, y_1)$; if $y_1 \geq y_2$ set $Ty = y$. Then T is positively homogeneous, isotonic, expectation invariant, idempotent, and distance reducing. But if $y = (-1, 1)$ then $(y, Ty) = (-1)(1)(1/2) + (1)(-1)(1/2) = -1$, so that T is not strictly monotonic at 0. Since T is not projection on its range, it is not a conditional expectation given a σ-lattice.

7.4 CONDITIONAL EXPECTATION GIVEN A σ-LATTICE AS A LEBESGUE–RADON–NIKODYM DERIVATIVE

The usual approach to conditional expectation given a σ-field is through the theory of the Lebesgue–Radon–Nikodym (LRN) derivative. A somewhat more general LRN theory yields the conditional expectation given a σ-lattice as an operator on L_1. It will be seen that it coincides in $L_1 \cap L_2$ with the projection operator studied in Section 7.3.

Again, the starting point is a measure space $(\Omega, \mathscr{S}, \lambda)$, in which Ω is an abstract space, $\mathscr{S}$ a σ-field of subsets of Ω, and λ a positive measure, not necessarily σ-finite. There is given also a σ-lattice $\mathscr{U} \subset \mathscr{S}$; the role of a σ-lattice in a study of isotonic regression is discussed in Section 7.2. Let R denote the set of real numbers, and let $R(\mathscr{U})$ denote the class of real valued $\mathscr{U}$-measurable functions on Ω. Define

$$\mathscr{L} = \mathscr{U}^c = \{L : L^c \in \mathscr{U}\}$$

and

$$\mathscr{F} = \{U \cap L : U \in \mathscr{U}, L \in \mathscr{L}\}.$$

The class $\mathscr{F}$ is a Boolean semi-algebra. Indeed, it is closed under countable intersection, and the complement of a set in $\mathscr{F}$ is a finite union of pairwise disjoint sets in $\mathscr{F}$: $(U \cap L)^c = U^c \cup L^c = U^c + U \cap L^c = (U^c \cap \Omega) + (U \cap L^c) \cap \Omega$; each term of this last expression is in $\mathscr{F}$. (Here and henceforth $+$ and Σ are used to denote *disjoint* unions of sets.)

Definition 7.4

A function v on $\mathscr{F}$ is a *measure* if

$$-\infty < v(F) \leq \infty \quad \text{for} \quad F \in \mathscr{F}$$
or
$$-\infty \leq v(F) < \infty \quad \text{for} \quad F \in \mathscr{F}; \quad (7.22)$$

$$v(\phi) = 0; \quad (7.23)$$

CONDITIONAL EXPECTATION AS A LRN DERIVATIVE

imply
$$\left.\begin{array}{c} F \in \mathscr{F}, F_i \in \mathscr{F}, \quad i = 1, 2, \ldots, \quad F = \sum_{i=1}^{\infty} F_i \\ v(F) = \sum_{i=1}^{\infty} v(F_i); \end{array}\right\} \quad (7.24)$$

v is a *positive measure* if in place of (7.22),

$$v(F) \geq 0 \quad \text{for} \quad F \in \mathscr{F}. \quad (7.25)$$

$E(y|\mathscr{U})$ will be obtained for $y \in L_1(\Omega, \mathscr{S}, \lambda)$ as Lebesgue–Radon–Nikodym derivative of φ with respect to λ given $\mathscr{U}$, where

$$\varphi(F) = \int_F y \, d\lambda \quad \text{for} \quad F \in \mathscr{F}.$$

In defining the Lebesgue–Radon–Nikodym (LRN) derivative of an arbitrary measure φ on $\mathscr{F}$ with respect to λ given $\mathscr{U}$, it is convenient to set

$$v_a = \varphi - a\lambda \quad \text{for} \quad a \in R.$$

It will always be assumed that either φ or λ is finite.

Definition 7.5

An extended real valued function f on Ω is a LRN *derivative of ϕ with respect to λ given $\mathscr{U}$* if

$$f \text{ is } \mathscr{U}\text{-measurable}; \quad (7.26)$$

$$v_b(U \cap [f < b]) \leq 0 \text{ for } b \in R, U \in \mathscr{U}; \quad (7.27)$$

$$v_a(L \cap [f > a]) \geq 0 \text{ for } a \in R, L \in \mathscr{L}. \quad (7.28)$$

We observe that f is LRN provided f satisfies (7.26) and satisfies (7.27) and (7.28) for a and b in a dense set D. For suppose that (7.28) holds for $a \in D$; if $a \notin D$, let $L \in \mathscr{L}$, $s_i \downarrow a$, $s_i \in D$. Then

$$v_a(L \cap [f > a]) = \sum_i v_a(L \cap [f \leq s_{i-1}] \cap [f > s_i]) \geq 0,$$

since $v_{s_i}(L \cap [f \leq s_{i-1}] \cap [f > s_i]) \geq 0$ for each i, and $v_{s_i} \geq 0$ implies $v_a \geq 0$. Thus (7.28) holds for $a \in R$. The proof that (7.27), for $b \in D$, implies (7.27) for $b \in R$ is similar.

Remark

If f is a LRN derivative of φ with respect to λ given $\mathscr{U}$ then $-f$ is a LRN derivative of $-\varphi$ with respect to λ given $\mathscr{L} = \mathscr{U}^c$. If $k > 0$ then kf is a LRN derivative of $k\varphi$ with respect to λ given $\mathscr{U}$.

Definition 7.6

If $y \in L_1(\Omega, \mathscr{S}, \lambda)$ and $\mathscr{U}$ is a sub-σ-lattice of $\mathscr{S}$, then $E(y|\mathscr{U})$, the *conditional expectation of y given $\mathscr{U}$*, denotes a LRN derivative of φ with respect to λ given $\mathscr{U}$, where

$$\varphi(F) = \int_F y \, d\lambda \quad \text{for} \quad F \in \mathscr{F}.$$

That is, $y^* = E(y|\mathscr{U})$ if and only if

$$y^* \in R(\mathscr{U}) \quad (y^* \text{ is } \mathscr{U}\text{-measurable}), \tag{7.29}$$

$$\int_{U \cap [y^* < b]} (y - b) \, d\lambda \leq 0 \quad \text{for} \quad b \in R, \, U \in \mathscr{U}, \tag{7.30}$$

$$\int_{L \cap [y^* > a]} (y - a) \, d\lambda \geq 0 \quad \text{for} \quad a \in R, \, L \in \mathscr{L}. \tag{7.31}$$

Again, $-y^* = E(-y|\mathscr{L})$; and if $k > 0$, $ky^* = E(ky|\mathscr{U})$.

Existence and representations

If either φ is finite or λ is finite the measures ν_a have the properties (7.32), (7.33) and (7.34) below; in (7.34) we can take $\alpha = 0$ if φ is finite, $\alpha = -\infty$ if $\lambda < \infty$, $\varphi < \infty$, $\alpha = \infty$ if $\lambda < \infty$, $\varphi > -\infty$.

For each real a, ν_a is a measure on $\mathscr{F}$. (7.32)

For $F \in \mathscr{F}$ and $a < b$,

$$\begin{aligned}&\nu_b(F) \geq 0 \quad \text{implies} \quad \nu_a(F) \geq 0 \\ \text{and} \quad &\nu_a(F) \leq 0 \quad \text{implies} \quad \nu_b(F) \leq 0;\end{aligned} \tag{7.33}$$

$\exists \alpha$, $-\infty \leq \alpha \leq \infty$, such that

$$\begin{aligned}&\nu_a(F) < \infty \quad \text{for} \quad a > \alpha, \, F \in \mathscr{F} \\ \text{and} \quad &\nu_a(F) > -\infty \quad \text{for} \quad a < \alpha, \, F \in \mathscr{F}.\end{aligned} \tag{7.34}$$

It is no more difficult to prove the existence of a LRN function for an arbitrary family $\{\nu_a, a \in R\}$ of measures on $\mathscr{F}$ satisfying (7.32), (7.33) and (7.34) than for the measures $\nu_a = \varphi - a\lambda$; this is done in Theorem 7.12. Applications of the more general theory are indicated in Brunk and Johansen (1970). Properties (7.29), (7.30) and (7.31) are taken as definition also of *a LRN function for $\{\nu_a, a \in R\}$ and $\mathscr{U}$*, when $\{\nu_a, a \in R\}$ is an arbitrary family of measures satisfying (7.32), (7.33) and (7.34).

Definition 7.7

If v is a measure on $\mathscr{F}$, a set $U \in \mathscr{U}$ is *positive* for v if $v(UL) \geq 0$ for all $L \in \mathscr{L}$; $L \in \mathscr{L}$ is *negative* for v if $v(UL) \leq 0$ for all $U \in \mathscr{U}$. $\mathscr{P}$ and $\mathscr{N}$ denote the classes of positive and negative sets:

$$\mathscr{P} = \{U \in \mathscr{U} : v(UL) \geq 0, \text{ for all } L \in \mathscr{L}\},$$
$$\mathscr{N} = \{L \in \mathscr{L} : v(UL) \leq 0, \text{ for all } U \in \mathscr{U}\}.$$

A positive set P is *maximal* if $v(P) \geq v(U)$ for all $U \in \mathscr{P}$; a negative set N is *minimal* if $v(N) \leq v(L)$ for all $L \in \mathscr{N}$.

Lemma A

$\mathscr{P}$ and $\mathscr{N}$ are closed under countable union. If P (N) is maximal (minimal) for v and $U \in \mathscr{P}$ $(L \in \mathscr{N})$ then $P \cup U$ $(N \cup L)$ is maximal (minimal). If $U_i \in \mathscr{P}$ $(L_i \in \mathscr{N})$, $i = 1, 2$, then $v(U_1 \cup U_2) \geq v(U_1) \vee v(U_2)$ $(v(L_1 \cup L_2) \leq v(L_1) \wedge v(L_2))$. If $P \in \mathscr{P}$ and $P^c \in \mathscr{N}$ then $v(U) \leq v(P)$ for $U \in \mathscr{U}$, $v(L) \geq v(P^c)$ for $L \in \mathscr{L}$, P is maximal, and P^c is minimal.

Proof. The first conclusion follows from the identity

$$\bigcup_{n=1}^{\infty} U_n = \sum_{n=1}^{\infty} U_n \cap \bigcap_{j=1}^{n-1} U_j^c$$

and the countable additivity of v. If $U_i \in \mathscr{P}$, $i = 1, 2$, then

$$v(U_1 \cup U_2) = v(U_1) + v(U_1^c \cap U_2) \geq v(U_1)$$

and similarly $v(U_1 \cup U_2) \geq v(U_2)$. If P is maximal and $U \in \mathscr{U}$ then $P \cup U \in \mathscr{P}$ by the first conclusion and $v(U \cup P) \geq v(P)$, so $U \cup P$ is maximal. If $P \in \mathscr{P}$, $P^c \in \mathscr{N}$ and $U \in \mathscr{U}$ then

$$v(U) \leq v(U) + v(U^c \cap P) = v(U \cup P) = v(P) + v(P^c \cap U) \leq v(P)$$

hence P is maximal. The proofs of the other conclusions are symmetric. ∎

Lemma B

Let $\{v_a, a \in R\}$ satisfy (7.32), (7.33) and (7.34). If $r < s$, the class $\mathscr{P}_r$ $(\mathscr{N}_r)$ of positive (negative) sets for v_r contains (is contained in) the class $\mathscr{P}_s$ $(\mathscr{N}_s)$ of positive (negative) sets for v_s. If $\alpha \leq r < s$, and if U_r and U_s are maximal and U_r^c and U_s^c are negative for v_r and v_s respectively, then $U_r \cap U_s$ is maximal for v_s. If $r < s \leq \alpha$ and if L_r and L_s are minimal and L_r^c and L_s^c are positive for v_r and v_s respectively, then $L_r \cap L_s$ is minimal for v_r.

Proof. The inclusions $\mathscr{P}_r \supset \mathscr{P}_s$ and $\mathscr{N}_r \subset \mathscr{N}_s$ for $r < s$ are immediate from (7.33). Now suppose $\alpha \leq r < s$. We show first that $\nu_s(U_s) = \nu_s(U_r \cap U_s)$. $U_r^c \in \mathscr{N}_r \subset \mathscr{N}_s$, hence $\nu_s(U_s \cap U_r^c) \leq 0$. Then

$$\nu_s(U_s) = \nu_s(U_s \cap U_r) + \nu_s(U_s \cap U_r^c) \leq \nu_s(U_s \cap U_r).$$

Since U_s is maximal and U_s^c is negative for ν_s, and $U_s \cap U_r \in \mathscr{U}$, the reverse inequality follows from Lemma A above, so that $\nu_s(U_s) = \nu_s(U_r \cap U_s)$. We now show that $U_r \cap U_s \in \mathscr{P}_s$. Let $L \in \mathscr{L}$. By Lemma A above, since U_s is maximal and U_s^c is negative for ν_s, we have $\nu_s(L^c \cap U_s \cap U_r) \leq \nu_s(U_s)$. Then

$$\nu_s(L \cap U_s \cap U_r) + \nu_s(U_s) \geq \nu_s(L \cap U_s \cap U_r) + \nu_s(L^c \cap U_s \cap U_r)$$
$$= \nu_s(U_s \cap U_r) = \nu_s(U_s).$$

Since $0 \leq \nu_s(U_s) < \infty$, $\nu_s(L \cap U_s \cap U_r) \geq 0$. Thus $U_s \cap U_r \in \mathscr{P}$. Since U_s is maximal for ν_s and $\nu_s(U_s) = \nu_s(U_s \cap U_r)$, also $U_s \cap U_r$ is maximal for ν_s. The proof that $L_r \cap L_s$ is minimal for ν_r, $r < s \leq \alpha$, is symmetric. ∎

Theorem 7.11

If ν is a measure on $\mathscr{F}$ it has a maximal set P such that P^c is minimal. If $\nu < \infty$ ($\nu > -\infty$) and if P is maximal (N is minimal) for ν, then P^c is minimal (N^c is maximal).

Proof. The proof is given for the case $\nu < \infty$; the proof when $\nu > -\infty$ is similar.

1^0. *There exists a maximal set P.*

Set
$$\gamma = \sup_{U \in \mathscr{P}} \nu(U).$$

Choose $U_n \in \mathscr{P}$, $n = 1, 2, \ldots$, such that $\nu(U_n) \uparrow \gamma$. By Lemma A above it may be assumed without loss of generality that $U_{n+1} \supset U_n$, $n = 1, 2, \ldots$. Set
$$P = \bigcup_n U_n \in \mathscr{P}$$

by Lemma A. Also
$$\nu(P) = \lim_n \nu(U_n) = \gamma.$$

2^0. *P maximal implies P^c is negative.*

We use Zorn's Lemma in the following form: If $(\mathscr{A}, \succ)$ is a partially ordered system, if $a_0 \in \mathscr{A}$, and if every chain has an upper bound in $\mathscr{A}$ then $\mathscr{A}$ has a maximal element $\bar{a}$ such that $\bar{a} \succ a_0$. Suppose P^c is not negative. Then there exists $U_0 \in \mathcal{U}$ such that $v(U_0 \cap P^c) > 0$. Assume without loss of generality that $U_0 \supset P$ (if not, use $U_0 \cup P$ in place of U_0). Set

$$\mathcal{A} = \{U \in \mathcal{U} : P \subset U \subset U_0\}$$

with a partial order "$\succ$" defined by:

$$U_1 \succ U_2 \text{ if } U_1 \subset U_2 \text{ and } v(U_1) \geq v(U_2).$$

Let $\mathcal{C}$ be an arbitrary chain in $\mathcal{A}$, and let

$$\gamma_\mathcal{C} = \sup_{U \in \mathcal{C}} v(U).$$

If there exists $U_1 \in \mathcal{C}$ such that $v(U_1) = \gamma_\mathcal{C}$ then U_1 is an upper bound in $\mathcal{A}$ of $\mathcal{C}$. If not, choose $U_n \in \mathcal{C}$, $n = 1, 2, \ldots$, such that $v(U_n) \uparrow \gamma_\mathcal{C}$. Since $\mathcal{C}$ is a chain and $v(U_{n+1}) > v(U_n)$, $U_{n+1} \succ U_n$ for $n = 1, 2, \ldots$. Set

$$\bar{U}_\mathcal{C} = \lim_n U_n = \bigcap_{n=1}^\infty U_n.$$

Then $v(\bar{U}_\mathcal{C}) = \gamma_\mathcal{C} > v(U)$ and $\bar{U}_\mathcal{C} \subset U$ for all $U \in \mathcal{C}$, so that $\bar{U}_\mathcal{C}$ is an upper bound in $\mathcal{A}$ of $\mathcal{C}$. By Zorn's Lemma there is a $\bar{U} \in \mathcal{A}$ such that $\bar{U} \succ U_0$ and $U \not\succ \bar{U}$ for all $U \in \mathcal{A}$. Therefore $v(\bar{U}) \geq v(U_0) = v(P) + v(U_0 \cap P^c) > 0$, and if $U \in \mathcal{U}$ and $P \subset U \subset \bar{U}$ then $v(U) \leq v(\bar{U})$.

We now show that $\bar{U} \in \mathcal{P}$. Suppose $L \in \mathcal{L}$, and suppose first that $L \subset P^c$. Then

$$v(\bar{U}) = v(\bar{U} \cap L) + v(\bar{U} \cap L^c).$$

Since $\bar{U} \cap L^c \in \mathcal{A}$ and $\bar{U} \cap L^c \subset \bar{U}$, and since $0 < v(\bar{U}) < \infty$, $v(\bar{U} \cap L^c) \leq v(\bar{U})$, hence $v(\bar{U} \cap L) \geq 0$. Now if L is an arbitrary set in $\mathcal{L}$,

$$v(\bar{U} \cap L) = v(\bar{U} \cap L \cap P^c) + v(P \cap L).$$

Also, since $P^c \supset L \cap P^c \in \mathcal{L}$, $v(\bar{U} \cap L \cap P^c) \geq 0$, and $v(P \cap L) \geq 0$ since P is positive. It follows that $\bar{U} \in \mathcal{P}$. But

$$v(\bar{U}) \geq v(U_0) = v(P) + v(U_0 \cap P^c) > v(P),$$

contradicting the maximality of P.

3°. *P^c is minimal for v.* This follows from Lemma A above. ∎

In the following lemma, R' denotes the set of rational numbers.

Lemma

Let $\{v_a, a \in R\}$ satisfy (7.32), (7.33) and (7.34). Then there are sets P_r maximal for v_r such that

$$N_r \stackrel{\text{def}}{=} P_r^c$$

is minimal for v_r, $r \in R'$, and such that $r, s \in R'$ and $r < s$ imply $P_r \supset P_s$.

Proof. For $r \in R' \cup \{\alpha\}$, let U_r be a maximal set for v_r such that

$$L_r \stackrel{\text{def}}{=} U_r^c$$

is minimal for v_r (Theorem 7.11). Then by Lemma B above, $U_r \cap U_\alpha$ is maximal for v_r for $r \geq \alpha$, $r \in R'$; and if $s \leq \alpha$, $s \in R'$, then $L_s \cap L_\alpha$ is minimal for v_s. Set

$$P_r = \cup \{U_s \cap U_\alpha : s \in R', s \geq r\}, \text{ for } r \geq \alpha, r \in R',$$

$$N_s = \cup \{L_r \cap L_\alpha : r \in R', r \leq s\}, \text{ for } s < \alpha, s \in R',$$

$$N_r = P_r^c, \text{ for } r \in R'.$$

By Lemma A above and Theorem 7.11, P_r is maximal and N_r is minimal for v_r if $r \in R'$, $r \geq \alpha$; and N_s is minimal and P_s is maximal for v_s if $s \in R'$, $s < \alpha$. Clearly also $P_r \supset P_s$ if $r, s \in R'$, $r < s$. ∎

In the following theorems, 1_A denotes the indicator function of the set $A: 1_A(\omega) = 1$ if $\omega \in A$, $1_A(\omega) = 0$ if $\omega \in A^c$.

Theorem 7.12

Let $\{v_a, a \in R\}$ satisfy (7.32), (7.33) and (7.34), and let $\{P_r, N_r, r \in R'\}$ have the properties given in the lemma preceding Theorem 7.12. Then the function

$$f \stackrel{\text{def}}{=} \sup_{r \in R'} (r 1_{P_r} - \infty 1_{N_r}) = \inf_{r \in R'} (r 1_{N_r} + \infty 1_{P_r})$$

is a LRN function for $\{v_a, a \in R\}$ and $\mathscr{U}$.

Proof. We show first that the two formulas coincide. Set

$$f_1 = \sup_{r \in R'} (r 1_{P_r} - \infty 1_{N_r}),$$

$$f_2 = \inf_{r \in R'} (r 1_{N_r} + \infty 1_{P_r}).$$

If $r, s \in R'$ then

$$r 1_{P_r} - \infty 1_{N_r} \leq s 1_{N_s} + \infty 1_{P_s},$$

hence $f_1 \leq f_2$. Set

$$N_{-\infty} = \cap_{r \in R'} N_r = \lim_{r \to -\infty} N_r \quad \text{and} \quad P_\infty = \cap_{r \in R'} P_r = \lim_{r \to \infty} P_r.$$

If $\omega \in N_{-\infty}$ then $f_1(\omega) = f_2(\omega) = -\infty$ while if $\omega \in P_\infty$ then $f_1(\omega) = f_2(\omega) = \infty$. If $\omega \notin N_{-\infty} \cup P_\infty$ then $|f_i(\omega)| < \infty$, $i = 1, 2$. Fix ω, and set $a = f_1(\omega)$; then $a = \sup\{r \in R' : \omega \in P_r\}$. For $r > a$, $\omega \in N_r$, i.e. $f_2(\omega) \leq r$; hence $f_2(\omega) \leq a$. Thus $f_2 \leq f_1$, whence $f_1 = f_2$.

Now if $t \in R$,

$$[f > t] = \cup \{P_r : r \in R', r > t\} \in \mathcal{U}.$$

Thus f is $\mathcal{U}$-measurable ($f \in R(\mathcal{U})$). Further, for $r > t$, $P_r \in \mathscr{P}_r \subset \mathscr{P}_t$ by Lemma B above, and from Lemma A above we have $[f > t] \in \mathscr{P}_t$. This verifies (7.28) in the definition of a LRN function for $\{v_a, a \in R\}$ and $\mathcal{U}$. The proof of (7.27) is symmetric. ∎

Theorem 7.13

Let $(\Omega, \mathscr{S}, \lambda)$ be a measure space and $\mathcal{U}$ a sub-σ-lattice of $\mathscr{S}$. If φ is a measure on $\mathscr{F}$ and if either φ or λ is finite then φ has a LRN derivative f with respect to λ given $\mathcal{U}$. Two LRN derivatives coincide a.e. (λ). If φ is finite then f is finite a.e. (λ).

Proof. The existence of a LRN derivative is a consequence of Theorem 7.12 with $v_a = \varphi - a\lambda$, $a \in R$. Suppose f and g are two such. To show they coincide it suffices to show that if $a, b \in R$, $a > b$, $A = \{f > a > b > g\}$, then $\lambda(A) = 0$. Since $[f > a] \in \mathcal{U}$ and $[g < b] \in \mathscr{L}$, from the defining properties of a LRN derivative it follows that

$$\varphi(A) = \varphi([f > a] \cap [g < b]) \geq a\lambda(A) \geq b\lambda(A) \geq \varphi(A).$$

Since either φ or λ is finite it follows that $\lambda(A) = 0$. Since $P_r \downarrow$ as $r \uparrow$, $\lambda(P_r) \downarrow$ as $r \uparrow$. If φ is finite then it follows from

$$\infty > |\varphi|(\Omega) \geq \varphi(P_r) \geq r\lambda(P_r)$$

that $\lambda(P_r) \downarrow 0$ as $r \uparrow \infty$. Let

$$P_\infty = \lim_{r \to \infty} P_r.$$

Then $\lambda([f = \infty]) = \lambda(P_\infty) = 0$. Similarly $\lambda([f = -\infty]) = 0$ if φ is finite. ∎

Corollary

If $y \in L_1(\Omega, \mathscr{S}, \lambda)$ and $\mathscr{U}$ is a sub-σ-lattice of $\mathscr{S}$, then y has an a.e. (λ) finite and unique conditional expectation given $\mathscr{U}$, $y^* = E(y|\mathscr{U})$.

When λ is finite and φ is absolutely continuous with respect to λ one can give formulas for the LRN derivative which specialize in the context of Chapter 1 to the max–min formulas given there. These formulas are particularly simple when φ also is finite.

Theorem 7.14

If $(\Omega, \mathscr{S}, \lambda)$ is a finite measure space, $\mathscr{U}$ is a sub-σ-lattice of $\mathscr{S}$, φ is a finite measure on $\mathscr{F}$ and if φ is absolutely continuous with respect to λ then

$$f \overset{\text{def}}{=} \underset{U \in \mathscr{U}}{\text{ess sup}}\, \{\inf\, [\varphi(UL)/\lambda(UL) : L \in \mathscr{L}, \lambda(UL) > 0]1_U - \infty 1_{U^c}\} \\ = \underset{L \in \mathscr{L}}{\text{ess inf}}\, \{\sup\, [\varphi(UL)/\lambda(UL) : U \in \mathscr{U}, \lambda(UL) > 0]1_L + \infty 1_{L^c}\} \quad (7.35)$$

is a LRN derivative of φ with respect to λ given $\mathscr{U}$.

Before proceeding with the proof it is remarked that in an alternative development for the case in which φ and λ are finite, Theorem 7.14 yields the existence of a LRN derivative without using Theorem 7.12 and the lemma which precedes it. (It is true that the proof of Theorem 7.14 uses sets P_r, $r \in R'$; but only the existence of such maximal sets, guaranteed by Theorem 7.11, is required; not the fact (lemma preceding Theorem 7.12) that they can be chosen so as to decrease as r increases).

Proof. Let f_1, f_2 be given by the first and second formulas respectively. There are countable subfamilies $\mathscr{U}_1, \mathscr{L}_1$ of $\mathscr{U}, \mathscr{L}$ respectively (cf. Neveu (1965, p. 44)) which may be supposed without loss of generality to contain $\{P_r, r \in R'\}$ and $\{N_r, r \in R'\}$ respectively, such that

$$f_1 = \underset{U \in \mathscr{U}_1}{\sup}\, \{\inf\, [\varphi(UL)/\lambda(UL) : L \in \mathscr{L}, \lambda(UL) > 0]1_U - \infty 1_{U^c}\}, \quad (7.36)$$

$$f_2 = \underset{L \in \mathscr{L}_1}{\inf}\, \{\sup\, [\varphi(UL)/\lambda(UL) : U \in \mathscr{U}, \lambda(UL) > 0]1_L + \infty 1_{L^c}\}. \quad (7.37)$$

Note first that since φ is absolutely continuous with respect to λ, if

$$\underset{L \in \mathscr{L}, \lambda(UL) > 0}{\inf} \varphi(UL)/\lambda(UL) = r$$

then $U \in \mathscr{P}_r$. Fix $t \in R$. Then

$$[f_1 > t] = \underset{r \in R', r > t}{\cup}\; \underset{U \in \mathscr{U}_1 \cap \mathscr{P}_r}{\cup}\; U \in \mathscr{U},$$

so that f_1 is $\mathcal{U}$-measurable ($f_1 \in R(\mathcal{U})$). So, similarly, is f_2. Also, by Lemmas A and B above, $[f_1 > t] \in \mathcal{P}_t$, and similarly $[f_2 < t] \in \mathcal{N}_t$. Let $r_n \uparrow t$, $r_n \in R'$, $n = 1, 2, \ldots$. For $L \in \mathcal{L}$,

$$\varphi(L \cap [f_1 \geq t]) - t\lambda(L \cap [f_1 \geq t])$$
$$= \lim_n \{\varphi(L \cap [f_1 > r_n]) - r_n \lambda(L \cap [f_1 > r_n])\} \geq 0.$$

Thus $[f_1 \geq t] \in \mathcal{P}_t$, and similarly $[f_2 \leq t] \in \mathcal{N}_t$.

It will now be shown that $f_1 \leq f_2$ a.e. (λ). It suffices to show that for a, $b \in R$, $a > b$, $\lambda[f_1 > a > b > f_2] = 0$. But since $[f_2 < b] \in \mathcal{N}_b$ and $[f_1 > a] \in \mathcal{P}_a$,

$$b\lambda[f_1 > a > b > f_2] \geq \varphi[f_1 > a > b > f_2] \geq a\lambda[f_1 > a > b > f_2]$$

so that $\lambda[f_1 > a > b > f_2] = 0$. Thus $f_1 \leq f_2$ a.e. (λ).

But because $P_r \in \mathcal{U}_1$ for $r \in R'$, from (7.36) $P_r \subset [f_1 \geq r]$ for $r \in R'$ and similarly $P_r^c = N_r \subset [f_2 \leq r]$ for $r \in R'$. Hence

$$[f_1 < r] \cap [f_2 > r] \subset N_r \cap P_r = \varphi, \text{ for } r \in R'$$

so that $f_1 \geq f_2$. Thus finally $f_1 = f_2$ a.e. (λ). Since $[f_1 > t] \in \mathcal{P}_t$ and $[f_2 < t] \in \mathcal{N}_t$ for $t \in R$, $f_1 = f_2$ (a.e. (λ)) is a LRN derivative of φ with respect to λ given $\mathcal{U}$. ∎

Corollary

If $y \in L_1(\Omega, \mathcal{S}, \lambda)$, $\lambda(\Omega) < \infty$, $\mathcal{U}$ is a sub-σ-lattice of $\mathcal{S}$, then $y^* = E(y|\mathcal{U})$ is given (a.e. (λ)) by (7.35).

We note in passing that if $\mathcal{U}$ is a σ-field then (7.35) with $\mathcal{L} = \mathcal{U}$ gives explicit formulas for the classical LRN derivative and conditional expectation.

Order properties

Theorem 7.15

Let φ_i be a measure on $\mathcal{F}$, $i = 1, 2$, and let λ be finite or else both φ_1 and φ_2 be finite. Let f_i be the LRN derivative of φ_i with respect to λ given $\mathcal{U}$, $i = 1, 2$. If $\varphi_2(F) \geq \varphi_1(F)$ for all $F \in \mathcal{F}$ then $f_2 \geq f_1$ a.e. (λ)).

Proof. Let r, s be rationals with $r < s$. Consider the set $F = [f_2 < r < s < f_1] \in \mathcal{F}$. From the defining properties of the LRN derivatives $\varphi_1(F) \geq s\lambda(F)$ and $\varphi_2(F) \leq r\lambda(F)$. Then $s\lambda(F) \leq \varphi_1(F) \leq \varphi_2(F) \leq r\lambda(F)$. If $r \geq 0$ then $s > 0$, while if $s \leq 0$ then $r < 0$, so that under the hypotheses $\lambda(F)$ is finite, and hence zero. Since this holds for all pairs of rationals, $\lambda[f_2 < f_1] = 0$, as was to be shown. ∎

Corollary

Let $(\Omega, \mathscr{S}, \lambda)$ be a measure space, $\mathscr{U}$ a sub-σ-lattice of $\mathscr{S}$, and $y_i \in L_1$, $i = 1, 2$. If $y_1 \leq y_2$ then $E(y_1|\mathscr{U}) \leq E(y_2|\mathscr{U})$ (a.e. (λ)).

Theorem 7.16

If $y \in L_1(\Omega, \mathscr{S}, \lambda)$, $\mathscr{U}$ is a sub-σ-lattice of $\mathscr{S}$, and $y^* = E(y|\mathscr{U})$, if $z \in R(\mathscr{U})$ and if $y \leq z$ ($y \geq z$) then $y^* \leq z$ ($y^* \geq z$) (a.e. (λ)).

Proof. It suffices to show that for arbitrary rational r, $\lambda[y^* > r > z] = 0$. By the definition of y^* as a LRN function,

$$\int_{[y^* > r > z]} (y - r) \, d\lambda \geq 0,$$

since $[z < r] \in \mathscr{L} = \mathscr{U}^c$. Since $y \leq z$, the integrand, $y - r$, is strictly negative on $[y^* > r > z]$, hence the integral is zero and $\lambda[y^* > r > z] = 0$. The proof of the parenthetical statement is symmetric. ∎

Integral properties

It has not yet been shown that the LRN derivative is integrable. We now do so, at the same time giving properties expressed by means of integrals which will be useful below.

Let $\mathscr{B}$ denote the class of Borel sets of real numbers.

Theorem 7.17

Let $(\Omega, \mathscr{S}, \lambda)$ be a measure space and $\mathscr{U}$ a sub-σ-lattice of $\mathscr{S}$. Let φ be a finite measure on $\mathscr{F}$ and let f be a LRN derivative of φ with respect to λ given $\mathscr{U}$. Then $f \in L_1$. Further, if $U \in \mathscr{U}$, $B \in \mathscr{B}$ and if either $\lambda(U) < \infty$ or $0 \notin B$ then

$$\varphi(U \cap [f \in B]) \leq \int_{U \cap [f \in B]} f \, d\lambda, \tag{7.38}$$

and if $L \in \mathscr{L}$, $B \in \mathscr{B}$, and either $\lambda(L) < \infty$ or $0 \notin B$ then

$$\varphi(L \cap [f \in B]) \geq \int_{L \cap [f \in B]} f \, d\lambda. \tag{7.39}$$

CONDITIONAL EXPECTATION AS A LRN DERIVATIVE

We remark that if $\lambda(\Omega) < \infty$, the phrases "and if either $\lambda(U) < \infty$ or $0 \notin B$" and "either $\lambda(L) < \infty$ or $0 \notin B$" can be deleted since $\lambda(U) < \infty$ and $\lambda(L) < \infty$ obtain for all $U \in \mathcal{U}, L \in \mathcal{L}$.

Proof. We first prove that $f \in L_1$ and that

$$\left. \begin{array}{c} \varphi(L \cap [a < f \leq b]) \geq \int_{L \cap [a < f \leq b]} f \, d\lambda \\ \text{for} \quad L \in \mathcal{L}, \quad 0 \leq a < b \leq \infty. \end{array} \right\} \quad (7.40)$$

First fix a, b and ρ, $0 \leq a < b < \infty$, $0 < \rho < 1$, and fix $L \in \mathcal{L}$. Set $\delta = b - a$. For $n = 1, 2, \ldots$, set

$$A_n = L \cap [a + \delta\rho^n < f \leq a + \delta\rho^{n-1}].$$

Since f is a LRN derivative of φ with respect to λ given $\mathcal{U}$,

$$\varphi(A_n) = \varphi(L \cap [f \leq a + \delta\rho^{n-1}] \cap [f > a + \delta\rho^n])$$
$$\geq (a + \delta\rho^n)\lambda(A_n) \geq 0.$$

Then

$$\varphi(A_n) \geq a\lambda(A_n) + \rho \int_{A_n} \delta\rho^{n-1} \, d\lambda$$
$$\geq a\lambda(A_n) + \rho \int_{A_n} (f - a) \, d\lambda$$
$$= (1 - \rho)a\lambda(A_n) + \rho \int_{A_n} f \, d\lambda.$$

Summing over n, for $n = 1, 2, \ldots$,

$$\infty > \varphi(L \cap [a < f \leq b]) \geq (1 - \rho)a\lambda(L \cap [a < f \leq b]) + \rho \int_{L \cap [a < f \leq b]} f \, d\lambda.$$

Since this implies $\lambda(L \cap [a < f \leq b]) < \infty$, on letting $\rho \uparrow 1$, (7.40) is verified. By Theorem 7.13, f is finite a.e. (λ) so that, on letting $b \uparrow \infty$, (7.40) is verified for $b = \infty$. Setting $L = \Omega$, $a = 0$, $b = \infty$ in (7.40) it is found that

$$\int f^+ \, d\lambda \leq \varphi[f > 0] < \infty.$$

Applying (7.40) with φ replaced by $-\varphi$, f by $-f$, and $\mathcal{U}$ by $\mathcal{L}$,

$$\varphi(U \cap [a \leq f < b]) \leq \int_{U \cap [a \leq f < b]} f \, d\lambda \quad (7.41)$$

for $U \in \mathcal{U}$, $-\infty \leq a < b \leq 0$, and

$$\int f^- \, d\lambda \leq -\varphi[f < 0] < \infty.$$

Thus $f \in L_1$.

Now fix again a, b and ρ, $0 \leq a < b \leq \infty$, $0 < \rho < 1$, fix $U \in \mathcal{U}$, and set $\delta = b - a$. Set

$$B_n = U \cap [a + \delta\rho^n \leq f < a + \delta\rho^{n-1}], \quad n = 1, 2, \ldots.$$

Since $f \in L_1$, $\lambda(B_n) < \infty$, $n = 1, 2, \ldots$. Also

$$\varphi(B_n) = \varphi(U \cap [f \geq a + \delta\rho^n] \cap [f < a + \delta\rho^{n-1}])$$

$$\leq (a + \delta\rho^{n-1})\lambda(B_n)$$

$$\leq a\lambda(B_n) + (1/\rho) \int_{B_n} \delta\rho^n \, d\lambda$$

$$\leq a\lambda(B_n) + (1/\rho) \int_{B_n} (f - a) \, d\lambda$$

$$= (1/\rho) \int_{B_n} f \, d\lambda - [(1/\rho) - 1]a\lambda(B_n).$$

Summing over n, we obtain

$$\varphi(U \cap [a < f < b])$$

$$\leq (1/\rho) \int_{U \cap [a < f < b]} f \, d\lambda - [(1/\rho) - 1]a\lambda(U \cap [a < f < b]).$$

Note that if $a > 0$ then $\lambda(U \cap [a < f < b]) < \infty$ since $f \in L_1$, while if $a = 0$, $a\lambda(U \cap [a < f < b]) = 0$. Letting $\rho \uparrow 1$ and $b \uparrow \infty$,

$$\varphi(U \cap [a < f < b]) \leq \int_{U \cap [a < f < b]} f \, d\lambda \qquad (7.42)$$

for $U \in \mathcal{U}$, $0 \leq a < b \leq \infty$. Applying this result with φ replaced by $-\varphi$, f by $-f$, $\mathcal{U}$ by $\mathcal{L}$,

$$\varphi(L \cap [a < f < b]) \geq \int_{L \cap [a < f < b]} f \, d\lambda \qquad (7.43)$$

for $L \in \mathcal{L}$, $-\infty \leq a < b \leq 0$.

Fix $U \in \mathcal{U}$. Let γ_U denote the measure on $\mathcal{B}$ defined by

$$\gamma_U(B) = \varphi(U \cap [f \in B] \cap [f \neq 0]) - \int_{U \cap [f \in B] \cap [f \neq 0]} f \, d\lambda.$$

By (7.41) and (7.42), γ_U is nonpositive at each interval. By the extension theorem for measures, γ_U is nonpositive on $\mathscr{B}$, verifying (7.38) in the case $0 \notin B$. A symmetric argument yields (7.39) under the hypothesis $0 \notin B$.

Now suppose $L \in \mathscr{L}$, $\lambda(L) < \infty$, $-\infty < a < b < \infty$. The argument used above for (7.40) yields the same inequality since again $\lambda(L \cap [a < f < b]) < \infty$. Since $f \in L_1$, letting $a \to -\infty$ and $b \to \infty$,

$$\varphi(L \cap [a < f < b]) \geq \int_{L \cap [a<f<b]} f \, d\lambda \tag{7.44}$$

for $L \in \mathscr{L}$, $\lambda(L) < \infty$, $-\infty \leq a < b \leq \infty$. Replacing φ by $-\varphi$, f by $-f$, and $\mathscr{U}$ by $\mathscr{L}$ gives

$$\varphi(U \cap [a < f < b]) \leq \int_{U \cap [a<f<b]} f \, d\lambda \tag{7.45}$$

for $U \in \mathscr{U}$, $\lambda(U) < \infty$, $-\infty \leq a < b \leq \infty$. Again the extension theorem for measures leads to (7.38) and (7.39) in the cases $\lambda(U) < \infty$, $\lambda(L) < \infty$. ∎

Theorem 7.18 is a converse of Theorem 7.17, and in part relaxes the requirements for a function to be a LRN derivative.

Theorem 7.18

Let $(\Omega, \mathscr{S}, \lambda)$ be a measure space and $\mathscr{U}$ a sub-σ-lattice of $\mathscr{S}$. If φ is a finite measure on $\mathscr{F}$, if $g \in R(\mathscr{U})$, and if g satisfies

$$\lambda([|g| > a]) < \infty \quad \text{for} \quad a > 0, \tag{7.46}$$

$$\varphi(U \cap [g < r]) \leq r\lambda(U \cap [g < r]) \quad \text{for} \quad U \in \mathscr{U} \tag{7.47}$$

and r rational such that either $\lambda(U) < \infty$ or $r < 0$, and

$$\varphi(L \cap [g > r]) \geq r\lambda(L \cap [g > r]) \quad \text{for} \quad L \in \mathscr{L} \tag{7.48}$$

and r rational such that either $\lambda(L) < \infty$ or $r > 0$, then g is a LRN derivative of φ with respect to λ given $\mathscr{U}$.

In particular, if $g \in L_1 \cap R(\mathscr{U})$ and satisfies

$$\varphi(U \cap [g < r]) \leq \int_{U \cap [g<r]} g \, d\lambda \quad \text{for} \quad U \in \mathscr{U} \tag{7.49}$$

and r rational such that either $\lambda(U) < \infty$ or $r < 0$, and

$$\varphi(L \cap [g > r]) \geq \int_{L \cap [g>r]} g \, d\lambda \quad \text{for} \quad L \in \mathscr{L} \tag{7.50}$$

and r rational such that either $\lambda(L) < \infty$ or $r > 0$, then g is a LRN *derivative of φ with respect to λ given $\mathcal{U}$.*

We remark that if $\lambda(\Omega) < \infty$, (7.46) and the hypotheses $\lambda(L) < \infty$, $\lambda(U) < \infty$, $r < 0$, $r > 0$ in (7.47), (7.48), (7.49) and (7.50) can be deleted since then $\lambda(U) < \infty$ and $\lambda(L) < \infty$ obtain for all $U \in \mathcal{U}, L \in \mathcal{L}$.

Proof. Let f denote the LRN derivative of φ with respect to λ given $\mathcal{U}$ guaranteed by Theorem 7.13. To show $f = g$ a.e. (λ) it suffices to show that $\lambda([f < r < s < g]) = 0$ and $\lambda([g < r < s < f]) = 0$ for rationals r, s such that either $r < s < 0$ or $0 < r < s$. Suppose first that $0 < r < s$. Then

$$\varphi([f < r < s < g]) \leq r\lambda([f < r < s < g])$$

since $[g > s] \in \mathcal{U}$ and f is LRN. But also $[f < r] \in \mathcal{L}$ and $s > 0$ so that by (7.48),

$$\varphi([f < r < s < g]) \geq s\lambda([f < r < s < g]).$$

Since $r < s$, $\lambda([f < r < s < g]) = 0$. Also

$$\varphi([g < r < s < f]) \geq s\lambda([g < r < s < f])$$

since f is LRN, and

$$\varphi([g < r < s < f]) \leq r\lambda([g < r < s < f])$$

by (7.47) since $\lambda([f > s]) < \infty$. Thus $\lambda([g < r < s < f]) = 0$. The proof for $r < s < 0$ is similar. The conclusion under the second set of hypotheses follows from the fact that $g \in L_1 \cap R(\mathcal{U})$ implies (7.46), (7.49) implies (7.47), and (7.50) implies (7.48). ∎

Corollary A

Let $(\Omega, \mathcal{S}, \lambda)$ be a finite measure space, let $\mathcal{U}$ be a sub-σ-lattice of $\mathcal{S}$, and let $y \in L_1$. If $g \in L_1 \cap R(\mathcal{U})$ and satisfies

$$\int_U (y - g) \, d\lambda \leq 0 \qquad \text{for} \qquad U \in \mathcal{U} \tag{7.51}$$

and

$$\int_{[a < g < b]} (y - g) \, d\lambda = 0 \qquad \text{for} \qquad -\infty \leq a < b \leq \infty \tag{7.52}$$

then $g = E(y|\mathcal{U})$.

Proof. It suffices to verify that g satisfies (7.49) and (7.50) with

$$\varphi(A) = \int_A y \, d\lambda, \qquad A \in \mathcal{F}.$$

Note first that if $L \in \mathscr{L}$ then (7.52) with $a = -\infty$, $b = \infty$ and (7.51) imply

$$\int_L (y - g) \, d\lambda \geq 0 \quad \text{for} \quad L \in \mathscr{L}; \tag{7.53}$$

for

$$\int_L (y - g) \, d\lambda = \int (y - g) \, d\lambda - \int_{L^c} (y - g) \, d\lambda \geq 0.$$

Now let $U \in \mathscr{U}$ and $r \in R$.

$$\varphi(U \cap [g < r]) = \int_{U \cap [g < r]} y \, d\lambda$$

$$= \int_{[g < r]} y \, d\lambda - \int_{U^c \cap [g < r]} y \, d\lambda$$

$$\leq \int_{[g < r]} g \, d\lambda - \int_{U^c \cap [g < r]} g \, d\lambda$$

$$= \int_{U \cap [g < r]} g \, d\lambda,$$

verifying (7.49). The proof of (7.50) is symmetric. ∎

Corollary B

Let $(\Omega, \mathscr{S}, \lambda)$ be a measure space and let $\mathscr{U}_1, \mathscr{U}$ be sub-σ-lattices of $\mathscr{S}$, with $\mathscr{U}_1 \supset \mathscr{U}$. Let φ be a finite measure on $\mathscr{S}$ and let f_1 be a LRN derivative of φ with respect to λ given $\mathscr{U}_1$. Set

$$\varphi_1(A) = \int_A f_1 \, d\lambda, \quad A \in \mathscr{F},$$

and let f_2 be a LRN derivative of φ_1 with respect to λ given $\mathscr{U}$. If either $\mathscr{U}$ or $\mathscr{U}_1$ is a σ-field then f_2 is a LRN derivative of φ with respect to λ given $\mathscr{U}$; in particular, if $y \in L_1$ then

$$E(y|\mathscr{U}) = E\{E(y|\mathscr{U}_1)|\mathscr{U}\}.$$

Proof. We verify equations (7.49) and (7.50). Let $U \in \mathscr{U}$ and $r \in R'$, and suppose either $\lambda(U) < \infty$ or $r < 0$. If $\mathscr{U}_1$ is a σ-field then, since $U \cap [f_2 < r] \in \mathscr{U}_1 = \mathscr{U}_1^c$, it follows from Theorem 7.17 that $\varphi(U \cap [f_2 < r]) = \varphi_1(U \cap [f_2 < r])$. But

$$\varphi_1(U \cap [f_2 < r]) \leq \int_{U \cap [f_2 < r]} f_2 \, d\lambda$$

by Theorem 7.17, verifying (7.49). Similarly f_2 satisfies (7.50) so that f_2 is a LRN derivative of φ with respect to λ given $\mathcal{U}$. Now suppose $\mathcal{U}$ is a σ-field but $\mathcal{U}_1$ may not be. Again let $U \in \mathcal{U}$, $r \in R'$, and suppose either $\lambda(U) < \infty$ or $r < 0$. By Theorem 7.17,

$$\int_{U \cap [f_2 < r]} f_1 \, d\lambda = \varphi_1(U \cap [f_2 < r]) = \int_{U \cap [f_2 < r]} f_2 \, d\lambda,$$

since $U \cap [f_2 < r] \in \mathcal{U} = \mathcal{L}$. Since $\mathcal{U} \subset \mathcal{U}_1$ and $\mathcal{U}^c \subset \mathcal{U}_1^c$, $U \cap [f_2 < r] \in \mathcal{U}_1$ and $U \cap [f_2 < r] \in \mathcal{U}_1^c$. Thus

$$\varphi(U \cap [f_2 < r]) = \int_{U \cap [f_2 < r]} f_1 \, d\lambda = \int_{U \cap [f_2 < r]} f_2 \, d\lambda$$

so that f_2 satisfies (7.49). Similarly f_2 satisfies (7.50) and again f_2 is a LRN derivative of φ with respect to λ given $\mathcal{U}$. ∎

Theorem 7.19

Let $(\Omega, \mathcal{S}, \lambda)$ be a measure space and let $\mathcal{U}$ be a sub-σ-lattice of $\mathcal{S}$. Let $y \in L_1$ and set $y^* = E(y|\mathcal{U})$. Let $z \in R(\mathcal{U})$ and let ρ be a non-negative Borel function on R. Let $(y - y^*)z\rho(y^*) \in L_1$. Then

$$\int (y - y^*)z\rho(y^*) \, d\lambda \leq 0 \qquad (7.54)$$

provided at least one of the following additional hypotheses is satisfied:

$$\rho(0) = 0 \qquad (7.55)$$

or

$$\lambda[|z| > a] < \infty \quad \text{for} \quad a > 0. \qquad (7.56)$$

Note that either $\lambda(\Omega) < \infty$ or $z \in L_p$, $p \geq 1$ suffices for (7.56), and that $(y - y^*)z\rho(y^*) \in L_1$ if $z\rho(y^*)$ is bounded.

Proof. To prove (7.54) successively more general cases are considered.

Case 1: $z = 1_U$, $U \in \mathcal{U}$, $\rho = 1_B$, $B \in \mathcal{B}$.

In this case (7.54) is (7.38) with

$$\varphi(\cdot) = \int_{(\cdot)} y \, d\lambda,$$

with $0 \notin B$ if (7.55) holds, and $\lambda(U) < \infty$ if (7.56) holds.

Case 2: $z = 1_U$, $U \in \mathcal{U}$.

For n a positive integer, set

$$\rho_n(t) = \sum_{k=1}^{n 2^n} (k/2^n) 1_{[k/2^n \leq \rho < (k+1)/2^n]}(t), \qquad t \in R.$$

Note that if $\rho(0) = 0$ then $\rho_n(0) = 0$, $n = 1, 2, \ldots$. By Case 1, (7.54) holds for $z = 1_U$ and ρ_n. By hypothesis, $|y - y^*|z\rho(y^*) \in L_1$, so that the dominated convergence theorem yields (7.54) for $z = 1_U$ and ρ.

Case 3: the general case. Consider first a non-negative simple function z in $R(\mathcal{U})$ whose positive values are $z_1 > z_2 > \ldots > z_k > 0$. Then

$$z = \sum_{i=1}^{k-1}(z_i - z_{i+1})1_{B_i} + z_k 1_{B_k},$$

where $B_i = [z \geq z_i]$, $i = 1, 2, \ldots, k$. For $i = 1, 2, \ldots, k-1$, $0 \leq (z - z_{i+1})1_{B_i} \leq z$, hence $|y - y^*|1_{B_i}\rho(y^*) \in L_1$. Case 2 now yields (7.54) for non-negative simple z in $R(\mathcal{U})$. For arbitrary non-negative $\mathcal{U}$-measurable z and for each positive integer n, set

$$z_n = \sum_{k=1}^{n2^n}(k/2^n)1_{[k/2^n \leq z < (k+1)/2^n]}.$$

If z satisfies (7.56), so does z_n for each n. Also

$$|y - y^*|z_n\rho(y^*) \leq |y - y^*|z\rho(y^*) \in L_1,$$

so that the dominated convergence theorem yields (7.54) for $z \geq 0$. Applying this result to the positive part, z^+, of z satisfying the hypotheses of the theorem gives

$$\int(y - y^*)z^+\rho(y^*)\,d\lambda \leq 0.$$

Also $z^- = -(z \wedge 0) \in R(\mathcal{L})$ and $-y^* = E(-y|\mathcal{L})$. Setting $\rho_1(t) = \rho(-t)$,

$$-\int(y - y^*)z^-\rho(y^*)\,d\lambda = \int(-y + y^*)z^-\rho_1(-y^*)\,d\lambda \leq 0.$$

Thus (7.54) obtains. ∎

Corollary A

Let $(\Omega, \mathcal{S}, \lambda)$ be a measure space and $\mathcal{U}$ a sub-σ-lattice of $\mathcal{S}$. Let $y \in L_1$ and $y^* = E(y|\mathcal{U})$. Let ψ be a Borel function on R such that $(y - y^*)\psi(y^*) \in L_1$. If either $\psi(0) = 0$ or $\lambda(\Omega) < \infty$, then

$$\int(y - y^*)\psi(y^*)\,d\lambda = 0. \tag{7.57}$$

Proof. From (7.54) with $z = 1$ and $z = -1$, $\int(y - y^*)\psi^+(y^*)\,d\lambda = 0$ and $\int(y - y^*)\psi^-(y^*)\,d\lambda = 0$. ∎

The following corollary of Theorems 7.18 and 7.19 gives conditions sufficient in order that the conditional expectation of the reciprocal be the reciprocal of the conditional expectation, with respect to a different measure.

For use here only, the measure is introduced explicitly in the notation: $E(y|\mathscr{U}, d\lambda)$. When $y > 0$ and the measure

$$\lambda_1(A) = \int_A y \, d\lambda, \qquad A \in \mathscr{F},$$

replaces λ, $E(z|\mathscr{U}, y d\lambda)$ is written for $E(z|\mathscr{U}, d\lambda_1)$.

Corollary B

Let $(\Omega, \mathscr{S}, \lambda)$ be a finite measure space and let $\mathscr{U}$ be a sub-σ-lattice of $\mathscr{S}$. Let $y \in L_1$. Set $y^* = E(y|\mathscr{U}, d\lambda)$. If $y > 0$ and $(y - y^*)/y^* \in L_1$ then

$$1/y^* = E(1/y|\mathscr{L}, y d\lambda).$$

Proof. Theorem 7.19 is applied to show that the hypotheses of Theorem 7.18 are satisfied, with $\mathscr{U}$ replaced by $\mathscr{L}$,

$$\varphi(A) = \int_A (1/y) y \, d\lambda = \lambda(A) \qquad \text{for} \qquad A \in \mathscr{F},$$

and $g = 1/y^*$. Then ϕ is finite and $g \in R(\mathscr{L})$. By the corollary to Theorem 7.15, $y^* \geq 0$ a.e. (λ). Since $(y - y^*)/y^* \in L_1$, $y^* > 0$ a.e. (λ). Consider first (7.49) in this instance: we wish to show that

$$\int_{L \cap [1/y^* < r]} (1/y - 1/y^*) y \, d\lambda \leq 0 \tag{7.58}$$

whenever either

$$\int_L y \, d\lambda < \infty$$

or $r < 0$. But since $y \in L_1$,

$$\int_L y \, d\lambda < \infty$$

for all $L \in \mathscr{L}$. If $r \leq 0$, the left member of (7.58) is zero. If $r > 0$,

$$\int_{L \cap [1/y^* < r]} (1/y - 1/y^*) y \, d\lambda = -\int_{L \cap [y^* > 1/r]} (y - y^*)(1/y^*) \, d\lambda.$$

In (7.54), set $z = -1_L$ and $\rho(t) = (1/t) 1_{(1/r, \infty)}(t)$ ($\rho(0) = 0$). By Theorem 7.19,

$$-\int_{L \cap [y^* > 1/r]} (y - y^*)(1/y^*) \, d\lambda \leq 0,$$

verifying (7.58). Now suppose $U \in \mathscr{U}$. Again

$$\int_U y \, d\lambda < \infty.$$

We wish to verify (7.50), which in this instance becomes

$$\int_{U \cap [1/y^* > r]} (1/y - 1/y^*) y \, d\lambda \geq 0, \qquad r \in R'. \tag{7.59}$$

If $r > 0$, the left member is

$$-\int_{U \cap [y^* > 1/r]} (y - y^*)(1/y^*) \, d\lambda \geq 0$$

by Theorem 7.19. If $r \leq 0$, the left member is

$$\int_U (1/y - 1/y^*) y \, d\lambda = -\int_U (y - y^*)(1/y^*) \, d\lambda.$$

Setting $\rho(t) = (1/t)1_{(0, \infty)}(t)$ ($\rho(0) = 0$) in Theorem 7.19, (7.59) is verified in this case also. ∎

Corollary C

Let $(\Omega, \mathscr{S}, \lambda)$ be a finite measure space and $\mathscr{U}$ a sub-σ-lattice of $\mathscr{S}$. Let $y \in L_1$ and set $y^* = E(y|\mathscr{U})$. Let ψ be a Borel function and ρ a non-negative Borel function such that $y\rho(y^*)$, $y^*\rho(y^*)$ and $\psi(y^*)$ all belong to L_1. Then

$$E[\rho(y^*)y + \psi(y^*)|\mathscr{U}] = \rho(y^*)y^* + \psi(y^*), \tag{7.60}$$

provided the right-hand member is $\mathscr{U}$-measurable.

Proof. We apply Corollary A to Theorem 7.18 with $y_1 = \rho(y^*)y + \psi(y^*)$ in place of y and with $g = \rho(y^*)y^* + \psi(y^*)$. If $U \in \mathscr{U}$,

$$\int_U (y_1 - g) \, d\lambda = \int_U (y - y^*)\rho(y^*) \, d\lambda \leq 0$$

by Theorem 7.19 with $z = 1_U$, verifying (7.51). Also

$$\int_{[a < g < b]} (y_1 - g) \, d\lambda = 0 \quad \text{for} \quad -\infty \leq a < b < \infty$$

follows from Theorem 7.19 with ρ replaced by the product of ρ and the indicator of the set $\{t : a < t\rho(t) + \psi(t) < b\}$, using first $z = 1$ then $z = -1$. ∎

Corollary C yields an inequality related to Jensen's. Let Φ be a convex function on an interval J and Φ' any determination of its derivative, so that

$$v = \Phi(u_0) + \Phi'(u_0)(u - u_0)$$

is a line of support in the (u, v) plane of the convex set $v \geq \Phi(u)$; that is,

$$\Phi(u) \geq \Phi(u_0) + \Phi'(u_0)(u - u_0) \text{ for } u_0, u \in J. \tag{7.61}$$

Corollary D

Let $(\Omega, \mathscr{S}, \lambda)$ be a finite measure space and $\mathscr{U}$ a sub-σ-lattice of $\mathscr{S}$. Let $y \in L_1$ and set $y^* = E(y|\mathscr{U})$. Let Φ be a nondecreasing convex function defined on an interval J containing the range of y, and Φ' a determination of its derivative. Suppose $y\Phi'(y^*)$, $y^*\Phi'(y^*)$ and $\Phi(y^*)$ all belong to L_1. Then $E(\Phi(y)|\mathscr{U}) \geq \Phi(y^*)$, i.e.

$$E(\Phi(y)|\mathscr{U}) \geq \Phi(E(y|\mathscr{U})). \tag{7.62}$$

Proof. It follows from Theorem 7.16 (using constant z) that J contains also the range of y^*. From (7.61),

$$\Phi(y) \geq y\Phi'(y^*) + \Phi(y^*) - y^*\Phi'(y^*);$$

note that $\Phi' \geq 0$ since Φ is nondecreasing. By the corollary to Theorem 7.15, the conditional expectation given $\mathscr{U}$ of the left member is not smaller than that of the right member, which is $\Phi(y^*)$ by Corollary C. ∎

Corollary A yields a direct generalization of Jensen's inequality in its usual form.

Corollary E

Let $(\Omega, \mathscr{S}, \lambda)$ be a measure space and $\mathscr{U}$ a sub-σ-lattice of $\mathscr{S}$. Let $y \in L_1$ and $y^* = E(y|\mathscr{U})$. Let Φ be a convex function defined on an interval J containing the range of y, and Φ' a determination of its derivative. Suppose $(y - y^*)\Phi'(y^*) \in L_1$. If either $\Phi'(0) = 0$ or $\lambda(\Omega) < \infty$ then

$$\int [\Phi(y) - \Phi(y^*)] \, d\lambda \geq 0.$$

Thus if $(\Omega, \mathscr{S}, \lambda)$ is a probability space,

$$E[\Phi(y)] \geq E(\Phi[E(y|\mathscr{U})]$$

if either member is defined. If $\mathscr{U}$ is the trivial σ-field, $\mathscr{U} = \{\phi, \Omega\}$, it is Jensen's inequality in its usual form:

$$E[\Phi(y)] \geq \Phi[E(y)].$$

Proof. From (7.61),
$$\Phi(y) - \Phi(y^*) \geq (y - y^*)\Phi'(y^*).$$
The conclusion is now immediate from Corollary A. ∎

Minimizing properties

We develop here extremizing properties of conditional expectation given a σ-lattice generalizing those of Chapter 1. These properties also permit the identification, in $L_1 \cap L_2$, of the conditional expectation given a σ-lattice defined as a LRN derivative with the conditional expectation given a σ-lattice defined in Section 7.3 as projection on $L_2 \cap R(\mathcal{U})$. The next lemma is the key to extremizing properties of conditional expectation given a σ-lattice. We recall that for real u,
$$\nu_u(A) = \varphi(A) - u\lambda(A), \quad A \in \mathcal{F}.$$

Lemma

Let $(\Omega, \mathcal{S}, \lambda)$ be a measure space and $\mathcal{U}$ a sub-σ-lattice of $\mathcal{S}$. Let φ be a finite measure on $\mathcal{F}$ and let f be a LRN derivative of φ with respect to λ given $\mathcal{U}$. If $U_n \in \mathcal{U}$, $\lambda(U_n) < \infty$ for $n = 1, 2, \ldots$, and if $U_n \uparrow U$ as $n \to \infty$, then for $u \in R$,
$$\nu_u(U \cap [f \leq u]) \leq 0 \tag{7.63}$$
and
$$\nu_u([f > u]) \geq \nu_u(U). \tag{7.64}$$
If $L_n \in \mathcal{L}$, $\lambda(L_n) < \infty$ for $n = 1, 2, \ldots$, and if $L_n \uparrow L$ as $n \to \infty$, then for $u \in R$,
$$\nu_u(L \cap [f \geq u]) \geq 0 \tag{7.65}$$
and
$$\nu_u([f < u]) \leq \nu_u(L). \tag{7.66}$$

Proof. A proof based directly on the definition of the LRN derivative can be given, or the theorem may be viewed as a consequence of Theorem 7.17. From (7.38),
$$\varphi(U_n \cap [f \leq u]) \leq \int_{U_n \cap [f \leq u]} f \, d\lambda, \quad n = 1, 2, \ldots.$$
Since φ is finite and $f \in L_1$, on passing to the limit we obtain
$$\varphi(U \cap [f \leq u]) \leq \int_{U \cap [f \leq u]} f \, d\lambda \leq u\lambda(U \cap [f \leq u]),$$

giving (7.63). Inequality (7.64) follows from (7.63), the identity

$$\nu_u([f > u]) + \nu_u(U \cap [f \le u]) = \nu_u(U \cup [f > u])$$
$$= \nu_u(U) + \nu_u(U^c \cap [f > u])$$

and the inequality $\nu_u(U^c \cap [f > u]) \ge 0$, one of the defining properties of f as a LRN derivative. The proofs of (7.65) and (7.66) are symmetric. ∎

Let Φ be finite and convex on an interval $J \subset R$, and let $\Phi'(t)$ denote its derivative from the left at t if $t \ge 0$, from the right if $t < 0$. For reals a, b in J define

$$\Delta(b, a) = \Phi(b) - \Phi(a) - (b - a)\Phi'(a). \qquad (7.67)$$

For real t, let $t^+ = t \lor 0 = \max(t, 0)$, and $t^- = -(t \land 0) = -\min(t, 0)$. It is easy to verify that

$$\Delta(b, a) = \int_{[a,\infty)} (b - u)^+ \, d\Phi'(u) + \int_{(-\infty,a)} (b - u)^- \, d\Phi'(u) \qquad (7.68)$$

for $a \ge 0$,

$$\Delta(b, a) = \int_{(a,\infty)} (b - u)^+ \, d\Phi'(u) + \int_{(-\infty,a]} (b - u)^- \, d\Phi'(u) \qquad (7.69)$$

if $a < 0$. For given $y \in L_1$ with range in J and an arbitrary $\mathscr{S}$-measurable function z on Ω whose range is in J, define

$$I(z) = \int \Delta(y, z) \, d\lambda; \qquad (7.70)$$

since $\Delta \ge 0$, $I(z)$ is defined, though not necessarily finite. It will be shown in Theorem 7.20 that y^* minimizes $I(z)$ in the class of functions $z \in R(\mathscr{U})$ with range in J which satisfy (7.56):

$$\lambda([|z| > a]) < \infty \qquad \text{for} \qquad a > 0.$$

From (7.30) and (7.31) applied to y^*, with $U = L = \Omega$ and $a > 0$:

$$\int_{[y^* < -a]} y \, d\lambda + a\lambda([y^* < -a]) \le 0$$

and

$$\int_{[y^* > a]} y \, d\lambda - a\lambda([y^* > a]) \ge 0.$$

Since $y \in L_1$, y^* satisfies (7.56). We remark also that every function which is in L_p for some positive finite p satisfies (7.56).

Theorem 7.20

Let $(\Omega, \mathscr{S}, \lambda)$ be a measure space, $\mathscr{U}$ be a sub-σ-lattice of $\mathscr{S}$, $y \in L_1$, $y^* = E(y|\mathscr{U})$. If the range of y is in J then y^* minimizes $I(z)$ in the class of functions z in $R(\mathscr{U})$ with range in J which satisfy (7.56).

Proof. That y^* is in this class follows from Theorem 7.16 (with z constant) and the remark above. In (7.68) and (7.69), set $a = z(\omega)$, $b = y(\omega)$, $\omega \in \Omega$. We shall interchange the order of integration in (7.70). For fixed $u \geq 0$, we shall integrate $[y(\omega) - u]^+$ with respect to $\lambda(d\omega)$ over the set $[0 \leq z \leq u] \cup [z < 0] = [z \leq u]$, and $[y(\omega) - u]^-$ over the set $[z > u] \cup \phi = [z > u]$. In so doing we obtain, as a first term of two whose sum is $I(z)$:

$$\int_{[0,\infty)} d\Phi'(u) \left\{ \int_{[z \leq u]} [y(\omega) - u]^+ \lambda(d\omega) + \int_{[z > u]} [y(\omega) - u]^- \lambda(d\omega) \right\}.$$

Considering $u < 0$, we obtain the second term:

$$\int_{(-\infty,0)} d\Phi'(u) \left\{ \int_{[z < u]} [y(\omega) - u]^+ \lambda(d\omega) + \int_{[z \geq u]} [y(\omega) - u]^- \lambda(d\omega) \right\}.$$

Set

$$\nu_u^+(A) = \int_A [y(\omega) - u]^+ \lambda(d\omega)$$

and

$$\nu_u^-(A) = \int_A [y(\omega) - u]^- \lambda(d\omega).$$

Then for $u \geq 0$, $\nu_u^+ < \infty$ and

$$\nu_u^+([z \leq u]) + \nu_u^-([z > u]) = \nu_u^+(\Omega) - \nu_u([z > u]).$$

Since z satisfies (7.56), we may let $[z > u]$ play the role of U in the lemma above. From (7.64),

$$\nu_u^+(\Omega) - \nu_u([z > u]) \geq \nu_u^+(\Omega) - \nu_u([y^* > u]).$$

Treating the second term similarly and repeating the calculations with z replaced by y^* it is found that $I(z) \geq I(y^*)$. ∎

Corollary

Let $(\Omega, \mathscr{S}, \lambda)$ be a measure space, let $y \in L_1 \cap L_2$, $\mathscr{U}$ be a sub-σ-lattice of $\mathscr{S}$, $y^* = E(y|\mathscr{U})$. Then $y^* \in L_2$ and y^* is the projection of y on $L_2(\mathscr{U})$, so that y^* coincides with $E(y|\mathscr{U})$ as defined in Section 7.3.

Proof. In Theorem 7.20, set $\Phi(t) = t^2$ and $z = 0$. Then $\Delta(b, a) = (b - a)^2$, and $\int (y - y^*)^2 \, d\lambda \leq \int (y - 0)^2 \, d\lambda < \infty$. Thus $y - y^* \in L_2$, hence $y^* \in L_2$. By Theorem 7.20, y^* minimizes $\int (y - z)^2 \, d\lambda$ for z in $L_2(\mathcal{U})$. Since projection on $L_2(\mathcal{U})$ is unique, this identifies y^* with $E(y|\mathcal{U})$ as defined in Section 7.3. ∎

Theorem 7.19 and its Corollary A permit a strengthening of Theorem 7.20 for appropriate convex Φ finite on an interval J. Define $\Delta(b, a)$ by (7.67) in which Φ' is an arbitrary determination of the derivative of Φ. Then for r, s, t in J,

$$\Delta(r, t) = \Delta(r, s) + \Delta(s, t) + (r - s)[\Phi'(s) - \Phi'(t)]. \tag{7.71}$$

Theorem 7.21

Let $(\Omega, \mathcal{S}, \lambda)$ be a measure space and $\mathcal{U}$ a sub-σ-lattice of $\mathcal{S}$. Let $y \in L_1$, $y^* = E(y|\mathcal{U})$, $z \in R(\mathcal{U})$, and suppose the range of each lies in an interval J. Let Φ be convex and finite in J and let Φ' be a determination of its derivative. Suppose $(y - y^*)\Phi'(y^*) \in L_1$ and $(y - y^*)\Phi'(z) \in L_1$. If either $\Phi'(0) = 0$ and $\Phi'(z)$ satisfies (7.56) or $\lambda(\Omega) < \infty$ then

$$\int \Delta(y, z) \, d\lambda \geq \int \Delta(y, y^*) \, d\lambda + \int \Delta(y^*, z) \, d\lambda. \tag{7.72}$$

The inequality in (7.72) is reversed if z is $\mathcal{L}$-measurable ($-z \in R(\mathcal{U})$). Equality may be obtained in (7.72) if in addition any of the following hold:

$$\mathcal{U} \text{ is a } \sigma\text{-field};$$

or

$$z = 0 \text{ and } \Phi'(0) = 0;$$

or

$$z \text{ is constant and } \lambda(\Omega) < \infty.$$

In particular we have again Corollary D to Theorem 7.8: if $(\Omega, \mathcal{S}, \lambda)$ is a probability space, then

$$\text{var } y = E(y - y^*)^2 + \text{var } y^*.$$

Proof. Since Φ' is nondecreasing, $\Phi'(z)$ is $\mathcal{U}$-measurable if z is. From (7.71),

$$\Delta(y, z) = \Delta(y, y^*) + \Delta(y^*, z) + (y - y^*)[\Phi'(y^*) - \Phi'(z)].$$

Also $\int (y - y^*)\Phi'(y^*) \, d\lambda = 0$ by Corollary A of Theorem 7.19, and $\int (y - y^*)\Phi'(z) \, d\lambda \leq 0$ by Theorem 7.19. ∎

Convergence properties

We prove two theorems giving conditions sufficient for

$$\lim_n E(y_n|\mathcal{U}) = E(\lim_n y_n|\mathcal{U}),$$

and a theorem of martingale type. A theorem on monotonic convergence can be obtained as a consequence of Theorem 7.23 on dominated convergence, but here Theorem 7.22 on monotonic convergence is used in proving Theorem 7.23.

Theorem 7.22

Let $(\Omega, \mathcal{S}, \lambda)$ be a measure space, $\mathcal{U}$ a sub-σ-lattice of $\mathcal{S}$. If φ, φ_n are finite measures on $\mathcal{F}$, $n = 1, 2, \ldots$, if $\varphi_n(A) \uparrow \varphi(A)$ for $A \in \mathcal{F}$ or if $\varphi_n(A) \downarrow \varphi(A)$ for $A \in \mathcal{F}$ and if f_n is a LRN derivative of φ_n with respect to λ given $\mathcal{U}$, $n = 1, 2, \ldots$, then $\lim f_n$ is a LRN derivative of φ with respect to λ given $\mathcal{U}$.

Proof. The proof is given under the hypothesis $\phi_n(A) \uparrow (\varphi A)$ for $A \in \mathcal{F}$. The proof under the alternative hypothesis $\varphi_n(A) \downarrow \varphi(A)$ for $A \in \mathcal{F}$ follows from application of this result to $-\varphi_n, -\varphi$ with $\mathcal{U}$ replaced by $\mathcal{L}$.

Let f be a LRN derivative of φ with respect to λ given $\mathcal{U}$. The sequence $\{f_n\}$ is nondecreasing (a.e. (λ)) by Theorem 7.15; let

$$g = \lim_n f_n.$$

Applying Theorem 7.15 again, $f \geq g$. It now suffices to prove that for arbitrary rationals r and s such that either $0 < r < s$ or $r < s < 0$, $\lambda(A) = 0$ is obtained, where

$$A = \{g \leq r < s < f\}.$$

Since f is a LRN derivative of φ with respect to λ given $\mathcal{U}$, and g is $\mathcal{U}$-measurable,

$$\varphi(A) \geq s\lambda(A). \tag{7.73}$$

Now set $A_n = \{f_n \leq r < s < f\}$ and consider

$$\varphi(A) - \varphi_n(A_n) = \varphi(A) - \varphi(A_n) + \varphi(A_n) - \varphi_n(A_n).$$

Since $[f_n \leq r] \downarrow [g \leq r]$ as $n \to \infty$,

$$\lim_n \varphi(A_n) = \varphi(A).$$

Also, $\varphi - \varphi_n$ is a positive measure on $\mathcal{F}$, so that

$$\varphi(A_n) - \varphi_n(A_n) \leq \varphi(\Omega) - \varphi_n(\Omega) \to 0 \text{ as } n \to \infty.$$

Thus
$$\varphi(A) = \lim_n \varphi_n(A_n).$$

Let $r_m \downarrow r$, and set
$$A_{n,m} = \{f_n < r_m < s < f\}.$$
Then
$$\varphi_n(A_{n,m}) \leq r_m \lambda([f_n < r_m < s < f])$$
since f_n is a LRN derivative of φ_n with respect to λ given $\mathcal{U}$. The right member is finite for $n = 1, 2, \ldots$, since if $r_m > 0$, $\lambda(A_{n,m}) \leq \lambda([f > s])$, while if $r_m < 0$,
$$r_m \lambda(A_{n,m}) \geq \int_{A_{n,m}} f_n \, d\lambda > -\infty$$
since $f_n \in L_1$. Thus
$$\lim_m \lambda(A_{n,m}) = \lambda(A_n),$$
and $\varphi_n(A_n) \leq r\lambda(A_n)$. Again $\lambda(A_n) < \infty$, and
$$\varphi(A) = \lim_n \varphi_n(A_n) \leq r\lambda(A). \tag{7.74}$$

Inequalities (7.73) and (7.74) imply that $\lambda(A) < \infty$ and hence that $\lambda(A) = 0$. ∎

Theorem 7.23

Let $(\Omega, \mathcal{S}, \lambda)$ be a measure space and $\mathcal{U}$ a sub-σ-lattice of $\mathcal{S}$. For $n = 1, 2, \ldots$, let y_n be an $\mathcal{S}$-measurable function on Ω, and let $y_n \to y$ as $n \to \infty$. Suppose there exists $y_0 \in L_1$ such that $|y_n| \leq y_0$, $n = 1, 2, \ldots$. Then $E(y_n|\mathcal{U}) \to E(y|\mathcal{U})$ as $n \to \infty$.

Proof. Set
$$z_n = \bigvee_{k \geq n} y_k, \quad n = 1, 2, \ldots.$$
Then $z_n \downarrow y$. Set
$$\varphi_n(A) = \int_A z_n \, d\lambda, \quad \varphi(A) = \int_A y \, d\lambda, \quad A \in \mathcal{F},$$
$n = 1, 2, \ldots$. By the dominated covergence theorem, $\varphi_n \downarrow \varphi$. By Theorem 7.22, $E(z_n|\mathcal{U}) \downarrow E(y|\mathcal{U})$. But
$$E(z_n|\mathcal{U}) \geq \bigvee_{k \geq n} E(y_i|\mathcal{U}),$$
hence
$$E(y|\mathcal{U}) \geq \limsup_n E(y_n|\mathcal{U}).$$

A symmetric argument yields

$$E(y|\mathcal{U}) \leq \liminf_n E(y_n|\mathcal{U}),$$

completing the proof of the theorem. ∎

This section is concluded with a theorem of martingale type.

Theorem 7.24

Let $(\Omega, \mathcal{S}, \lambda)$ be a finite measure space. Let $y \in L_1(\Omega, \mathcal{S}, \lambda)$. Let $\{\mathcal{U}_n\}$ be an increasing sequence of sub-σ-lattices of $\mathcal{S}$: $\mathcal{U}_n \subset \mathcal{U}_{n+1}$, $n = 1, 2, \ldots$. Let $\mathcal{U}$ be the σ-lattice generated by $\{\mathcal{U}_n\}$. Let $y_n^* = E(y|\mathcal{U}_n)$ and $y^* = E(y|\mathcal{U})$. Then $y_n^* \to y^*$ a.e. (λ). The same conclusion obtains if $\{\mathcal{U}_n\}$ is a decreasing sequence of sub-σ-lattices of $\mathcal{S}$, with

$$\mathcal{U} = \bigcap_{n \geq 1} \mathcal{U}_n.$$

Proof. Let

$$\bar{y} = \limsup_n y_n^*, \qquad \underline{y} = \liminf_n y_n^*.$$

For real a and b and for $\varepsilon_n \downarrow 0$,

$$[\bar{y} \geq b] = \bigcap_{n \geq 1} \bigcap_{p \geq n} [y_p^* > b - \varepsilon_n] \in \mathcal{U}$$

and

$$[\underline{y} \leq a] = \bigcap_{n \geq 1} \bigcup_{p \geq n} [y_p^* < a + \varepsilon_n] \in \mathcal{L} = \mathcal{U}^c,$$

so that $\underline{y}$ and $\bar{y}$ are $\mathcal{U}$-measurable.

For $p \geq n$ set

$$H_{n,p} = [y_n^* \leq b - \varepsilon_n, \ldots, y_{p-1}^* \leq b - \varepsilon_n, y_p^* > b - \varepsilon_n],$$

and for $n = 1, 2, \ldots,$

$$H_n = \bigcup_{p \geq n} H_{n,p} = \bigcup_{p \geq n} [y_p^* > b - \varepsilon_n];$$

then $H_n \downarrow [\bar{y} \geq b]$.

For real a and $A \in \mathcal{S}$,

$$\nu_a(A) = \int_A (y - a) \, d\lambda;$$

for a positive integer n let $v_{a,n}$ denote the contraction of v_a to $\mathscr{F}_n$. If $L \in \mathscr{L}_m$ then $L \in \mathscr{L}_n$ for $n \geq m$; also $H_{n,p} \in \mathscr{F}_p$ for $p \geq n$. Then

$$v_{b-\varepsilon_n}(L \cap H_{n,p}) = v_{b-\varepsilon_n,p}(L \cap H_{n,p}) \geq 0$$

by the defining properties of y_p^* as a LRN derivative. Then

$$v_{b-\varepsilon_n}(L \cap H_n) = \sum_{p=n}^{\infty} v_{b-\varepsilon_n}(L \cap H_{n,p}) \geq 0.$$

The left hand member is

$$\int_{L \cap H_n} (y - b + \varepsilon_n) \, d\lambda.$$

Taking the limit as $n \to \infty$ it is found that

$$v_b(L \cap [\bar{y} \geq b]) = \int_{L \cap [\bar{y} \geq b]} (y - b) \, d\lambda \geq 0. \qquad (7.75)$$

This holds for $L \in \mathscr{L}_m$ for each positive integer m, and by the monotone class argument it can be shown to hold for all $L \in \mathscr{L}$. Similarly

$$v_a(U \cap [\underline{y} \leq a]) \leq 0, \qquad U \in \mathscr{U}. \qquad (7.76)$$

Set $A = [\underline{y} \leq a < b \leq \bar{y}]$. Since $[\underline{y} < a] \in \mathscr{L}$, we have $v_b(A) \geq 0$. Since $[\bar{y} \geq b] \in \mathscr{U}$, also $v_a(A) \leq 0$. Since $v_a \geq v_b$ for $a < b$, we conclude that $v_a(A) = v_b(A) = 0$, that is

$$\int_A (y - a) \, d\lambda = \int_A (y - b) \, d\lambda = 0,$$

which implies $\lambda(A) = 0$. Since this holds for each pair of reals a, b with $a < b$, it follows that $\underline{y} = \bar{y}$ a.e. (λ). But now (7.75) and (7.76) identify them with y^*.

The above argument with minor modifications suffices also for the second statement of the theorem; the monotone class argument is not required. ∎

7.5 COMPLEMENTS

In Section 7.2 it was observed that a quasi-order on X generates a σ-lattice $\mathscr{U}$ of sets such that a function f on X is isotonic if and only if it is $\mathscr{U}$-measurable. Further comments on the relationship between partial orders and σ-lattices may be found in Brunk (1961, Section 6). It is remarked in Brunk (1965, Lemma 3.1) that if $\mathscr{U}$ is a σ-lattice, if $z \in R(\mathscr{U})$, and if φ is nondecreasing, then the composition $\varphi(z)$ is in $R(\mathscr{U})$. Necessary and sufficient conditions for a family of random variables to be $R(\mathcal{U})$ for some σ-lattice $\mathcal{U}$, and necessary and sufficient conditions for a family in $L_2(\Omega, \mathcal{S}, \lambda)$ to be $L_2(\mathcal{U})$ for some σ-lattice $\mathcal{U}$ are given in Brunk (1963). However, there is an error in the theorem giving the latter conditions. This error is corrected and an improved version of the theorem is given in Dykstra (1970b).

Probably the explication of "conditional expectation given a σ-field" most frequently used is that it is a Radon–Nikodym derivative. The corresponding theory for σ-lattices has been developed in Section 7.4. It may be remarked however, that von Neumann (1940) used a Hilbert-space approach to the Radon–Nikodym theory, and that Blackwell (1947) commented that for random variables possessing finite variance, conditional expectation given a σ-field may be regarded as projection in Hilbert space. For the case of a totally finite measure space, conditional expectation given a σ-lattice was introduced in Brunk (1961) as projection on a closed convex cone in Hilbert space. The requirement that the measure be finite was relaxed in Brunk (1963), and an extension to L_1 was given. Further results in L_2 appear in Brunk (1961).

Theorems 7.2 and 7.3 on projection on a closed convex subset of a Hilbert space H are well known (cf., e.g., Luenberger (1969, p. 69)). Another interesting property of projection on a closed convex set in H, related to Corollary D to Theorem 7.8, is found in Brunk (1965): if A_i is a closed convex cone, $i = 1, 2$, and if $A_1 \subset A_2$, then

$$||P(y|A_2)||^2 - ||P(y|A_1)||^2 = (P(y|A_2) - P(y|A_1), y)$$
$$\geq ||P(y|A_2) - P(y|A_1)||^2.$$

Monotonic operators have been studied by Browder (1965) and by Minty (1965) (see these papers for further references). Theorem 7.7 appears in Nashed (1968, p. 784).

Corollary A to Theorem 7.8 generalizes Theorem 1.4.

In a recent paper on efficiency-robust estimation of location, van Eeden (1970) has occasion to consider the following problem: minimize

$$\int (y - u)^2 \, d\lambda$$

in the class of nondecreasing functions $u \in L_2$ on $[0, 1]$. Here y is a given function on $[0, 1]$, $\Omega = [0, 1]$, $\mathcal{S}$ is the class of Borel subsets of $[0, 1]$, and λ is Lebesgue measure. The class of nondecreasing functions in L_2 on $[0, 1]$ is $L_2(\mathcal{U})$, where $\mathcal{U}$ is the class of intervals $(a, 1]$ or $[a, 1]$, $0 \leq a \leq 1$, plus the empty set ϕ. The solution is $E(y|\mathcal{U})$.

A problem discussed by W. T. Reid (1968, p. 373) is more general in that restrictions bounding the difference quotient both above and below are admitted.

Corollary E to Theorem 7.8 generalizes remarks of Brillinger (1966) and Rao (1965, p. 221) to the effect that the regression function $E(y|x)$ is the function of x most highly correlated with y; and of Rao (1965, p. 223) that the linear regression function of y on x is the linear function of x most highly correlated with y.

Corollary A to Theorem 7.9 generalizes Theorem 1.6, as do also Theorem 7.15, its corollary, and Theorem 7.16.

The approach to conditional expectation given a σ-lattice followed in Section 7.4 is due to Johansen (1967). In particular Definition 7.5 is Johansen's adaptation of a standard definition for the case of a σ-field (see Hewitt and Stromberg (1965, p. 367)); and Theorem 7.11 is Johansen's. A theorem of this kind in a special setting may be found also in Brunk, Ewing and Utz (1957).

Darst (1970) has extended the theory of the Lebesgue decomposition and the Radon–Nikodym derivative to encompass additive set functions defined on lattices of sets.

Theorem 7.14 is a special case of a theorem of Brunk and Johansen (1970) which is of interest also when $\mathcal{U}$ is a σ-field. It is a general formulation of the max–min formulas of Section 1.2. The first special case known to the authors appears in Ayer and coworkers (1955). Other special cases occur in Brunk, Ewing and Utz (1957), van Eeden (1957c), van Eeden (1958) and Marshall and Proschan (1965).

When applied in the context of Chapter 1 (X finite), Corollary A of Theorem 7.18 shows that condition (1.21) and the equation in Theorem 1.7 are not only necessary but also sufficient in order that g^* be the isotonic regression of g.

Corollary B to Theorem 7.18 expresses a smoothing property of conditional expectation given a σ-lattice, well known when both $\mathcal{U}_1$ and $\mathcal{U}$ are σ-fields. In the case of a totally finite measure space, this corollary is due to Robertson (1968). The finite measure case of Theorem 7.19 is also a remark of Robertson (1966). Corollary A to Theorem 7.19 generalizes Theorem 1.7.

An application of Theorem 7.19 extends an inequality due to Levin and Stečkin (1960). Let $\Omega = (-1/2, 1/2)$ and let λ be Lebesgue measure on the class $\mathcal{S}$ of Borel subsets of $(-1/2, 1/2)$. Let $\mathcal{L}$ be the σ-lattice of subintervals of $(-1/2, 1/2)$ containing the origin, and set $\mathcal{U} = \mathcal{L}^c$. A function p is termed *unimodal* on $(-1/2, 1/2)$ if p is $\mathcal{L}$-measurable; $p \in R(\mathcal{L})$, or $-p \in R(\mathcal{U})$. We remark first that if p is integrable on $(-1/2, 1/2)$ and unimodal and if

$$\int_{-\frac{1}{2}}^{0} p(x)\, dx = \int_{0}^{\frac{1}{2}} p(x)\, dx$$

then
$$E(p|\mathscr{U}) = \int_{-\frac{1}{2}}^{\frac{1}{2}} p(x)\, dx.$$

This follows from Corollary A to Theorem 7.18. To see this, set $y = p$ in that corollary,
$$g = \int_{-\frac{1}{2}}^{\frac{1}{2}} p(x)\, dx = 2\int_{0}^{\frac{1}{2}} p(x)\, dx.$$

Set $p_1(x) = p(x) - g$; then p_1 is unimodal and
$$\int_{-\frac{1}{2}}^{0} p_1(x)\, dx = \int_{0}^{\frac{1}{2}} p_1(x)\, dx = 0.$$

For real a, b such that $-1/2 < b < 0 < a < 1/2$,
$$\int_{a}^{\frac{1}{2}} p_1(x)\, dx \leq \int_{0}^{\frac{1}{2}} p_1(x)\, dx = 0$$
and
$$\int_{-\frac{1}{2}}^{b} p_1(x)\, dx \leq \int_{-\frac{1}{2}}^{0} p_1(x)\, dx = 0.$$

Thus
$$\int_{U} [p(x) - g]\, dx \leq 0,$$

where $U = (-1/2, b) \cup (a, 1/2)$, so that (7.51) is satisfied. Equation (7.52) is immediate from the fact that g is constant.

Now let Φ be a function on $(-1/2, 1/2)$ such that $-\Phi$ is unimodal, i.e., Φ is $\mathscr{U}$-measurable, and such that Φ and $p\Phi$ are integrable. In (7.54) we set $y = p$, $\rho \equiv 1$, $z = \Phi$,
$$y^* = \int_{-\frac{1}{2}}^{\frac{1}{2}} p(x)\, dx,$$
and find that
$$\int_{-\frac{1}{2}}^{\frac{1}{2}} p(x)\Phi(x)\, dx \leq \left[\int_{-\frac{1}{2}}^{\frac{1}{2}} p(x)\, dx\right]\int_{-\frac{1}{2}}^{\frac{1}{2}} \Phi(x)\, dx. \tag{7.77}$$

Now suppose p is integrable and unimodal on $(-1/2, 1/2)$,
$$\int_{-\frac{1}{2}}^{0} p(x)\, dx = \int_{0}^{\frac{1}{2}} p(x)\, dx.$$

Suppose that Ψ is convex on $(-1/2, 1/2)$, and that Ψ and $p\Psi$ are integrable. Let m be a determination of the derivative of Ψ at 0; i.e., $\Psi''(0-) \leq m$

$\leq \Psi''(0+)$. Set $\Phi(x) = \Psi(x) - mx$; then Φ satisfies the hypotheses leading to (7.77). If

$$m \int_{-\frac{1}{2}}^{\frac{1}{2}} xp(x) \, dx \leq 0,$$

this gives

$$\int_{-\frac{1}{2}}^{\frac{1}{2}} p(x)\Psi(x) \, dx \leq \left[\int_{-\frac{1}{2}}^{\frac{1}{2}} p(x) \, dx\right] \int_{-\frac{1}{2}}^{\frac{1}{2}} \Psi(x) \, dx,$$

the inequality of Levin and Stečkin, under less restrictive hypotheses on p.

Corollary B to Theorem 7.19 is due to Robertson (1965), who applied it to the generalized isotonic regression problem, and in particular to the problem of maximizing a product (Example 1.10).

Corollary C to Theorem 7.19 expresses a very restricted additive property of conditional expectation given a σ-lattice. Additivity of conditional expectation given a σ-lattice has been studied by Kuenzi (1969).

The treatment of extremum problems in Section 7.4 is based on that in Brunk and Johansen (1970). Theorem 7.21 generalizes Theorem 1.10. The special case $\Phi(s) = s^2$ is given in Ayer and coworkers (1955). The first special case known to the authors which involves arbitrary convex Φ appears for special σ-lattices in Brunk, Ewing and Utz (1957) and was discovered independently by W. T. Reid. For general σ-lattices, a theorem of this kind is given in Brunk (1961).

If $\Phi(s) = s \log s$ then

$$\Delta(y, z) = y \log (y/z) - (y - z).$$

If y, z are probability densities with respect to λ, then

$$\int \Delta(y, z) \, d\lambda = \int y \log (y/z) \, d\lambda.$$

This may be interpreted as the mean information per observation from a distribution with density y for discriminating in favour of y against z. By Theorem 7.21, this is minimized for z in the class of functions measurable with respect to a σ-lattice $\mathscr{U}$ by $y^* = E(y|\mathscr{U})$; and the mean information per observation from y for discriminating in favour of y against $\mathscr{U}$-measurable z is at least as large as that for discriminating in favour of y against y^* plus the mean information per observation from y^* for discriminating in favour of y^* against z.

Theorem 7.24 is a special case of a theorem in Brunk and Johansen (1970), proved using methods of Sparre Andersen and Jessen (1948). Theorems of martingale convergence type in L_2 are given in Brunk (1965).

Appendix: Tables

These tables are provided to facilitate the application of the likelihood ratio tests described in Chapter 3. They cover some of the cases most likely to be encountered in practice, and comprise four tables of critical values of the test statistics, all relating to the simple order alternative, and two tables of the probabilities $P(l, k)$ when the weights are equal, for the simple order and simple tree alternatives. Some of the tables are reproduced, with permission, from previously published work; the others extend earlier tables.

TABLES A.1 AND A.2. CRITICAL VALUES OF THE $\bar{\chi}_3^2$ AND $\bar{\chi}_4^2$ DISTRIBUTIONS FOR GENERAL WEIGHTS

These tables give the upper 10%, 5%, 2.5%, 1% and 0.5% points of the null hypothesis distribution of the $\bar{\chi}_k^2$ statistic for testing against the simple order alternative, when $k = 3$ and $k = 4$. These critical values are tabulated against the values of the correlation coefficients ρ_{12} and (in Table A.2) ρ_{23}, where, from (3.15),

$$-\rho_{12} = \left[\frac{w_1 w_3}{(w_1 + w_2)(w_2 + w_3)}\right]^{\frac{1}{2}}$$

and

$$-\rho_{23} = \left[\frac{w_2 w_4}{(w_2 + w_3)(w_3 + w_4)}\right]^{\frac{1}{2}}.$$

In Table A.2 the entries correspond only to values of these correlation coefficients satisfying $\rho_{12}^2 + \rho_{23}^2 \leq 1$. It can readily be seen from the form of the above expressions that the only possible values of ρ_{12} and ρ_{23} are negative and satisfy the strict form of the above inequality; the entries in the table for $\rho_{12} = 0$ are included for interpolation purposes. Further reduction in the size of Table A.2 has been achieved by omitting entries for $\rho_{23} > \rho_{12}$; nothing is lost through this omission, as the probability integral of the $\bar{\chi}_4^2$ distribution is symmetrical in the two correlation coefficients.

Table A.1 is reproduced from Bartholomew (1959b), and Table A.2 extends, and makes minor corrections to, tables of Bartholomew (1959a, 1959b). If greater accuracy is required than rough interpolation in these tables can give, recourse must be had to direct calculation of tail probabilities using Theorem 3.1, with the probabilities $P(l, k)$ given by the appropriate expressions chosen from (3.24)–(3.28).

TABLES A.3 AND A.4. CRITICAL VALUES OF THE $\bar{\chi}^2$ AND $\bar{E}^2$ DISTRIBUTIONS FOR EQUAL WEIGHTS

The production of tables corresponding to Tables A.1 and A.2 for values of k greater than 4 would involve two major problems. One is that, as we have remarked in Chapter 3, closed expressions for the probabilities $P(l, k)$, $k > 4$ do not exist, and recourse must be had to approximations or summation of series; the other is that, as tabulation must be in terms of $k - 2$ correlation coefficients, the size of any such tables would increase very rapidly with k. Thus the only table of critical values of $\bar{\chi}_k^2$ for $k > 4$ is Table A.3, which is for equal weights and gives the upper 10%, 5%, 2.5%, 1% and 0.5% points for $k = 3\,(1)\,12$. The source is Bartholomew (1959b), and calculations are based on Theorem 3.1 with the probabilities $P(l, k)$ given by Corollary A to Theorem 3.3. We also reproduce, from Chacko (1963), a table giving the upper 5% and 1% points of the null hypothesis distribution of the $\bar{E}_k^2$ statistic, for $k = 3, 4, 5, 6$. This table, appearing as Table A.4, relates to the most common "equal weights" case, when the error variance is constant and the same number, n, of observations is made in each of the k populations; in the table $n = 2\,(1)\,8, 10, 16$. Under these conditions the null hypothesis distribution of $\bar{E}_k^2$ is given by Theorem 3.2, with $N = nk$ and the probabilities $P(l, k)$ as in Corollary A to Theorem 3.3.

TABLES A.5 AND A.6. THE PROBABILITIES $P(l, k)$ FOR EQUAL WEIGHTS

These tables are provided to facilitate calculations of tail area probabilities using Theorems 3.1 and 3.2, and give the necessary probabilities $P(l, k)$ for equal weights and $k = 2\,(1)\,12$. Table A.5 relates to the simple order alternative. Here the probabilities have been calculated, using Corollary A to Theorem 3.3, from Table 24.3 of Abramowitz and Stegun (1965), which gives the Stirling numbers $\{S_{k}^{l}\}$ for $k \leq 25$. Table A.6 is for the simple tree alternative, and is reproduced from Bartholomew (1961a). The entries in this table are derived from Ruben's (1954) table, using the expression (3.40).

Table A.1 Critical Values of the $\bar{\chi}_3^2$ Statistic for Testing Against the Simple Order Alternative

$-\rho_{12}$		0.0	0.1	0.2	0.3	0.4	0.5	0.6	0.7	0.8	0.9	1.0
	0.1	2.952	2.885	2.816	2.742	2.664	2.580	2.486	2.379	2.251	2.080	1.642
	0.05	4.231	4.158	4.081	4.001	3.914	3.820	3.715	3.593	3.446	3.245	2.706
α	0.025	5.537	5.459	5.378	5.292	5.200	5.098	4.985	4.852	4.689	4.465	3.841
	0.01	7.289	7.208	7.122	7.030	6.932	6.822	6.700	6.556	6.377	6.130	5.413
	0.005	8.628	8.543	8.455	8.360	8.258	8.146	8.016	7.865	7.677	7.413	6.635

Significance level = α. For definition of ρ_{12}, see text.

APPENDIX: TABLES

Table A.2 *Critical Values of the $\overline{\chi}_4^2$ Statistic for Testing Against the Simple Order Alternative*

$-\rho_{23}$	α	\multicolumn{8}{c}{$-\rho_{12}$}							
		0.0	0.1	0.2	0.3	0.4	0.5	0.6	0.7
0.0	0.1	4.010							
	0.05	5.435							
	0.025	6.861							
	0.01	8.746							
	0.005	10.171							
0.1	0.1	3.952	3.891						
	0.05	5.372	5.305						
	0.025	6.794	6.724						
	0.01	8.676	8.601						
	0.005	10.098	10.020						
0.2	0.1	3.893	3.827	3.758					
	0.05	5.307	5.235	5.160					
	0.025	6.725	6.649	6.570					
	0.01	8.602	8.522	8.437					
	0.005	10.022	9.939	9.851					
0.3	0.1	3.831	3.760	3.685	3.606				
	0.05	5.239	5.162	5.080	4.993				
	0.025	6.653	6.571	6.484	6.391				
	0.01	8.525	8.438	8.346	8.246				
	0.005	9.942	9.852	9.756	9.653				
0.4	0.1	3.765	3.688	3.607	3.519	3.423			
	0.05	5.166	5.083	4.994	4.898	4.791			
	0.025	6.575	6.486	6.392	6.289	6.174			
	0.01	8.442	8.348	8.247	8.137	8.014			
	0.005	9.855	9.758	9.653	9.539	9.411			
0.5	0.1	3.695	3.610	3.521	3.423	3.313	3.187		
	0.05	5.088	4.997	4.898	4.791	4.670	4.528		
	0.025	6.491	6.394	6.289	6.173	6.043	5.891		
	0.01	8.352	8.248	8.136	8.013	7.873	7.709		
	0.005	9.761	9.654	9.537	9.409	9.264	9.092		
0.6	0.1	3.617	3.523	3.422	3.310	3.183	3.031	2.837	
	0.05	5.002	4.900	4.789	4.665	4.524	4.354	4.135	
	0.025	6.398	6.289	6.170	6.038	5.886	5.702	5.462	
	0.01	8.251	8.135	8.008	7.867	7.703	7.504	7.244	
	0.005	9.656	9.535	9.404	9.256	9.085	8.877	8.604	
0.7	0.1	3.530	3.422	3.305	3.172	3.017	2.822	2.550	1.987
	0.05	4.904	4.787	4.657	4.510	4.337	4.118	3.805	3.137
	0.025	6.291	6.166	6.027	5.870	5.682	5.443	5.100	4.346
	0.01	8.135	8.002	7.854	7.684	7.482	7.223	6.846	6.000
	0.005	9.534	9.395	9.242	9.065	8.853	8.581	8.183	7.279
0.8	0.1	3.427	3.296	3.151	2.981	2.770	2.473	1.642	
	0.05	4.787	4.644	4.483	4.294	4.056	3.715	2.706	
	0.025	6.163	6.011	5.838	5.634	5.375	4.999	3.841	
	0.01	7.994	7.832	7.647	7.427	7.146	6.734	5.412	
	0.005	9.385	9.217	9.025	8.795	8.500	8.064	6.635	
0.9	0.1	3.291	3.110	2.897	2.621	2.166			
	0.05	4.631	4.432	4.195	3.883	3.353			
	0.025	5.990	5.778	5.523	5.182	4.591			
	0.01	7.804	7.577	7.303	6.933	6.277			
	0.005	9.183	8.948	8.661	8.273	7.576			
1.0	0.1	2.952							
	0.05	4.231							
	0.025	5.537							
	0.01	7.289							
	0.005	8.628							

Significance level = α. For definitions of ρ_{12}, ρ_{23}, see text.

Table A.3 Critical Values of the $\overline{\chi}_k^2$ Statistic for Testing Against the Simple Order Alternative, Equal Weights

k	3	4	5	6	7	8	9	10	11	12
α 0.1	2.580	3.187	3.636	3.994	4.289	4.542	4.761	4.956	5.130	5.288
0.05	3.820	4.528	5.049	5.460	5.800	6.088	6.339	6.560	6.758	6.937
0.025	5.098	5.891	6.471	6.928	7.304	7.624	7.901	8.145	8.363	8.561
0.01	6.822	7.709	8.356	8.865	9.284	9.639	9.946	10.216	10.458	10.676
0.005	8.146	9.092	9.784	10.327	10.774	11.153	11.480	11.767	12.025	12.257

Significance level = α.

Table A.4 Critical Values of the $\overline{E}_k^2$ Statistic for Testing Against the Simple Order Alternative, Equal Weights (i.e. Constant Error Variance and Equal Sample Sizes)

n	α	k			
		3	4	5	6
2	0.05	0.687	0.590	0.518	0.461
	0.01	0.878	0.787	0.708	0.641
3	0.05	0.455	0.392	0.345	0.308
	0.01	0.665	0.575	0.506	0.453
4	0.05	0.337	0.292	0.258	0.231
	0.01	0.522	0.447	0.391	0.348
5	0.05	0.267	0.233	0.206	0.184
	0.01	0.427	0.364	0.318	0.282
6	0.05	0.221	0.193	0.170	0.153
	0.01	0.361	0.307	0.268	0.237
7	0.05	0.189	0.165	0.146	0.131
	0.01	0.312	0.265	0.231	0.205
8	0.05	0.164	0.144	0.128	0.115
	0.01	0.287	0.223	0.203	0.180
10	0.05	0.131	0.113	0.102	0.092
	0.01	0.222	0.188	0.164	0.145
16	0.05	0.081	0.071	0.064	0.057
	0.01	0.140	0.119	0.103	0.091

Significance level = α. Common sample size = n.

Table A.5 The Probabilities $P(l, k)$ for Equal Weights (Testing Against the Simple Order Alternative)

l	\multicolumn{11}{c}{k}
	2	3	4	5	6	7	8	9	10	11	12
1	0.50000	0.33333	0.25000	0.20000	0.16667	0.14286	0.12500	0.11111	0.10000	0.09091	0.08333
2	0.50000	0.50000	0.45833	0.41667	0.38056	0.35000	0.32411	0.30198	0.28290	0.26627	0.25166
3		0.16667	0.25000	0.29167	0.31250	0.32222	0.32569	0.32552	0.32316	0.31950	0.31507
4			0.04167	0.08333	0.11806	0.14583	0.16788	0.18542	0.19943	0.21068	0.21974
5				0.00833	0.02083	0.03472	0.04861	0.06186	0.07422	0.08560	0.09602
6					0.00139	0.00417	0.00799	0.01250	0.01744	0.02260	0.02785
7						0.00020	0.00069	0.00150	0.00260	0.00395	0.00551
8							0.00002	0.00010	0.00024	0.00045	0.00075
9								0.00000	0.00001	0.00003	0.00007
10									0.00000	0.00000	0.00000
11										0.00000	0.00000
12											0.00000

Table A.6 The Probabilities $P(l, k)$ for Equal Weights (Testing Against the Simple Tree Alternative)

l	\multicolumn{11}{c}{k}
	2	3	4	5	6	7	8	9	10	11	12
1	0.50000	0.16667	0.04387	0.00978	0.00192	0.00034	0.00006	0.00001	0.00000	0.00000	0.00000
2	0.50000	0.50000	0.25000	0.08774	0.02446	0.00577	0.00119	0.00022	0.00004	0.00001	0.00000
3		0.33333	0.45613	0.29021	0.12373	0.04060	0.01101	0.00258	0.00053	0.00010	0.00002
4			0.25000	0.41226	0.30887	0.15185	0.05650	0.01714	0.00443	0.00101	0.00021
5				0.20000	0.37434	0.31620	0.17352	0.07139	0.02374	0.00669	0.00164
6					0.16667	0.34238	0.31731	0.19018	0.08499	0.03054	0.00925
7						0.14286	0.31541	0.31492	0.20299	0.09725	0.03734
8							0.12500	0.29246	0.31054	0.21282	0.10824
9								0.11111	0.27274	0.30505	0.22035
10									0.10000	0.25563	0.29898
11										0.09091	0.24064
12											0.08333

Bibliography and Author Index

[Pages on which authors are cited are given in brackets]

Abelson, R. P., and J. W. Tukey (1963). "Efficient utilization of non-numerical information in quantitative analysis: general theory and the case of simple order." *Ann. math. Statist.*, **34**, 1347–1369. [186, 187, 190, 211, 215, 219]

Abrahamson, I. G. (1964). "Orthant probabilities for the quadrivariate normal distribution." *Ann. math. Statist.*, **35**, 1685–1703. [137, 141]

Abramowitz, M., and I. S. Stegun (1965). *Handbook of Mathematical Functions*, Dover, New York. [360]

Alexander, M. J. (1970). "An algorithm for obtaining maximum-likelihood estimates of a set of partially ordered parameters." *Rocketdyne Technical Paper*. [81]

Andrews, F. C. (1954). "Asymptotic behavior of some rank tests for analysis of variance." *Ann. math. Statist.*, **25**, 724–735. [202]

Armitage, P. (1955). "Tests for linear trends in proportions and frequencies." *Biometrics*, **11**, 375–386. [184, 194]

Arnold, B. C. (1970). "An alternative derivation of a result due to Srivastava and Bancroft." *Jl R. statist. Soc. (B)*, **32**, 265–267.

Arrow, K. J., S. Karlin and H. Scarf (1958). *Studies in the Mathematical Theory of Inventory and Production*, Stanford University Press. [61]

Ayer, M., H. D. Brunk, G. M. Ewing, W. T. Reid and E. Silverman (1955). "An empirical distribution function for sampling with incomplete information." *Ann. math. Statist.*, **26**, 641–647. [3, 55, 112, 113, 114, 356, 358]

Bancroft, T. A. *See* Srivastava and Bancroft (1967).

Barlow, R. E. (1968a). "Likelihood ratio tests for restricted families of probability distributions." *Ann. math. Statist.*, **39**, 547–560. [255, 305, 306]

Barlow, R. E. (1968b). "Some recent developments in reliability theory." In *Selected Statistical Papers*, 2, Mathematical Centre, Amsterdam, 49–66.

Barlow, R. E., and H. D. Brunk (1972). "The isotonic regression problem and its dual." *J. Am. statist. Ass.*, **67**, to appear. [115]

Barlow, R. E., and K. Doksum (1972). "Isotonic tests for convex orderings." *Proc. 6th Berkeley Symp. math. Statist. Probab.*, **I**, 293–323. [264, 305]

Barlow, R. E., A. W. Marshall and F. Proschan (1963). "Properties of probability distributions with monotone hazard rate." *Ann. math. Statist.*, **34**, 375–389.

Barlow, R. E., and F. Proschan (1966). "Inequalities for linear combinations of order statistics from restricted families." *Ann. math. Statist.*, **37**, 1574–1592. [305]

Barlow, R. E., and F. Proschan (1967). "Exponential life test procedures when the distribution has monotone failure rate." *J. Am. statist. Ass.*, **62**, 548–560.

Barlow, R. E., and F. Proschan (1969). "A note on tests for monotone failure rate based on incomplete data." *Ann. math. Statist.*, **40**, 595–600. [306]

Barlow, R. E., F. Proschan and E. M. Scheuer (1971). "A system debugging model." In D. Grouchko (Ed.), *Operations Research and Reliability*, Gordon and Breach, New York, 401–420. [305]

Barlow, R. E., and E. M. Scheuer (1966). "Reliability growth during a development testing program." *Technometrics*, **8**, 53–60. [4]

Barlow, R. E., and E. M. Scheuer (1971). "Estimation from accelerated life tests." *Technometrics*, **13**, 145–159. [262]

Barlow, R. E., and V. A. Ubhaya (1971). "Isotonic approximation." In J. S. Rustagi (Ed.), *Optimizing Methods in Statistics*, Academic Press, New York.

Barlow, R. E., and W. R. van Zwet (1969). "Asymptotic properties of isotonic estimators for the generalized failure rate function. Part II: asymptotic distributions." *Operations Research Center Report ORC* 69-10, University of California, Berkeley. [252, 254, 262]

Barlow, R. E., and W. R. van Zwet (1970). "Asymptotic properties of isotonic estimators for the generalized failure rate function. Part I: strong consistency." In M. L. Puri (Ed.), *Nonparametric Techniques in Statistical Inference*, Cambridge University Press, 159–173. [228, 262]

Barlow, R. E. *See also* Proschan, Barlow, Madansky and Scheuer (1968).

Bartholomew, D. J. (1956). "Tests for randomness in a series of events when the alternative is a trend." *Jl R. statist. Soc.* (*B*), **18**, 234–239. [196, 300]

Bartholomew, D. J. (1959a). "A test of homogeneity for ordered alternatives." *Biometrika*, **46**, 36–48. [2, 55, 114, 122, 127, 170, 171, 172, 193, 359]

Bartholomew, D. J. (1959b). "A test of homogeneity for ordered alternatives II." *Biometrika*, **46**, 328–335. [137, 359, 360]

Bartholomew, D. J. (1961a). "A test of homogeneity of means under restricted alternatives" (with discussion). *Jl R. statist. Soc.* (*B*), **23**, 239–281. [127, 153, 159, 169, 170, 171, 172, 360]

Bartholomew, D. J. (1961b). "Ordered tests in the analysis of variance." *Biometrika*, **48**, 325–332. [171, 206, 219]

Bartholomew, D. J. (1963). "On Chassan's test for order." *Biometrics*, **19**, 188–191.

Barton, D. E., and C. L. Mallows (1961). "The randomization bases of the problem of the amalgamation of weighted means." *Jl R. statist. Soc.* (*B*), **23**, 423–433. [172]

Basu, A. P. (1967). "On two k-sample rank tests for censored data." *Ann. math. Statist.*, **38**, 1520–1535.

Bennett, B. M. (1962). "On an exact test for trend in binomial trials and its power function." *Metrika*, **5**, 49–53. [194, 220]

Bhattacharyya, G. K., and R. A. Johnson (1970). "A layer rank test for ordered bivariate alternatives." *Ann. math. Statist.*, **41**, 1296–1310.

Bhattacharyya, G. K., and J. H. Klotz (1966). "The bivariate trend of Lake Mendota." *Technical Report No.* 98, Department of Statistics, University of Wisconsin, Madison, Wisconsin. [5, 15]

Bickel, P. J. (1969). "Tests for monotone failure rate II." *Ann. math. Statist.*, **40**, 1250–1260. [287, 289, 305]

Bickel, P. J., and K. Doksum (1969). "Tests for monotone failure rate based on normalized spacings." *Ann. math. Statist.*, **40**, 1216–1235. [267, 283, 287, 297, 305]

Birnbaum, Z. W. (1953). "On the power of a one-sided test of fit for continuous probability functions." *Ann. math. Statist.*, **24**, 484–489. [280]

Birnbaum, Z. W., and F. H. Tingey (1951). "One-sided confidence contours for probability distribution functions." *Ann. math. Statist.*, **22**, 592–596. [196]

Birnbaum, Z. W., J. D. Esary and A. W. Marshall (1966). "Stochastic characterization of wear-out for components and systems." *Ann. math. Statist.*, **37**, 816–825. [262]

Blackwell, D. (1947). "Conditional expectation and unbiased estimation." *Ann. math. Statist.*, **18**, 105–110. [355]

Blumenthal, S., and A. Cohen (1968). "Estimation of two ordered translation parameters." *Ann. math. Statist.*, **39**, 517–530.

Boswell, M. T. (1966). "Estimating and testing trend in a stochastic process of Poisson type." *Ann. math. Statist.*, **37**, 1564–1573. [5, 306]

Boswell, M. T., and H. D. Brunk (1969). "Distribution of likelihood ratio in testing against trend." *Ann. math. Statist.*, **40**, 371–380. [191, 195]

Bray, T. A., G. B. Crawford and F. Proschan (1967). "Maximum likelihood estimation of a U-shaped failure rate function." *Math. Note No. 534*, Math. Res. Lab., Boeing Sci. Res. Labs. [232]

Bremner, J. M. (1967). "Problems of estimation and testing arising from ordered hypotheses concerning normal means." *Unpublished M.Sc. dissertation*, University of Wales. [177]

Brillinger, D. R. (1966). "An extremal property of the conditional expectation." *Biometrika*, **53**, 594–595. [356]

British Association Mathematical Tables, Vol. 1, 3rd ed., Cambridge University Press, 1951. [151]

Browder, F. E. (1965). "Nonlinear monotone operators and convex sets in Banach spaces." *Bull. Am. math. Soc.*, **71**, 780–785. [355]

Brunk, H. D. (1955). "Maximum likelihood estimates of monotone parameters." *Ann. math. Statist.*, **26**, 607–616. [112, 113, 114]

Brunk, H. D. (1956). "On an inequality for convex functions." *Proc. Am. math. Soc.*, **7**, 817–824. [55]

Brunk, H. D. (1958). "On the estimation of parameters restricted by inequalities." *Ann. math. Statist.*, **29**, 437–454. [112, 114]

Brunk, H. D. (1960). "On a theorem of E. Sparre Andersen and its application to tests against trend." *Mathematica scand.*, **8**, 305–326. [200]

Brunk, H. D. (1961). "Best fit to a random variable by a random variable measurable with respect to a σ-lattice." *Pacif. J. Math.*, **11**, 785–802. [354, 355, 358]

Brunk, H. D. (1963). "On an extension of the concept conditional expectation." *Proc. Am. math. Soc.*, **14**, 298–304. [355]

Brunk, H. D. (1964). "A generalization of Spitzer's combinatorial lemma." *Z. Wahrscheinlichkeitstheorie verw. Geb.*, **2**, 395–405.

Brunk, H. D. (1965). "Conditional expectation given a σ-lattice and applications." *Ann. math. Statist.*, **36**, 1339–1350. [56, 354, 355, 358]

Brunk, H. D. (1970). "Estimation of isotonic regression" (with discussion by Ronald Pyke). In M. L. Puri (Ed.), *Nonparametric Techniques in Statistical Inference*, Cambridge University Press, 177–197. [70, 111, 112, 114]

Brunk, H. D., G. M. Ewing and W. R. Utz (1957). "Minimizing integrals in certain classes of monotone functions." *Pacif. J. Math.*, **7**, 833–847. [2, 56, 113, 114, 356, 358]

Brunk, H. D., W. E. Franck, D. L. Hanson and R. V. Hogg (1966). "Maximum likelihood estimation of the distributions of two stochastically ordered random variables." *J. Am. statist. Ass.*, **61**, 1067–1080. [115]

Brunk, H. D., and S. Johansen (1970). "A generalized Radon–Nikodym derivative." *Pacif. J. Math.*, **34**, 585–617. [57, 328, 356, 358]

Brunk, H. D. *See also* Ayer, Brunk, Ewing, Reid and Silverman (1955); Barlow and Brunk (1972); Boswell and Brunk (1969); *and* Lombard and Brunk (1963).

Burr, E. J. (1960). "The distribution of Kendall's score S for a pair of tied rankings." *Biometrika*, **47**, 151–171.

Chacko, V. J. (1963). "Testing homogeneity against ordered alternatives." *Ann. math. Statist.*, **34**, 945–956. [170, 174, 201, 202, 221, 360]

Chacko, V. J. (1966). "Modified chi-square test for ordered alternatives." *Sankhyā (B)*, **28**, 185–190. [111]

Chapman, D. G. (1958). "A comparative study of several one-sided goodness-of-fit tests." *Ann. math. Statist.*, **29**, 655–674. [280]

Chassan, J. B. (1960.) "On a test for order." *Biometrics*, **16**, 119–121. [194, 220]

Chassan, J. B. (1962). "An extension of a test for order." *Biometrics*, **18**, 245–247. [194, 220]

Chernoff, H. (1954). "On the distribution of the likelihood ratio." *Ann. math. Statist.*, **25**, 573–578. [214]

Childs, D. R. (1967). "Reduction of the multivariate normal integral to characteristic form." *Biometrika*, **54**, 293–300. [136, 137]

Chung, K. L. (1968). *A Course in Probability Theory*. Harcourt, Brace and World, New York. [97]

Ciesielski, A. (1958). "A note on some inequalities of Jensen's type." *Annls. pol. math.*, **4**, 269–274. [56]

Cochran, W. G. (1954). "Some methods for strengthening the common χ^2 tests." *Biometrics*, **10**, 417–451.

Cochran, W. G. (1955.) "A test of a linear function of the deviations between observed and expected numbers." *J. Am. statist. Ass.*, **50**, 377–397.

Cochran, W. G. (1966). "Analyse des classifications d'ordre." *Revue Statist. appl.*, **14**, 5–17.

Cohen, A., and H. B. Sackrowitz (1970). "Estimation of the last mean of a monotone sequence." *Ann. math. Statist.*, **41**, 2021–2034.

Cohen, A. *See also* Blumenthal and Cohen (1968).

Conover, W. J. (1967). "A k-sample extension of the one-sided two sample Smirnov test statistic." *Ann. math. Statist.*, **38**, 1726–1730.

Cox, D. R. (1955). "Some statistical methods connected with series of events" (with discussion). *Jl R. statist. Soc. (B)*, **17**, 129–164. [300, 305]

Cox, D. R., and P. A. W. Lewis (1966). *The Statistical Analysis of Series of Events*, Methuen, London. [305]

Crawford, G. B. (1967). "Maximum likelihood estimation: a practical theorem on consistency of the nonparametric maximum likelihood estimates with applications." *Math. Note No. 503*, Math. Res. Lab., Boeing Sci. Res. Labs.

Crawford, G. B. *See also* Bray, Crawford and Proschan (1967).

Daly, C. (1962). "A simple test for trends in a contingency table." *Biometrics*, **18**, 114–119.

Dantzig, G. B. (1971). "A control problem of Bellman." *Management Science (Theory)*, **17**, 542–546. [61]

Darst, R. B. (1970). "The Lebesgue decomposition, Radon–Nikodym derivative, conditional expectation, and martingale convergence for lattices of sets." *Pacif. J. Math.*, **35**, 581–600. [356]

Davis, H. T. (1933, 1935). *Tables of the Higher Mathematical Functions*, Vols. 1 and 2, Principia Press, Bloomington, Indiana. [151]

Denier van der Gon, J. J. (1963). "Comparison of two tests against trend for a number of probabilities" (in Dutch). *Statist. Neerlandica*, **17**, 105–112.

Doksum, K. (1966). "Asymptotically minimax distribution-free procedures." *Ann. math. Statist.*, **37**, 619–628. [280]

Doksum, K. (1967a). "Robust procedures for some linear models with one observation per cell." *Ann. math. Statist.*, **38**, 878–883. [211, 212]

Doksum, K. (1967b). "Asymptotically optimal statistics in some models with increasing failure rate averages." *Ann. math. Statist.*, **38**, 1731–1739 (for corrections to the above paper, see *Ann. math. Statist.*, **39**, 684–685).

Doksum, K. (1969a). "Starshaped transformations and the power of rank tests." *Ann. math. Statist.*, **40**, 1167–1176.

Doksum, K. (1969b). "Minimax results for IFRA scale alternatives." *Ann. math. Statist.*, **40**, 1778–1783.

Doksum, K. *See also* Barlow and Doksum (1972); *and* Bickel and Doksum (1969).

Doran, J. E. *See* Hodson, Sneath and Doran (1966).

Dudewicz, E. J. (1969). "Estimation of ordered parameters." *Technical Report*, Department of Operations Research, Cornell University.

Dunnett, C. W., and M. Sobel (1955). "Approximations to the probability integral and certain percentage points of a multivariate analogue of Student's t-distribution." *Biometrika*, **42**, 258–260. [146]

Dykstra, R. L. (1970a). "A characterization of a conditional expectation with respect to a σ-lattice." *Ann. math. Statist.*, **41**, 698–701. [322]

Dykstra, R. L. (1970b). "A note on a theorem by H. D. Brunk." *Proc. Am. math. Soc.*, **24**, 171–174. [355]

Eaton, M. L. (1970). "A complete class theorem for multidimensional one-sided alternatives." *Ann. math. Statist.*, **41**, 1884–1888.

Eeden, C. van (1956). "Maximum likelihood estimation of ordered probabilities." *Proc. K. ned. Akad. Wet. (A)*, **59**/*Indag. math.*, **18**, 444–455. [3, 55, 113]

Eeden, C. van (1957a). "Maximum likelihood estimation of partially or completely ordered parameters. I." *Proc. K. ned. Akad. Wet. (A)*, **60**/*Indag. math.*, **19**, 128–136. [3, 56, 57, 88, 113, 114]

Eeden, C. van (1957b). "Maximum likelihood estimation of partially or completely ordered parameters. II." *Proc. K. ned. Akad. Wet. (A)*, **60**/*Indag. math.*, **19**, 201–211. [3, 56, 57, 88, 113]

Eeden, C. van (1957c). "Note on two methods for estimating ordered parameters of probability distributions." *Proc. K. ned. Akad. Wet. (A)*, **60**/*Indag. math.*, **19**, 506–512. [356]

Eeden, C. van (1957d). "A least squares inequality for maximum likelihood estimates of ordered parameters." *Proc. K. ned. Akad. Wet. (A)*, **60**/*Indag. math.*, **19**, 513–521.

Eeden, C. van (1958). "Testing and estimating ordered parameters of probability distributions," (*Doctoral Dissertation, University of Amsterdam*), Studentendrukkerij Poortpers, Amsterdam. [56, 57, 88, 113, 170, 356]

Eeden, C. van (1960). "On distribution-free bio-assay." *Proc. Symp. Quantitative Methods in Pharmacology, Leiden*, 206–210. [113]

Eeden, C. van (1970). "Efficiency-robust estimation of location." *Ann. math. Statist.*, **41**, 172–181. [355]

Eeden, C. van, and J. Hemelrijk (1955a). "A test for the equality of probabilities against a class of specified alternative hypotheses, including trend. I." *Proc. K. ned. Akad. Wet. (A)*, **58**/*Indag. math.*, **17**, 191–198.

Eeden, C. van, and J. Hemelrijk (1955b). "A test for the equality of probabilities against a class of specified alternative hypotheses, including trend. II." *Proc. K. ned. Akad. Wet. (A)*, **58**/*Indag. math.*, **17**, 301–308.

Eilbott, J. M. *See* Nadler and Eilbott (1967).

Elfving, G., and J. H. Whitlock (1950). "A simple trend test with application to erythrocyte data." *Biometrics*, **6**, 282–288.

Epstein, B. (1960a). "Tests for the validity of the assumption that the underlying distribution of life is exponential. Part I." *Technometrics*, **2**, 83–101. [305]

Epstein, B. (1960b). "Tests for the validity of the assumption that the underlying distribution of life is exponential. Part II." *Technometrics*, **2**, 167–183.

Esary, J. D. *See* Birnbaum, Esary and Marshall (1966).

Ewing, G. M. *See* Ayer, Brunk, Ewing, Reid and Silverman (1955); *and* Brunk, Ewing and Utz (1957).

Franck, W. E. *See* Brunk, Franck, Hanson and Hogg (1966).

Friedman, M. (1937). "The use of ranks to avoid the assumption of normality implicit in the analysis of variance." *J. Am. statist. Ass.*, **32**, 675–701. [207]

Fujisawa, H. *See* Kudô and Fujisawa (1964).

Gebhardt, F. (1970). "An algorithm for monotone regression with one or more independent variables." *Biometrika*, **57**, 263–271. [113]

Gove, W. R. *See* Leik and Gove (1969).

Govindarajulu, Z., and H. S. Haller (1968). "c-sample tests for homogeneity against ordered alternatives" (abstract). *Ann. math. Statist.*, **39**, 1089.

Grenander, U. (1956). "On the theory of mortality measurement. Part II." *Skand. Akt.*, **39**, 125–153. [55, 261]

Gupta, S. S. (1963). "Bibliography on the multivariate normal integrals and related topics." *Ann. math. Statist.*, **34**, 829–838. [136]

Hájek, J., and Z. Šidák (1967). *Theory of Rank Tests*, Academic Press, New York. [282]

Haller, H. S. *See* Govindarajulu and Haller (1968).

Hanson, D. L. *See* Brunk, Franck, Hanson and Hogg (1966).

Hardy, G. H., J. E. Littlewood and G. Pólya (1952). *Inequalities*, 2nd ed., Cambridge University Press. [56]

Hartigan, J. A. (1967a). "Distribution of the residual sum of squares in fitting inequalities." *Biometrika*, **54**, 69–84.

Hartigan, J. A. (1967b). "Representation of similarity matrices by trees." *J. Am. statist. Ass.*, **62**, 1140–1158. [30, 31, 32]

Hemelrijk, J. (1958). "Distribution-free tests against trend and maximum likelihood estimates of ordered parameters." *Bull. Inst. int. Statist.*, **36**, 15–25.

Hemelrijk, J. *See also* van Eeden and Hemelrijk (1955a, 1955b).

Herbach, L. H. (1959). "Properties of model II-type analysis of variance tests, A: optimum nature of the F-test for model II in the balanced case." *Ann. math. Statist.*, **30**, 939–959. [100]

Hewitt, E., and K. Stromberg (1965). *Real and Abstract Analysis*, Springer–Verlag, Berlin. [356]

Hoadley, B. (1971). "The southern Bell left-in station study I: statistical methods for data analysis." *Bell Laboratories Memorandum.* [113]

Hodges, J. L., Jr., and E. L. Lehmann (1956). "The efficiency of some nonparametric competitors of the t-test." *Ann. math. Statist.*, **27**, 324–335. [204]

Hodson, F. R., P. H. A. Sneath and J. E. Doran (1966). "Some experiments in the numerical analysis of archaeological data." *Biometrika*, **53**, 311–324. [22]

Hogg, R. V. (1965). "On models and hypotheses with restricted alternatives." *J. Am. statist. Ass.*, **60**, 1153–1162. [219]

Hogg, R. V. *See also* Brunk, Franck, Hanson and Hogg (1966).
Hollander, M. (1967). "Rank tests for randomized blocks when the alternatives have an *a priori* ordering." *Ann. math. Statist.*, **38**, 867–877. [210, 211, 212, 219, 220]
Hollander, M. *See also* Pirie and Hollander (1971).
Horan, C. B. (1969). "Multidimensional scaling: combining observations when individuals have different perceptual structures." *Psychometrika*, **34**, 139–165. [56]
Jessen, B. *See* Sparre Andersen and Jessen (1948).
Johansen, S. (1967). "The descriptive approach to the derivative of a set function with respect to a σ-lattice." *Pacif. J. Math.*, **21**, 49–58. [356]
Johansen, S. *See also* Brunk and Johansen (1970).
Johnson, R. A. *See* Bhattacharyya and Johnson (1970).
Jonckheere, A. R. (1954a). "A distribution-free k-sample test against ordered alternatives." *Biometrika*, **41**, 133–145. [204]
Jonckheere, A. R. (1954b). "A test for significance of the relation between m rankings and k ranked categories." *Br. J. statist. Psychol.*, **7**, 93–100. [209]
Karlin, S. *See* Arrow, Karlin and Scarf (1958).
Katz, M. W. (1963). "Estimating ordered probabilities." *Ann. math. Statist.*, **34**, 967–972.
Kendall, M. G. (1941). "Proof of relations connected with the tetrachoric series and its generalization." *Biometrika*, **32**, 196–198. [137]
Kendall, M. G., and A. Stuart (1958). *The Advanced Theory of Statistics, Vol. 1*, Griffin, London. [129]
Klahr, D. (1969). "A Monte Carlo investigation of the statistical significance of Kruskal's nonmetric scaling procedure." *Psychometrika*, **34**, 319–333. [56]
Klotz, J. H. *See* Bhattacharyya and Klotz (1966).
Knoll, R. L. *See* Stenson and Knoll (1969).
Konkin, P. R. *See* May and Konkin (1970).
Kruskal, J. B. (1964a). "Multidimensional scaling by optimizing goodness of fit to a nonmetric hypothesis." *Psychometrika*, **29**, 1–27. [22, 23]
Kruskal, J. B. (1964b). "Nonmetric multidimensional scaling: a numerical method." *Psychometrika*, **29**, 115–129. [15, 72, 113]
Kruskal, J. B. (1965). "Analysis of factorial experiments by estimating monotone transformations of the data." *Jl R. statist. Soc. (B)*, **27**, 251–263. [56]
Kruskal, W. H. (1952). "A nonparametric test for the several sample problem." *Ann. math. Statist.*, **23**, 525–540. [202]
Kudô, A. (1963). "A multivariate analogue of the one-sided test." *Biometrika*, **50**, 403–418. [170, 176, 177, 213]
Kudô, A., and H. Fujisawa (1964). "A bivariate normal test with two sided alternative." *Mem. Fac. Sci. Kyushu Univ. (A)*, **18**, 104–108. [150]
Kuenzi, N. J. (1969). "An investigation of some problems concerning the additivity of conditional expectation with respect to a σ-lattice." *Unpublished Ph.D. thesis*, University of Iowa. [358]
Leadbetter, M. R. *See* Watson and Leadbetter (1964).
LeCam, L. (1966). "Likelihood functions for large numbers of independent observations." In Γ. N. David (Ed.), *Research Papers in Statistics* (Festschrift for J. Neyman), Wiley, New York, 167–187. [282]
Lehmann, E. L. (1959). *Testing Statistical Hypotheses*, Wiley, New York. [219]
Lehmann, E. L. (1964). "Asymptotically nonparametric inference in some linear models with one observation per cell." *Ann. math. Statist.*, **35**, 726–734. [212]

Lehmann, E. L. *See also* Hodges and Lehmann (1956).
Leik, R. K., and W. R. Gove (1969). "The conception and measurement of asymmetric monotonic relationships in sociology." *Am. J. Sociology*, **74**, 696–709. [56]
Levin, V. I., and S. B. Stečkin (1960). "Inequalities." *Am. math. Soc. Translations* (2), **14**, 1–29. [356, 358]
Lewis, P. A. W. *See* Cox and Lewis (1966).
Littlewood, J. E. *See* Hardy, Littlewood and Pólya (1952).
Lombard, P. B., and H. D. Brunk (1963). "Evaluating the relation of juice composition of mandarin oranges to percent acceptance of a taste panel." *Fd. Technol.*, **17**, 113–115. [56, 67]
Lubin, A. (1957). "A rank order test for trend in correlated means" (abstract). *Ann. math. Statist.*, **28**, 524.
Luenberger, D. G. (1969). *Optimization by Vector Space Methods*, Wiley, New York. [355]
Lyerly, S. B. (1952). "The average Spearman rank correlation coefficient." *Psychometrika*, **17**, 421–428.
McFadden, J. A. (1960). "Two expansions for the quadrivariate normal integral." *Biometrika*, **47**, 325–333. [137]
Madansky, A. *See* Proschan, Barlow, Madansky and Scheuer (1968).
Mallows, C. L. *See* Barton and Mallows (1961).
Mann, H. B. (1945). "Nonparametric tests against trend." *Econometrica*, **13**, 245–259.
Marshall, A. W. (1970). Discussion on Barlow and van Zwet's paper. In M. L. Puri (Ed.), *Nonparametric Techniques in Statistical Inference*, Cambridge University Press, 175–176. [111]
Marshall, A. W., and F. Proschan (1965). "Maximum likelihood estimation for distributions with monotone failure rate." *Ann. math. Statist.*, **36**, 69–77. [114, 258, 261, 305, 356]
Marshall, A. W. *See also* Barlow, Marshall and Proschan (1963); *and* Birnbaum, Esary and Marshall (1966).
May, R. B., and P. R. Konkin (1970). "A nonparametric test of an ordered hypothesis for k independent samples." *Educ. psychol. Measur.*, **30**, 251–257.
Miles, R. E. (1959). "The complete amalgamation into blocks, by weighted means, of a finite set of real numbers." *Biometrika*, **46**, 317–327. [55, 113, 161, 172, 174]
Minty, G. J. (1965). "A theorem on maximal monotonic sets in Hilbert space." *J. math. Analysis Applic.*, **11**, 434–439. [355]
Moore, D. S. (1968). "An elementary proof of asymptotic normality of linear functions of order statistics." *Ann. math. Statist.*, **39**, 263–265. [277, 278, 287, 294]
Moore, J. R. *See* Thompson and Moore (1963).
Moran, P. A. P. (1948). "Rank correlation and product moment correlation." *Biometrika*, **35**, 203–206. [137]
Moran, P. A. P. (1966). "Estimation from inequalities." *Aust. J. Statist.*, **8**, 1–8. [113]
Nadler, J., and J. M. Eilbott (1967). "Testing for monotone failure rate." *Unpublished.* [279, 305]
Nashed, M. Z. (1968). "A decomposition relative to convex sets." *Proc. Am. math Soc.*, **19**, 782–788. [355]
Neumann, J. von (1940). "On rings of operators, III." *Ann. Math.*, **41**, 94–161. [355]
Neveu, J. (1965). *Mathematical Foundations of the Calculus of Probability* (translated by Amiel Feinstein), Holden-Day, San Francisco. [334]

Nüesch, P. E. (1964). "Multivariate tests of location for restricted alternatives" (*Doctoral dissertation, Swiss Federal Institute of Technology*), Juris-Verlag, Zürich. [177, 213]

Nüesch, P. E. (1966). "On the problem of testing location in multivariate populations for restricted alternatives." *Ann. math. Statist.*, 37, 113–119. [176, 177, 213]

Nüesch, P. E. (1970). "Estimation of monotone parameters and the Kuhn–Tucker conditions" (abstract). *Ann. math. Statist.*, 41, 1800.

Odeh, R. E. (1971). "On Jonckheere's k-sample test against ordered alternatives." *Technometrics*, 13, 912–918.

Odeh, R. E. (1972). "On the power of Jonckheere's k-sample test against ordered alternatives." *Biometrika*, 59, to appear.

Oosterhoff, J. *See* van Zwet and Oosterhoff (1967).

Owen, D. B. (1962). *Handbook of Statistical Tables*, Addison–Wesley, Reading, Mass. [196]

Page, E. B. (1963). "Order hypotheses for multiple treatments: a significance test for linear ranks." *J. Am. statist. Ass.*, 58, 216–230. [209]

Parzen, E. (1962). "On estimation of a probability density and mode." *Ann. math. Statist.*, 33, 1065–1076. [262]

Patnaik, P. B. (1949). "The non-central χ^2- and F-distributions and their application." *Biometrika*, 36, 202–232. [157]

Pearson, K. (1909). "On a new method of determining correlation between a character A, and a character B, of which only the percentage of cases wherein B exceeds (or falls short of) a given intensity is recorded for each grade of A.' *Biometrika*, 7, 96–105.

Pearson, K. (1910). "On a new method of determining correlation, when one variable is given by an alternative and the other by multiple categories." *Biometrika*, 7, 248–257.

Pearson, K. (1934). *Tables of the Incomplete Γ-Function*, Cambridge University Press. [153]

Perlman, M. D. (1969). "One-sided problems in multivariate analysis." *Ann. math. Statist.*, 40, 549–567 (for corrections to the above paper, see *Ann. math. Statist.*, 42, 1777). [213, 214]

Perron, O. (1910). "Über das Verhalten der Integrale linearer Differenzengleichungen im Unendlichen." *Jber. dt. MatVerein*, 19, 129–137. [10]

Pfanzagl, J. (1960a). "Tests und Konfidenzintervalle für exponentielle Verteilungen und deren Anwendung auf einige diskrete Verteilungen." *Metrika*, 3, 1–25.

Pfanzagl, J. (1960b). "Über lokal optimale Rang-Tests." *Metrika*, 3, 143–150.

Pirie, W. R., and M. Hollander (1971). "A distribution-free normal scores test for ordered alternatives in the randomized block design." *Florida State University Statistics Report M213*, submitted for publication.

Plackett, R. L. (1954). "A reduction formula for normal multivariate integrals." *Biometrika*, 41, 351–360. [137, 141]

Pólya, G. *See* Hardy, Littlewood and Pólya (1952).

Prakasa Rao, B. L. S. (1966). "Asymptotic distributions in some non-regular statistical problems." *Tech. report No. 9*, Department of Statistics and Probability, Michigan State University.

Prakasa Rao, B. L. S. (1969). "Estimation of a unimodal density." *Sankhyā (A)*, 31, 23–36. [228, 261]

Prakasa Rao, B. L. S. (1970). "Estimation for distributions with monotone failure rate." *Ann. math. Statist.*, **41**, 507–519. [240, 253]

Proschan, F. (1963). "Theoretical explanation of observed decreasing failure rate." *Technometrics*, **5**, 375–383. [305]

Proschan, F. (1966). "Reliability estimation under plausible assumptions." *Bull. Inst. int. Statist.*, **41**, 143–155.

Proschan, F., R. E. Barlow, A. Madansky and E. M. Scheuer (1968). "Statistical estimation procedures for the 'burn-in' process." *Technometrics*, **10**, 51–62.

Proschan, F., and R. Pyke (1967). "Tests for monotone failure rate." *Proc. 5th Berkeley Symp. math. Statist. Probab.*, **III,** 293–312. [267, 305]

Proschan, F. *See also* Barlow, Marshall and Proschan (1963); Barlow and Proschan (1966, 1967, 1969); Barlow, Proschan and Scheuer (1971); Bray, Crawford and Proschan (1967); *and* Marshall and Proschan (1965).

Puri, M. L. (1965). "Some distribution-free k-sample rank tests of homogeneity against ordered alternatives." *Communs pure appl. Math.*, **18**, 51–63. [205, 206]

Puri, M. L., and P. K. Sen (1968). "On Chernoff–Savage tests for ordered alternatives in randomized blocks." *Ann. math. Statist.*, **39**, 967–972.

Pyke, R. *See* Brunk (1970); *and* Proschan and Pyke (1967).

Rao, C. R. (1965). *Linear Statistical Inference and its Applications*, Wiley, New York. [356]

Reid, W. T. (1968). "A simple optimal control problem involving approximation by monotone functions." *J. Optimization Theory and Applications*, **2,** 365–377. [58, 355]

Reid, W. T. *See also* Ayer, Brunk, Ewing, Reid and Silverman (1955).

Robertson, T. (1965). "A note on the reciprocal of the conditional expectation of a positive random variable." *Ann. math. Statist.*, **36**, 1302–1305. [37, 61, 358]

Robertson, T. (1966). "A representation for conditional expectations given σ-lattices." *Ann. math. Statist.*, **37**, 1279–1283. [356]

Robertson, T. (1967). "On estimating a density which is measurable with respect to a σ-lattice." *Ann. math. Statist.*, **38**, 482–493. [261, 262]

Robertson, T. (1968). "A smoothing property for conditional expectations given σ-lattices." *Am. math. Mon.*, **75**, 515–518. [356]

Robertson, T., and P. Waltman (1968). "On estimating monotone parameters." *Ann. math. Statist.*, **39**, 1030–1039. [56]

Rockafellar, R. T. (1970). *Convex Analysis*, Princeton University Press. [62]

Rosner, N. (1961). "System analysis: non-linear estimation techniques." In *Proceedings of the Seventh National Symposium on Reliability and Quality Control*, 203–207. [298]

Ruben, H. (1954). "On the moments of order statistics in samples from normal populations." *Biometrika*, **41**, 200–227. [146, 360]

Sackrowitz, H. B. (1970). "Estimation for ordered parameter sequences: the discrete case." *Ann. math. Statist.*, **41**, 609–620.

Sackrowitz, H. B. *See also* Cohen and Sackrowitz (1970).

Savage, I. R. (1957). "Contributions to the theory of rank order statistics—the 'trend' case." *Ann. math. Statist.*, **28**, 968–977.

Scarf, H. *See* Arrow, Karlin and Scarf (1958).

Schaafsma, W. (1966). "Hypothesis testing problems with the alternative restricted by a number of inequalities" (*Doctoral dissertation, University of Groningen*), Noordhoff, Groningen. [190, 214, 219]

Schaafsma, W. (1968). "A comparison of the most stringent and the most stringent somewhere most powerful test for certain problems with restricted alternative." *Ann. math. Statist.*, **39**, 531–546.

Schaafsma, W. (1970). "Most stringent and maximin tests as solutions of linear programming problems." *Z. Wahrscheinlichkeitstheorie verw. Geb.*, **14**, 290–307.

Schaafsma, W., and L. J. Smid (1966). "Most stringent somewhere most powerful tests against alternatives restricted by a number of linear inequalities." *Ann. math. Statist.*, **37**, 1161–1172. [190]

Scheffé, H. (1959). *The Analysis of Variance*, Wiley, New York. [182, 199]

Scheuer, E. M. *See* Barlow, Proschan and Scheuer (1971); Barlow and Scheuer (1966, 1971); *and* Proschan, Barlow, Madansky and Scheuer (1968).

Sen, P. K. (1968). "On a class of rank order tests in two-way layouts." *Ann. math. Statist.*, **39**, 1115–1124.

Sen, P. K. *See also* Puri and Sen (1968).

Shepard, R. N. (1966). "Metric structures in ordinal data." *J. Math. Psych.*, **3**, 287–315. [56]

Shorack, G. R. (1967). "Testing against ordered alternatives in model I analysis of variance; normal theory and nonparametric." *Ann. math. Statist.*, **38**, 1740–1753. [123, 124, 127, 171, 172, 193, 201, 207, 213]

Šidák, Z. *See* Hájek and Šidák (1967).

Silverman, E. *See* Ayer, Brunk, Ewing, Reid and Silverman (1955).

Smid, L. J. *See* Schaafsma and Smid (1966).

Sneath, P. H. A. *See* Hodson, Sneath and Doran (1966).

Sobel, M. *See* Dunnett and Sobel (1955).

Sparre Andersen, E. (1953a). "On sums of symmetrically dependent random variables." *Skand. Akt.*, **36**, 123–138.

Sparre Andersen, E. (1953b). "On the fluctuations of sums of random variables." *Mathematica scand.*, **1**, 263–285 (for corrections to the above paper, see *Mathematica scand.*, **2**, 193–194).

Sparre Andersen, E. (1954). "On the fluctuations of sums of random variables II." *Mathematica scand.*, **2**, 195–223. [172, 174, 200]

Sparre Andersen, E., and B. Jessen (1948). "Some limit theorems on set-functions." *Math.-fys. Meddr.*, **25**, No. 5, 8pp. [358]

Srivastava, S. R., and T. A. Bancroft (1967). "Inferences concerning a population correlation coefficient from one or possibly two samples subsequent to a preliminary test of significance." *Jl R. statist. Soc. (B)*, **29**, 282–291.

Stečkin, S. B. *See* Levin and Stečkin (1960).

Stegun, I. S. *See* Abramowitz and Stegun (1965).

Stenson, H. H., and R. L. Knoll (1969). "Goodness of fit for random rankings in Kruskal's nonmetric scaling procedure." *Psych. Bull.*, **71**, 122–126. [56]

Stromberg, K. *See* Hewitt and Stromberg (1965).

Stuart, A. (1963). "Calculation of Spearman's Rho for ordered 2-way classifications." *Am. Statistn.*, **17**, 23–24. [194]

Stuart, A. *See also* Kendall and Stuart (1958).

Terpstra, T. J. (1952). "The asymptotic normality and consistency of Kendall's test against trend when ties are present in one ranking." *Proc. Sect. Sci. K. ned. Akad. Wet. (A)*, **55**/*Indag. math.*, **14**, 327–333. [204]

Terpstra, T. J. (1953). "The exact probability distribution of the T statistic for testing against trend and its normal approximation." *Proc. Sect. Sci. K. ned. Akad. Wet. (A)*, **56**/*Indag. math.*, **15**, 433–437.

Terpstra, T. J. (1955). "A generalization of Kendall's rank correlation statistic. I." *Proc. K. ned. Akad. Wet. (A)*, **58**/*Indag. math.*, **17**, 690–696. [205]

Terpstra, T. J. (1956). "A generalization of Kendall's rank correlation statistic. II." *Proc. K. ned. Akad. Wet. (A)*, **59**/*Indag. math.*, **18**, 59–66. [205]

Thompson, W. A., Jr. (1962). "The problem of negative estimates of variance components." *Ann. math. Statist.*, **33**, 273–289. [3, 73, 74, 113]

Thompson, W. A., Jr., and J. R. Moore (1963). "Non-negative estimates of variance components." *Technometrics*, **5**, 441–449.

Tingey, F. H. *See* Birnbaum and Tingey (1951).

Tryon, P. V. (1970). "Nonparametric tests of homogeneity against restricted alternatives in a one-way classification." *Unpublished Ph.D. thesis*, Pennsylvania State University.

Tukey, J. W. *See* Abelson and Tukey (1963).

Ubhaya, V. A. *See* Barlow and Ubhaya (1971).

Ury, H. K. (1968). "The behavior of some tests for ordered alternatives under interior slippage." In *Selected Statistical Papers*, 2, Mathematical Centre, Amsterdam, 15–25.

Utz, W. R. *See* Brunk, Ewing and Utz (1957).

Veinott, A. F., Jr. (1971). "Least d-majorized network flows with inventory and statistical applications." *Management Science (Theory)*, **17**, 547–567. [61]

Vincent, S. E. (1961). "A test of homogeneity for ordered variances." *Jl R. statist. Soc. (B)*, **23**, 195–206. [198]

Waltman, P. *See* Robertson and Waltman (1968).

Watson, G. S., and M. R. Leadbetter (1964). "Hazard analysis II." *Sankhyā*, **26**, 101–116. [262]

Wegman, E. J. (1969a). "Maximum likelihood histograms." *University of North Carolina Mimeo Series 629.*

Wegman, E. J. (1969b). "A note on estimating a unimodal density." *Ann. math. Statist.*, **40**, 1661–1667.

Wegman, E. J. (1970a). "Maximum likelihood estimation of a unimodal density function." *Ann. math. Statist.*, **41**, 457–471. [226, 229, 231]

Wegman, E. J. (1970b). "Maximum likelihood estimation of a unimodal density, II." *Ann. math. Statist.*, **41**, 2169–2174. [226, 231]

Weiss, L., and J. Wolfowitz (1967). "Estimators of a density function at a point." *Z. Wahrscheinlichkeitstheorie verw. Geb.*, **7**, 327–335. [262]

Welch, B. L. (1937). "On the z-test in randomized blocks and Latin squares." *Biometrika*, **29**, 21–52. [121]

Whitlock, J. H. *See* Elfving and Whitlock (1950).

Williams, D. A. (1971). "A test for differences between treatment means when several dose levels are compared with a zero dose control." *Biometrics*, **27**, 103–117.

Wolfowitz, J. *See* Weiss and Wolfowitz (1967).

Yates, F. (1948). "The analysis of contingency tables with groupings based on quantitative characters." *Biometrika*, **35**, 176–181.

Zwet, W. R. van (1964). *Convex Transformations of Random Variables*, Mathematical Centre, Amsterdam. [242, 262]

Zwet, W. R. van, and Oosterhoff, J. (1967). "On the combination of independent test statistics." *Ann. math. Statist.*, **38**, 659–680.

Zwet, W. R. van. *See also* Barlow and van Zwet (1969, 1970).

Subject Index

Active block, 72
Actuarial methods of estimation, 4, 222
Air conditioning equipment failures, 269–270
Algorithms for calculation of isotonic regression, 72–91, 113
  for bounded isotonic regression problem, 57, 113
  Maximum Upper Sets, 77, 113
  Maximum Violator, 75
  Minimax Order, 81–85, 87–88
  Minimum Lower Sets, 57, 76–77, 80, 86, 113, 130, 167
  Minimum Violator, 57, 74–75, 79, 113
  Pool-Adjacent-Violators, 13–20, 55, 57, 79, 113, 120, 127, 160, 171
  Up-and-Down Blocks, 72–73, 79, 113
  van Eeden's, 90–91, 113
Analysis of variance, Model II, 2–3, 100–101
Antitonic function, 20
Antitonic regression, 21, 22, 37, 70–72
Archaeology, 22
Arcsin transformation, 193
Av, definitions, 9, 34

Bayesian estimation of ordered parameters, 94–95
  examples, 98, 102–103
  see also Estimation of ordered parameters
Bernoulli distribution, estimation of ordered parameters, 68
Beta distribution, truncated, 103

Binomial distribution
  estimation of ordered parameters, 38–41, 102–105, 114
  testing against trend, 192–194
Bioassay, 38, 55, 103, 113
  see also Dosage-mortality experiment, Stimulus-response curve
Blocks, 13, 72
  active, 72
  down-satisfied, 72
  for partial order represented by rooted tree, 75
  generalization to level sets, 75
  independent, 82
  solution, 13, 82
    average size of, 161
  up-satisfied, 72

$\bar{\chi}^2$ test for general linear model, 180
$\bar{\chi}^2$ test for multivariate normal mean, 177, 213, 214
$\bar{\chi}^2$ test for two-way classification, 124
$\bar{\chi}^2$ test of equality of normal means, 117–120, 126, 170
  comparison with tests based on scores, 186, 190, 219
  consistency, 165–167
  critical values, 360–362
  distribution-free counterpart, see $\bar{\chi}^2_{\text{rank}}$
  distribution of test statistic when means equal, 126 134, 171, 172
  approximation to, 150–153
  as approximation for non-normal distributions, 191, 193, 202

$\bar{\chi}^2$ test (*contd.*)—
  distribution (*contd.*)—
    asymptotic normality, 152
    characteristic function, 151
    conditional on l, 134, 171
    cumulants, 151
    of two-sided version, 149
  distribution of test statistic when means unequal, 153–157, 221
    approximations, 160–164
    for $k = 2$, 161–162
  numerical example, 120
  optimality, 169–170
  power function, 153–168
    approximations, 160–165
    comparisons with conditional test, 170
    comparisons with tests based on scores, 188–189, 220
    effect of prior information on, 157, 158
    expressions for $k = 3$, 155–157
    maxima and minima for given Δ^2, 156, 157, 159, 160, 162, 164
    numerical values, 158, 159, 162, 164, 165, 168
    outside alternative hypothesis region, 165–168, 220
  robustness, 198–199
  two-sided version, 126, 149
$\bar{\chi}^2$ test of equality of proportions, 193
$\bar{\chi}^2_{\text{rank}}$ test, 201–204, 221
  asymptotic distribution of test statistic, 202
  asymptotic relative efficiency relative to $\bar{\chi}^2$, 204
  variant for use with randomized blocks, 208
Censored data, 245, 301–305, 306
Cluster analysis, 22–24, 30–34
Coherent systems, 255, 262
Computer failure data, 298
Conditional expectation given a σ-field, 307, 355
Conditional expectation given a σ-lattice as a Lebesgue–Radon–Nikodym derivative, 326–354, 356–358
  additivity, 345, 358
  convergence, 351–353, 358
    dominated, 352

Conditional expectation (*contd.*)—
  convergence (*contd.*)—
    martingale, 353, 358
    monotonic, 351
  definition, 328
  existence, 334
  expressions for, 334, 335
  identification with definition in terms of projection, 349
  integrability, 336
  integral characterizations, 339, 340
  minimizing properties, 349
  order properties, 336
  reciprocal as conditional expectation, 344
  uniqueness, 334
Conditional expectation given a σ-lattice as projection in L_2, 314–326, 349, 354–355
  characterizations, 318, 325
    in terms of variances and covariances, 319, 320
  definition, 315
  existence, 315
  identification with definition as a Lebesgue–Radon–Nikodym derivative, 349
  uniqueness, 315
Conditional test of equality of normal means, 170
Cone, 27
  class of isotonic functions, 27
  convex, 48
  dual, 49
Conjugate family of distributions, 94
Consistency
  of empirical distribution as estimator of distribution, 227
  of isotonic estimators, 65, 67, 70, 111, 112, 114
    of decreasing density, 227–228
    of IFR distribution, 240, 261
    of increasing failure rate, 239–240
    of star-ordered distribution, 259–260
    of unimodal density, 231
    when basic estimator inconsistent, 223, 226–228, 245, 255, 258
    window estimators for generalized failure rate, 245, 249–251

Consistency (*contd.*)—
  of tests,
    $\bar{\chi}^2$ test of equality of means, 165–167
    distribution-free tests against ordered alternatives, 205, 210, 219–220
    tests for trend in proportions, 220
    tests, based on scores, of equality of means, 219
Contiguous alternatives, 282–283
Contrasts, tests of equality of normal means based on, 183–190
  from regression arguments, 184
  see also Scores
Convergence properties, poor, of maximum likelihood estimators of densities and failure rates, 228, 240, 241
Convergence properties of conditional expectation given a σ-lattice, 351–353, 358
Convex cone, 27–28, 48
  class of isotonic functions, 27
  class of functions measurable with respect to a σ-lattice, 311
  dual, 49
  see also Projection
Convex ordering, 242, 262
  and star-ordering, 254, 273
  conversion to stochastic ordering by total time on test transformation, 285
Convex set, 48, 314
  class of isotonic functions, 25
  see also Projection
c-ordering *see* Convex ordering
Correlation ratio, 56
Cumulative sum diagram (CSD), 9
  as basis of geometrical construction for isotonic regression, 10
  examples, 11, 12, 17, 84
Cumulative total time on test statistic, 236, 267
  and linear spacings tests, 305
  and uniform scores test, 284
  asymptotic distribution, 268, 276–279
  asymptotic minimax property, 297
  distribution under exponentiality, 268–302
  for censored data, 302

Cumulative total time on test (*contd.*)—
  for series of events, 300
  generalizations, 284
  in terms of regression of normalized spacings, 267, 305
  isotonic power with respect to star-ordering, 276, 303
  optimality with respect to parametric alternatives, 305
  originally proposed by Laplace, 305
  unbiasedness against IFRA alternatives, 285, 303

Debugging of systems, 3–4, 43, 297–298
Decomposable alternative, 163
  approximate power of $\bar{\chi}^2$ test against, 163–164
  isotonic regression calculations for, 74, 113
  $P(l, k)$s for, 148
Density, as case of generalized failure rate function, 242
Density, estimation of
  consistency, 227–228, 231, 261
  decreasing, 223–229
  unimodal, 223–231
    as conditional expectation given a σ-lattice, 261
    known mode, 223, 261
    unknown mode, 229–231
  see also Window estimators
DFR *see* Distributions, decreasing failure rate
Dissimilarity matrices *see* Similarity matrices
Distance reducing operator, 322
Distribution-free tests against ordered alternatives, 194, 198–213
  asymptotic relative efficiency, 204, 207, 208, 210, 212, 221
  consistency, 205, 210, 219–220
  for one observation per group, 200
  for partial order alternatives, 204
  for randomized blocks, 207–213
  for several observations per group, 200–207
Distributions
  convex ordered, 242
  decreasing failure rate, 241

SUBJECT INDEX

Distributions (*contd.*)—
  increasing failure rate, 232
    and convex ordering, 242
  increasing failure rate average, 255
    and coherent systems, 255, 262
    and star-ordering, 255
  infinitely divisible, 92, 97
  star-ordered, 255
  unimodal, 223
  see also Estimation of distributions
Dominated convergence property of conditional expectation given a σ-lattice, 352
Dosage-mortality experiment, 3, 38, 103
  see also Bioassay, Stimulus-response curve
Down-satisfied, 72
Dual cone, 49

$\bar{E}^2$ test for general linear model, 180–181
$\bar{E}^2$ test for multivariate normal mean, 178
$\bar{E}^2$ test of equality of normal means, 120–123, 127
  comparisons with $\bar{F}$ test, 122, 171
  critical values, 362
  distribution of test statistic when means equal, 127–134
    conditional on l, 134
  numerical example, 122
  robustness, 198–199
  two-sided version, 126
$\bar{E}^2$ tests for two-way classifications, 125, 181–182
Essential restrictions, 91
Estimation
  of distributions, 222–262
    consistency, 227–228, 231, 239–240, 249–251, 259, 260, 261
    convex ordered, 242–254, 262
    decreasing density, 223–229
    IFRA, 255, 258
    inconsistency, 257–258
    monotone failure rate, 104–105, 114, 231–242, 261–262
    of safe distance in traffic, 103–104
    star-ordered, 254–260, 262
    starshaped, 255–258
    stochastically ordered, 105–110, 115

Estimation (*contd.*)—
  of distributions (*contd.*)—
    unimodal, 223, 229–231, 261
    Ushaped failure rate, 232
  of location, efficiency-robust, 355
  of multivariate normal mean, restricted, 176–177
  of ordered parameters, 91–95, 113–115
    Bernoulli distribution, 68
    binomial distribution, 38–41, 102–105, 114
    consistency, 65, 114
    exponential distribution, 45, 99–100
    gamma distribution, 45, 99–100
    geometric distribution, 42–43
    multinomial distribution, 45, 65–66, 111, 114–115
    normal distribution, 5–6, 92, 98–99, 100, 114
    Poisson distribution, 43–45
    two-way classification, 124
  of regression, 66–72, 111–113
  of stimulus-response curve, 3, 38, 103, 113
  of variance components, 2–3, 100–101
Excess, 82
Expectation invariant operator, 323
Exponential distribution
  estimation of ordered means, 45, 99–100
  goodness of fit tests for, 263–306
    against DFR alternatives, 271–297
    against IFR alternatives, 267–268, 302, 305
    against IFRA alternatives, 273–276, 303
    against parametric families, 287, 305
    based on incomplete data, 301–303
    see also Cumulative total time on test statistic
  mixtures, 241
  tests for trend, 194–198, 299–300, 306
Exponential families of distributions, 91–98
  estimation of ordered parameters, 93–95
  testing against trend, asymptotic result, 191
Exponential scores test, 284

SUBJECT INDEX

Extremum problems solved by isotonic regression, 38–55

$\bar{F}$ test, 122, 171
Failure rate function, 231
  maximum likelihood estimators
    assumed decreasing, 241–242
    assumed increasing, 104–105, 232–234
    assumed U shaped, 232
    asymptotic distributions, 240
    consistency, 239–240
  see also Generalized failure rate function
Freezing of Lake Mendota, 5, 15–18

g^*-simple order, 81
  construction of, 82–84
  existence, 87
  minimax property, 88
Gamma distribution
  as approximation to distribution of χ^2 and $\bar{\chi}^2$, 152, 157, 161, 164
  convex ordering property, 242
  estimation of ordered parameters, 45, 99–100
  tests for trend, 194, 197–198
GCM see Greatest convex minorant
Generalized failure rate function, 242
  estimation see Window estimators
General linear model, 179–181
Geometric distribution, estimation of ordered parameters, 42–43
Goodness of fit tests, 218
  see also Exponential distribution
Graphical estimator for generalized failure rate function, 245
Greatest convex minorant, 9
  and taut string problem, 10
  as basis of graphical construction for isotonic regression, 10, 12, 19
  examples, 11, 17
  in estimation of cumulative regression function, 69
  in estimation of distributions, 230
Grid
  dense, 245
  subexponential in the right tail, 246
  see also Windows

Hilbert space, 314
  see also Operators, Projection, Riesz space

IFR see Distributions, increasing failure rate
IFRA see Distributions, increasing failure rate average
Incomplete data, 245, 301–305, 306
Inconsistency
  of maximum likelihood estimators for starshaped and star-ordered distributions, 257–258
  of naive estimator for a density, 226
Independent blocks, 82
Inequalities
  due to Levin and Stečkin, extension of, 356–358
  Jensen's, 346
  Tchebycheff's, 56
Infinitely divisible distributions, 92, 97
Inventory control, 53–55
Isotonic function
  class as convex cone and lattice, 25, 27, 29
  with respect to quasi-order, 24
  with respect to simple order, 9
Isotonic operator, 323
Isotonic regression, 1–63
  as projection in L_2, 310–314
  averaging property, 34
  bounded, 57, 113
  bounds on, 29
  characterizations, 25, 27, 36, 56, 356
  definitions,
    for simple order, 9
    for quasi-order, 25
  dual problems, 46–55
  existence, 12, 30
  generalized, 38–46
  graphical construction for simple order, 9–12, 20
  in estimation of ordered parameters, 91–95, 113–115
    consistency, 65
    examples, 5–6, 38–45, 65–66, 98–105, 114–115
  in estimation of regression, 66–72, 111–113

Isotonic regression (*contd.*)—
  in estimation of regression (*contd.*)—
    consistency, 67, 111–113
  in estimation of stimulus-response curve, 3, 38, 103, 113
  in estimation of variance components, 2–3, 100–101
  in inventory control, 53–55
  in maximizing a product, 45–46
  in multidimensional scaling, 24
  in production planning, 52–53
  in representation of similarity matrices by trees, 30
  loss-reducing property, 64, 110–111
  max–min and min–max formulas, 19, 20, 55, 80
  minimizing properties, 20, 22, 41–42
  numerical examples, 7–8, 14, 15–16, 18, 20, 84–85, 120, 122
  reciprocal as antitonic regression, 37
  uniqueness, 12, 25
  with respect to quasi-order, 24–38
  with respect to reverse simple order, 20
  with respect to simple order, 5–24
  see also Algorithms, Estimation of distributions, Estimation of ordered parameters
Isotonic test, 273, 276, 284, 300
Isotonic window estimators *see* Window estimators

Jensen's inequality, 346

Kendall's τ, 194, 205, 209
Kolmogorov distance, 280
  as means of separating distributions, 282
Kruskal–Wallis test, 201

Latin square, 123
Lattice, 29, 356
  class of functions measurable with respect to a σ-lattice, 311
  class of isotonic functions, 29
  vector, 320
  see also σ-lattice
Least concave majorant (LCM)
  as basis of graphical construction for isotonic regression, 20
  example, 225

Least concave majorant (*contd.*)—
  in estimation of distributions, 225, 230, 237, 238
Lebesgue decomposition, 356
Lebesgue–Radon–Nikodym derivative
  given a σ-lattice, 327, 356
  as definition of conditional expectation given a σ-lattice, 328
  existence, 332
  uniqueness, 333
Level set, 13, 75–76
Likelihood ratio tests against restricted alternatives
  asymptotic result for testing against trend in exponential family, 191
  for binomial distribution, 192
  for exponential distribution, 194–195
  for exponentiality against IFR alternatives, 306
  for gamma distribution, 198
  for general linear model, 179–181
  for multivariate normal mean, 176–178, 213–214
  for normal distribution
    means *see* $\bar{\chi}^2$ test, $\bar{E}^2$ test
    variances, 197
  for trend in series of events, 306
  for two-way classification, 124–125, 181–182
  usual asymptotic results invalid, 117, 127, 169
Linear spacings test, 305
Lower set, 75
LRN derivative *see* Lebesgue–Radon–Nikodym derivative

Mann–Whitney statistic, 205, 206
Martingale convergence property of conditional expectation given a σ-lattice, 353, 358
Maximal set, 329
Maximum likelihood estimation
  generalized, 255
  inconsistency for starshaped and IFRA distributions, 257–258
  of densities, 223–226, 229–231, 244, 261
  of distributions, 103–104, 105–110, 115, 240, 255–257, 261–262

SUBJECT INDEX

Maximum likelihood (*contd.*)—
  of failure rates, 104–105, 232–234, 241, 244, 261
  of multivariate normal mean, restricted, 176–177
  of ordered parameters, 93
    examples, 5–6, 38–45, 65–66, 98–105, 114–115
  of stimulus-response curve, 103
  of variance components, 100–101
  see also Estimation of distributions, Estimation of ordered parameters
Maximum Upper Sets algorithm, 77, 113
Maximum Violator algorithm, 75
Max–min and min–max formulas
  for bounded isotonic regression problem, 57
  for estimation of
    densities, 226, 230
    failure rate, 234, 237, 241
    generalized failure rate, 243, 244, 245, 249
    star-ordered distributions, 258, 260
  for isotonic regression with respect to
    partial order, 80
    reverse simple order, 20
    simple order, 19, 55
  for Poisson extremum problem, 45
  generalizations, 334, 356
Measurable function, with respect to a σ-lattice, 310
  class as convex cone and σ-lattice, 311
Measure, 326
  positive, 327
  transition, 309
Mills' ratio, 232
Minimal set, 329
Minimax order algorithm, 81–85, 87–88
  example, 84–85
Minimax property, asymptotic, of cumulative total time on test statistic, 293–297, 305
Minimax test, 280
Minimum Lower Sets algorithm, 57, 76–77, 80, 86, 113, 130, 167
Minimum Violator algorithm, 57, 74–75, 79, 113
Min–max formulas *see* Max–min and min–max formulas

Modal interval, 229
Monotonic convergence property of conditional expectation given a σ-lattice, 351
Monotonic test, 275
Most stringent somewhere most powerful (MSSMP) test, 215
Multidimensional scaling, 22–24, 56
Multinomial distribution, estimation of ordered parameters, 45, 65–66, 111, 114–115
Multivariate normal distribution
  as limiting distribution of vector of mean ranks, 202, 206, 208
  estimation of restricted mean, 176–177
  probability that all elements positive, 136–138, 146
  tests for restricted mean, 176–183, 213–215
    covariance matrix known, 177, 213, 214–215
    covariance matrix proportional to known matrix, 178, 213, 214–215
    covariance matrix unknown, 213–214
    likelihood ratio, 177–178, 213–214
    regression approach, 190, 214–215
  use for testing equality of normal means, 175–176
  use for testing in general linear model, 175–176
  truncated, 98

Negative conjugate cone, 49
Negative set, 329
Newton–Puiseux polygon, 10
Non-additivity, tests for, 218
Noncentral $\bar{\chi}_2^2$ distribution, 161
Normal distribution
  estimation of ordered parameters
    coefficients of variation, 99
    means, 5–6, 92, 98–99, 100, 114
    variances, 100
  tests of equality of ordered parameters
    means *see* $\bar{\chi}^2$ test, $\bar{E}^2$ test, Scores
    use with non-normal distributions, 117, 191, 193
    variances, 197, 198

Normalized spacings, 264
  as basis for cumulative total time on test statistic, 267, 305
  properties for monotone failure rate alternatives, 265–266, 305
Normal scores test, 284

Operators
  on Hilbert space, 316
    distance reducing, 322
    positively homogeneous, 322
    strictly monotonic, 317, 355
    unitary, 316
  on L_2
    expectation invariant, 323
    isotonic, 323
Oranges, mandarin, 67
Orthogonal designs, tests of equality of ordered parameters in, 123

$P(l, k)$s, 126
  conjecture regarding, 174
  distribution-free property for simple order and equal weights, 136, 139, 172, 192, 200
  for decomposable alternative, 148
  for equal weights, 142–145, 146, 148, 172–174, 363
  for simple loop alternative, 146–148
  for simple order alternative, 140–145, 172–174, 363
  for simple tree alternative, 145–146, 363
  lack of general expressions for $k > 4$, 141
  methods for finding, 134–139
  recurrence relations, 139, 142, 145, 172
  relationship to Stirling numbers, 143–144, 173–174
  tables, 363
  test for trend based on, 200
Partial order, 24
  and σ-lattice, 354
Percentile estimators for star-ordered distributions, 259–260, 262
  consistency, 260
Poisson distribution, 115, 234
  estimation of ordered means, 43–45

Poisson process, 5, 194–197, 297–300, 305–306
Polar cone, 49
Political tree for the U.S.A., 32
Poolable elements in a partially ordered set, 78
Pool–Adjacent–Violators algorithm, 13–15, 55, 57, 79, 113, 127, 160, 171
  examples, 14–18, 21, 120
  for reverse order, 20
  see also Up-and-Down Blocks algorithm
Positive dependence, tests for, 219
Positively homogeneous operator, 322
Positive set, 329
Power of tests see $\bar{\chi}^2$ test, Scores
Precision of distribution in exponential family, 93–94
Predecessor, immediate, of element in partially ordered set, 73
Product, maximization of, 45–46, 226, 358
Production planning, 52–53, 61
Projection on a closed convex set in Hilbert space, 314–322, 355
  as a distance reducing operator, 317
  as a strictly monotonic operator, 317
  characterizations, 315, 318, 322
  existence, 314
  on a closed convex cone, 318
  on a closed convex cone and lattice, 321
  uniqueness, 314
Proportions, testing equality of against trend, 117, 192–194, 220

Quasi-order, 24
  and σ-lattice, 310, 354

Randomized blocks, 178, 207–213
Rank correlation coefficients, 194, 200, 205, 206, 209
Reduced isotonic regression problem, 78, 90
Regression
  as projection in L_2, 307–310
  isotonic estimation, 66–72, 111–113
  tests for normal and multivariate normal means see Scores

Tests of equality (*contd.*)—
  asymptotic result for exponential families, 191
  binomial distribution, 117, 192–194, 220
  exponential distribution, 194–197
  gamma distribution, 197–198
  normal distribution
    means *see* $\bar{\chi}^2$ test, $\bar{E}^2$ test, Scores
    variances, 197–198
  see also Distribution-free tests
Total time on test statistic *see* Cumulative total time on test statistic
Total time on test transformation, 235, 271, 280, 285, 305
Total time on test weights, 244
Traffic, estimation of safe distance distribution in, 103–104, 113
Transition measure, 309
Tree
  as representation of a partial order, 74
  as representation of a similarity matrix, 30–34
Trend, tests for
  in series of events, 194–197, 299–300, 306
  see also Distribution-free tests, Tests of equality of parameters against restricted alternatives

Two-way classification, 123–126, 181–182, 216–218

Uniform scores test, 284
Unitary operator, 316
Up-and-Down Blocks algorithm, 72–73, 79, 113
  see also Pool-Adjacent-Violators algorithm
Upper set, 75
Up-satisfied, 72

Van Eeden's algorithm, 90–91, 113
Variance components, estimation of, 2–3, 100–101

Window estimators, isotonic, for generalized failure rate function, 223, 228–229, 240, 243–245, 254, 262
  asymptotic distributions, 251–254
  choice of window size, 254, 262
  consistency, 249–251
Windows
  narrow, 245, 252
  wide, 252
  see also Grid

SUBJECT INDEX

Riesz space, 320
Robustness of normal-theory tests against trend, 198–199

σ-lattice, 310
  functions measurable with respect to, 310–311
  generated by partial and quasi-orders, 310, 354
Scores, tests for equality of normal means based on, 185–190
  comparisons with $\bar{\chi}^2$, 186, 190, 219
  consistency, 219
  derivation from regression arguments, 185
  distribution-free counterparts, 204–207
  linear-2 and linear-2-4 scores for, 219
  optimum scores for, 185–190, 206, 211, 215–219
    simple order, 187, 215
    simple tree, 188
    two-way classification, 217, 218
  power, 185–190
    comparisons with $\bar{\chi}^2$, 188–189, 220
  robustness, 198–199
  use for testing equality of proportions, 193
Scores, tests for multivariate normal mean based on, 214
Scores tests for exponentiality against IFR alternatives, 283–284
  asymptotic distribution of test statistic, 292–293
  consistency, 286
  isotonic power, 284
Series of events, tests for trend in, 194–197, 299–300, 306
Shortcoming of a test, 215
Similarity and dissimilarity matrices
  representation by trees, 30–34
  representation in terms of distances, 22–24
Simple loop alternative, 135
Simple order, 24
  alternative, 118
Simple tree alternative, 145
Solution block, 13, 82

Solution block (*contd.*)—
  average size of for simple order, 161
  distribution of number of *see* $P(l, k)$s
Solution partition, 82
Spearman's ρ, 194, 206, 209
Star-ordered distributions, estimators for, 258–260, 262
  consistency, 259–260
  inconsistency of maximum likelihood estimators, 258, 262
Star-ordering
  and IFRA distributions, 255
  implied by convex ordering, 254, 273
  on distributions, 255, 263
  on mean functions of Poisson processes, 299
Starshaped distributions, estimation of, 255–257
  inconsistency of maximum likelihood estimator, 257–258
  see also Star-ordered distributions
Starshaped function, 254
Stimulus-response curve, 38, 64, 66, 103, 113
  see also Bioassay, Dosage-mortality experiment
Stirling numbers, 143, 173, 174
Strictly monotonic operator, 317, 355
Symmetry, tests of, 219

Taut string problem, 10, 47–51
Tchebycheff's inequality, 56
Tests
  combination of, 215–216, 284
  for positive dependence, 219
  for restricted multivariate normal mean, 175–179, 190, 213–215
  for trend in series of events, 194–197, 299–300, 306
  goodness of fit, 218
    see also Exponential distribution
  isotonic, 273, 276, 284, 300
  monotonic, 275
  most stringent somewhere most powerful, 215
  of symmetry, 219
  see also Distribution-free tests
Tests of equality of parameters against restricted alternatives